Science Curriculum Topic Study

Second Edition

This book is dedicated to all our hardworking colleagues who have informed our work and provided valuable feedback as we updated and improved the CTS process.

We especially want to dedicate this book to our good friend and colleague, Ted Willard, a true "standard-bearer" who lives and breathes standards with a combination of deep knowledge and delightful humor. We can think of no single individual who has greater insight into standards and has pushed our thinking more than Ted. He is our treasure trove of knowledge about standards and their history, and we, as well as thousands of science educators, will be forever grateful to have known, worked with, and learned from him.

Science Curriculum Topic Study

Bridging the Gap Between Three-Dimensional Standards, Research, and Practice

Second Edition

Page Keeley and Joyce Tugel

A Joint Publication with the National Science Teachers Association

FOR INFORMATION:

Corwin
A SAGE Company
2455 Teller Road
Thousand Oaks, California 91320
(800) 233-9936
www.corwin.com

SAGE Publications Ltd.
1 Oliver's Yard
55 City Road
London EC1Y 1SP
United Kingdom

SAGE Publications India Pvt. Ltd.
B 1/I 1 Mohan Cooperative Industrial Area
Mathura Road, New Delhi 110 044
India

SAGE Publications Asia-Pacific Pte. Ltd.
18 Cross Street #10-10/11/12
China Square Central
Singapore 048423

Program Director: Jessica Allan
Content Development Editor: Lucas Schleicher
Senior Editorial Assistant: Mia Rodriguez
Production Editor: Tori Mirsadjadi
Copy Editor: Deanna Noga
Typesetter: C&M Digitals (P) Ltd.
Proofreader: Lawrence Baker
Indexer: Sheila Bodell
Cover Designer: Anupama Krishnan
Marketing Manager: Deena Meyer

Copyright © 2020 by Corwin Press, Inc.

All rights reserved. Except as permitted by U.S. copyright law, no part of this work may be reproduced or distributed in any form or by any means, or stored in a database or retrieval system, without permission in writing from the publisher.

When forms and sample documents appearing in this work are intended for reproduction, they will be marked as such. Reproduction of their use is authorized for educational use by educators, local school sites, and/or noncommercial or nonprofit entities that have purchased the book.

All third-party trademarks referenced or depicted herein are included solely for the purpose of illustration and are the property of their respective owners. Reference to these trademarks in no way indicates any relationship with, or endorsement by, the trademark owner.

Printed in the United States of America

Library of Congress Cataloging-in-Publication Data

Names: Keeley, Page, author. | Tugel, Joyce, author.

Title: Science curriculum topic study : bridging the gap between three-dimensional standards, research, and practice / Page Keeley and Joyce Tugel.

Description: Second edition. | Thousand Oaks, California : National Science Teachers Association ; Corwin, [2020] | Revised edition of: Science curriculum topic study : bridging the gap between standards and practice / Page Keeley ; foreword by Harold Pratt. c2005. | Includes bibliographical references and index.

Identifiers: LCCN 2019020880 | ISBN 9781452244648 (paperback)

Subjects: LCSH: Science—Study and teaching (Secondary)—United States. | Curriculum planning—United States.

Classification: LCC Q183.3.A1 K44 2020 | DDC 507.1/2073—dc23
LC record available at https://lccn.loc.gov/2019020880

This book is printed on acid-free paper.

19 20 21 22 23 10 9 8 7 6 5 4 3 2 1

DISCLAIMER: This book may direct you to access third-party content via web links, QR codes, or other scannable technologies, which are provided for your reference by the author(s). Corwin makes no guarantee that such third-party content will be available for your use and encourages you to review the terms and conditions of such third-party content. Corwin takes no responsibility and assumes no liability for your use of any third-party content, nor does Corwin approve, sponsor, endorse, verify, or certify such third-party content.

Contents

LIST OF CURRICULUM TOPIC STUDY GUIDES	vii
PREFACE	xiii
ACKNOWLEDGMENTS	xxi
ABOUT THE AUTHORS	xxiii

CHAPTER 1 INTRODUCTION TO CURRICULUM TOPIC STUDY (CTS)	**1**
What Is Curriculum Topic Study (CTS)?	2
Why Study a Curriculum Topic?	6
Why Focus on Topics?	9
CTS Impacts Beliefs About Teaching and Learning	11

CHAPTER 2 THE CURRICULUM TOPIC STUDY GUIDE AND ITS COMPONENTS	**13**
The CTS Study Guide	13
Sections and Outcomes Column on the Left of Each Guide	17
Selected Readings Column on the Right of Each Guide	18
Adding Your Own Supplementary Material	20
CTS Website	21
Collective Resources Used in the CTS Guides	21
Building a Professional Collection: Experts at Your Fingertips 24/7!	21
Descriptions of the Common Resources Used in CTS	22

CHAPTER 3 USING CURRICULUM TOPIC STUDY	**37**
Guiding the CTS Process	37
Deciding Which Resources to Use	37
Accessing the Resources That Are Free Online	37
Gathering and Managing CTS Resources	38
Managing CTS Resources When Working With a Group	39
Selecting a Guide, Defining Your Purpose, and Choosing Your Outcomes	41
Getting Started	53
Reading the CTS Sections	53
Guiding Questions for Getting Started with CTS	54

CHAPTER 4 USING AND APPLYING CURRICULUM TOPIC STUDY   61

   What CTS Is Not   61

   Individual Use of CTS   62

   Group Use of CTS   64

      Assigning Jigsaw Readings   65

      Other Group Strategies   66

   Utilizing Curriculum Topic Study for Different Purposes   66

      Using CTS to Enhance Content Knowledge   66

      Using CTS to Inform Curriculum   68

      Using CTS to Inform Instruction   76

      Using CTS to Inform Classroom Assessment   79

   Other Applications of CTS   84

CHAPTER 5 THE CURRICULUM TOPIC STUDY GUIDES   85

   Category A: Life Science Guides   86

   Category B: Physical Science Guides   142

   Category C: Earth and Space Science Guides   200

   Category D: Scientific and Engineering Practices Guides   247

   Category E: Crosscutting Concepts Guides   271

   Category F: STEM Connections Guides   289

REFERENCES   317

INDEX   319

 Visit www.curriculumtopicstudy2.org for downloadable resources.

List of Curriculum Topic Study Guides

CATEGORY A: LIFE SCIENCE GUIDES: BIOLOGICAL STRUCTURE AND FUNCTION

- Behavior, Senses, Feedback, and Response
- Brain and the Nervous System
- Cell Division and Differentiation
- Cells and Biomolecules: Structure and Function
- Characteristics of Living Things
- Energy Extraction, Food, and Nutrition
- Macroscopic Structure and Function of Organisms
- Organs and Systems of the Human Body
- Photosynthesis and Respiration at the Organism Level

CATEGORY A: LIFE SCIENCE GUIDES: ECOSYSTEMS AND ECOLOGICAL RELATIONSHIPS

- Cycling of Matter and Flow of Energy in Ecosystems
- Cycling of Matter in Ecosystems
- Decomposition and Decay
- Ecosystem Stability, Disruption, and Change
- Food Chains and Food Webs
- Group Behaviors and Social Interactions in Ecosystems
- Interdependency in Ecosystems
- Transfer of Energy in Ecosystems

Category A: Life Science Guides: Life's Continuity, Change, and Diversity

Adaptation

Biodiversity and Human Impact

Biological Evolution

Diversity of Species and Evidence of Common Ancestry

DNA, Genes, and Proteins

Inheritance of Traits

Natural Selection

Reproduction, Growth, and Development

Variation of Traits

Category B: Physical Science Guides: Matter

Atoms and Molecules

Behavior and Characteristics of Gases

Chemical Bonding

Chemical Reactions

Conservation of Matter

Elements, Compounds, and the Periodic Table

Mixtures and Solutions

Nuclear Processes

Particulate Nature of Matter

Properties of Matter

States of Matter

Category B: Physical Science Guides: Force, Motion, and Energy

Concept of Energy

Conservation of Energy

Electric Charge and Current

Energy in Chemical Processes and Everyday Life

Force and Motion

Forces Between Objects

Gravitational Force

Kinetic and Potential Energy

Magnetism

Nuclear Energy

Relationship Between Energy and Forces

Sound

Transfer of Energy

Visible Light and Electromagnetic Radiation

Waves and Information Technologies

Waves and Wave Properties

CATEGORY C: EARTH AND SPACE SCIENCE GUIDES: EARTH STRUCTURE, MATERIALS, AND SYSTEMS

Earth's Materials and Systems

Earth's Natural Resources

Global Climate Change

Human Impact on Earth Systems

Structure of the Solid Earth

Water Cycle and Distribution

Water in the Earth System

Weather and Climate

CATEGORY C: EARTH AND SPACE SCIENCE GUIDES: EARTH HISTORY AND PROCESSES THAT CHANGE THE EARTH

Biogeology

Earthquakes and Volcanoes

Earth's History

Natural Hazards

Plate Tectonics

Weathering, Erosion, and Deposition

Category C: Earth and Space Science Guides: Earth in Space, Solar System, and the Universe

Earth and Our Solar System

Earth, Moon, Sun System

Earth-Sun System

Formation of the Earth, Solar System, and the Universe

Phases of the Moon

Seasons and Seasonal Patterns in the Sky

Stars and Galaxies

Category D: Scientific and Engineering Practices Guides

Asking Questions in Science

Defining Problems in Engineering

Developing and Using Models

Planning and Carrying Out Investigations

Analyzing and Interpreting Data

Using Mathematics and Computational Thinking

Constructing Scientific Investigations

Designing Solutions in Engineering

Argumentation

Obtaining, Evaluating, and Communicating Information

Category E: Crosscutting Concepts Guides

Patterns

Cause and Effect

Scale, Proportion, and Quantity

Systems and System Models

Energy and Matter: Flows, Cycles, and Conservation

Structure and Function

Stability and Change

CATEGORY F: STEM CONNECTIONS GUIDES: ENGINEERING, TECHNOLOGY, AND APPLICATIONS OF SCIENCE

Defining and Delimiting an Engineering Problem

Developing Possible Design Solutions

Engineering Design

Improving Proposed Designs

Influence of Science, Technology, and Engineering on Society and the Natural World

Interdependence of Science, Engineering, and Technology

CATEGORY F: STEM CONNECTIONS GUIDES: NATURE OF SCIENCE

Hypotheses, Theories, and Laws

Methods of Scientific Investigations

Science Addresses Questions

Science as a Human Endeavor

Science as a Way of Knowing

Scientific Knowledge Demands Evidence

Preface

OVERVIEW

To become an effective science teacher in today's classroom environment requires a specific set of abilities and knowledge that combine a deep understanding of science concepts and ideas, science pedagogy, and familiarity with standards and research on learning. Accomplished teachers recognize that achieving scientific literacy for all students requires substantive changes in what is taught and how it is structured and delivered. A new vision for science education (which includes engineering) actively engages students in using scientific and engineering practices and applying crosscutting concepts to understand core disciplinary ideas in the life, physical, earth, and space sciences and engineering. This new vision guides teachers and students through a process of continually building upon ideas and ways of doing science and engineering throughout multiple years of school, culminating in a scientifically based and coherent view of the natural and designed world. However, to accomplish this vision, teachers need tools, resources, and professional learning processes to help them transition to new or more effective ways of teaching and learning. Curriculum Topic Study (CTS) is a tool that combines resources and professional learning processes to help teachers and those who support teachers move toward this new vision.

THE HISTORY OF CTS: FROM PAST TO PRESENT

This book is the second edition of Curriculum Topic Study, which was first released in 2005 as *Science Curriculum Topic Study: Bridging the Gap Between Standards and Practice* (Keeley, 2005). The CTS process was originally adapted from the Project 2061 study of a benchmark from *Benchmarks for Science Literacy* (American Association for the Advancement of Science [AAAS], 2009). Studying a benchmark was a powerful approach to understanding the meaning and intent of a learning goal. In our work with teachers and teacher educators, we realized a process was needed to examine learning goals more comprehensively rather than focusing a study on a single benchmark. We also realized that teachers benefitted from using a range of different standards-, content-, and research-based resources. Based on our experiences in teacher professional development, we expanded the study process to focus on topics as a collection of related learning goals and incorporated both the AAAS (2009) *Benchmarks for Science Literacy* and the *National Science Education Standards* (National Research Council [NRC], 1996) in the first edition of CTS.

In 2002, a report was released by the National Research Council (NRC), *Investigating the Influence of Standards*, which showed that although standards had been around for almost a decade, they had not made a significant impact where they matter the most—the classroom. Around that time, the NRC also released *How People Learn* (Bransford, Brown, & Cocking, 2000), which raised awareness among science educators of the need to understand the ideas students bring to their learning.

Expert Teachers

- Know the structure of knowledge in their discipline.
- Know the conceptual barriers that are likely to hinder learning.
- Have a well-organized knowledge of concepts, inquiry procedures, and problem solving (based on pedagogical content knowledge).

SOURCE: Bransford et al. (2000). *How People Learn*.

CTS—by virtue of its focus on science content knowledge, research on commonly held conceptions, clarification of learning goals, and pedagogical implications—is consistent with the findings from *How People Learn*, that distinguish expert teachers from novices as shown in the chart above. There is a strong link between teacher expertise, which involves both content and pedagogical content knowledge, and student achievement. Because teacher expertise has such a demonstrated impact on student learning, it stands to reason that processes that support the development of science teachers' content and pedagogical knowledge, such as CTS, are a sound investment in improving student achievement in science.

The National Science Foundation awarded a $1,410,885 grant to Principal Investigator Page Keeley at the Maine Mathematics and Science Alliance, for Curriculum Topic Study: A Systematic Approach to Utilizing National Standards and Cognitive Research. The project ran from 2003 to 2009 and developed tools and resources for using CTS in both science and mathematics. The first book, *Science Curriculum Topic Study: Bridging the Gap Between Standards and Practice* (Keeley, 2005), was widely used by science teachers, science specialists, and preservice instructors for almost a decade. Results from the project evaluation showed positive effects on teachers' knowledge, attitudes, and beliefs toward teaching science as well as positive impacts on classroom practice.

Fast forward now to a new decade. With the release of *A Framework for K–12 Science Education* (NRC, 2012), the *Next Generation Science Standards* (NGSS Lead States, 2013), new research on learning, and revisions to state standards, CTS needed to be updated with new standards- and research-based resources. The study process is essentially the same, but much of the source material has been replaced or expanded with current resources driving science education today.

National, State, and Local Standards and STEM Integration

In CTS the term *standards*, sometimes referred to as benchmarks, learning goals, performance indicators, frameworks, and so on, refers to a set of outcomes that identify the specific concepts, ideas, processes, and practices that provide the knowledge and ways of doing and thinking about science that all students should have the opportunity to learn and use. Standards, by themselves, are not a curriculum. Standards can be thought of as what students should know and be able to do by the end of instruction, but they do not describe how students get there. Curricular decisions at the state, district, and classroom level are informed by standards, but there are a variety of ways to organize the concepts, ideas, and practices in standards into a coherent storyline that becomes the curriculum and is enacted through effective instruction.

The new standards-related resources used in this second edition of CTS include *A Framework for K–12 Science Education* (NRC, 2012) and the *Next Generation Science Standards* (NGSS Lead States, 2013). These two resources replace the *Benchmarks for Science Literacy* (AAAS, 2009) and the *National Science Education Standards* (NRC, 1996) that were used in the first edition of CTS and were heavily drawn upon, in addition to *Science for All Americans* (AAAS, 1989), to develop *A Framework for K–12 Science Education* (*Framework*) and the *Next Generation Science Standards* (*NGSS*).

Every *NGSS* standard, referred to as a *performance expectation*, has three dimensions: a disciplinary core idea, a science and engineering practice, and a crosscutting concept. The integration of content and application reflects how science and engineering are practiced in the real world. While CTS topics provide a deep study of a particular concept, idea, or practice, it should be noted that curriculum, instruction, and assessment would include a multidimensional approach where content (core ideas and crosscutting concepts) and practices are interwoven (sometimes referred to as *three-dimensional teaching and learning*).

The *NGSS* are based on *A Framework for K–12 Science Education: Practices, Crosscutting Concepts, and Core Ideas* (NRC, 2012), a product of the National Research Council's Committee on a Conceptual Framework for New K–12 Science Education Standards (Helen R. Quinn, Chair). The Committee recognized that while existing national science standards encompassed science content for grades K–12, there was a growing body of research on learning and teaching in science that could inform a revision of the standards and revitalize science education (NRC, 2012, p. ix). The *Framework* builds on the foundation of *Science for All Americans* (1989) and *Benchmarks for Science Literacy* (2009), developed by the American Association for the Advancement of Science (AAAS), and *National Science Education Standards* (1996), developed by the National Research Council (NRC), and incorporates what is now known about effective teaching and learning.

Several states have adopted the *NGSS* as their state standards. For states and districts that have not, most are choosing to adapt components of the three dimensions (disciplinary core ideas, practices, and crosscutting concepts) and embrace the vision of the *Framework*. Private K–12 schools that are not under the mandate of state testing have developed their own standards, often based on the *Framework* or the *NGSS*. These resources are described further in Chapter 2.

The current interest in STEM as an integration of closely related and interdependent disciplines (science, technology, engineering, and mathematics) is also reflected in new standards. The *Framework* and *NGSS* not only define the disciplinary content of science, engineering, and technology, but also provide clarification regarding practices at the K–2, 3–5, 6–8, and 9–12 grade spans. Science, technology, engineering, and mathematics practices are explicitly linked to content in the *NGSS* standards, and an examination of the developmentally appropriate content and practices for a particular grade can clarify the expectations and boundaries. In addition, the nature of science has been added to the *NGSS*.

While some states and districts have established a separate set of engineering standards, many are choosing to integrate engineering expectations within a district's science program. The *Framework* and *NGSS* provide the guidance for a variety of curriculum designs that integrate science and engineering.

Each time new standards are released by states there is a flurry of activity in aligning curriculum, instruction, and assessment to their new standards. Often interpretation of the new standards is left up to the teachers and little time or resources have been spent on helping them understand the content, curricular and instructional implications, and assessment contexts. As a result, consistency and coherency—and what counts as "alignment"—vary across classrooms, districts, and states. CTS provides a process and identifies a set of resources to help educators reliably analyze and interpret their standards and translate them into classroom practice. Using a science analogy, Copernicus did not change the motions of Earth and the sun; he just provided a more accurate interpretation of those motions. CTS can provide a more accurate and consistent interpretation of the standards that make up a curriculum topic, resulting in teaching and learning that is consistent with the current vision of science education reflected in the standards.

Examining the research into the commonly held ideas students bring to their learning is another important feature of CTS. Teachers typically do not have time to access and read a twenty-five page primary research paper, sifting through the methodology to discover findings related to common misconceptions or difficulty students have in learning a concept or idea. CTS locates the research and vets it so that teachers can read short summaries to build their understanding of the ideas their students are likely to bring to their learning and how those ideas are shaped or affected during the learning process. Understanding students' ideas before planning or teaching a lesson, developing curriculum, or looking at student work provides deeper insight into what constitutes effective teaching and learning.

The research literature on students' science conceptions has grown considerably since the first edition of CTS was released. Recent research summaries from the NSTA Press *Uncovering Student Ideas in Science* series are included in the new CTS guides. In addition, much of the older research that was conducted from 1980 to 1995 is still applicable and useful today and includes two of the original CTS sources in the guides—*The Benchmarks for Science Literacy:* Chapter 15 The Research Base (AAAS, 2009) and *Making Sense of Secondary Science: Research Into Children's Ideas* (Driver, Squires, Rushworth, & Wood-Robinson, 1994). *Making Sense of Secondary Science: Research Into Children's Ideas* (Driver et al., 1994) is considered a landmark reference for exploring children's alternative conceptions in science. These resources are described further in Chapter 2.

Effective science teaching requires an understanding not only of the content of the discipline of science teachers teach but also the structure of that content—how ideas interconnect and build on one another and how the practices of science support an understanding of scientific ideas. Teachers with deep knowledge of the content they teach and how the content forms a network of connected ideas as practices are utilized are better able to guide student learning, diagnose and address difficulties, and assess fairly and effectively. The former *Atlas of Science Literacy* (AAAS, 2001) has been replaced with maps that represent the three dimensions of learning, *The NSTA Atlas of the Three Dimensions* (Willard, 2019).

CTS can help teachers identify and learn more about the content they need to understand to provide appropriate and effective learning experiences for their students. It can be used to refresh previously learned content that teachers may have not used for some time, fill gaps in teachers' existing content knowledge, provide insight into phenomena and appropriate contexts used for teaching the content, or expand a teacher's existing content knowledge. This edition of CTS guides includes two of the resources for enhancing teachers' content knowledge that were used in the first edition of CTS: *Science for All Americans* (AAAS, 1989) and *Science Matters* (Hazen & Trefil, 1991, 2009). In addition, content readings from the *Framework for K–12 Science Education* (NSTA, 2012) narratives are included, which are also repeated in two NSTA Press publications that have been added, *The NSTA Quick Reference-Guide to the NGSS K–12* (Willard, 2015) and *The NSTA Atlas of the Three Dimensions* (Willard, 2019). All these resources are described in more detail in Chapter 2.

SHIFTS IN INSTRUCTIONAL PRACTICES

The conceptual shifts reflected in the new standards require today's teachers to make changes to their instructional practice as well. Prior to the development of new standards, the NRC's Committee on Science Learning, Grades K–8, was charged with examining how science is learned and how science should be taught in K–8 classrooms.

> Effective teaching requires more than simple mastery. Quality instruction entails strategically designing student encounters with science that take place in real time and over a period of months and years (e.g., learning progressions). Teachers draw on their knowledge of science, of their students, and of pedagogy to plan and enact instruction. Thus, in addition to understanding the science content itself, effective teachers need to understand learners and pedagogy design and need to monitor students' science learning experiences. (NRC, 2007)

CTS can help teachers understand the importance of and what they need to do to make these instructional shifts.

A range of instructional approaches will help teachers implement the vision of new standards. But how will teachers know which instructional practices support the development of disciplinary core ideas, scientific and engineering practices, and crosscutting concepts? CTS provides the missing link by helping educators identify topic-specific strategies that support student learning in science. These strategies may include use of particular models and representations, observable phenomena, investigations, teacher talk and questioning, collaborative activities, or science discussions. CTS links instructional suggestions to the important ideas and practices students need to learn and use in science. New resources that have been added to CTS that address instruction include two NSTA Press publications, *Disciplinary Core Ideas: Reshaping Teaching and Learning* (Duncan, Krajcik, & Rivet, 2017) and *Helping Students Make Sense of the World Using Next Generation Science and Engineering Practices* (Schwartz, Passmore, & Reiser, 2017).

In addition to curricular and instructional implications, this second edition of CTS has added formative assessment. CTS identifies formative assessment probes from the *Uncovering Student Ideas in Science* series, linked to each topic, that teachers can use to understand students' thinking and provide diagnostic feedback to the teacher prior to or during instruction that becomes formative when teachers use that feedback to inform instruction. The resource also includes instructional suggestions that may be used to address students' alternative conceptions or difficulties associated with a core idea, concept, or use of a practice. These resources used for examining curriculum, instruction, and formative assessment are described in more detail in Chapter 2.

If your goal is to attain the current vision of science education, we hope this new edition of CTS will help you reach that goal. This is not a how-to book about curriculum, instruction, and assessment. Rather than providing the answers, CTS promotes intellectual inquiry in discovering new knowledge or deepening existing knowledge about teaching and learning connected to important science topics. While selected readings have been identified on each of the CTS guides, there is much more to be learned from the collective set of new and old resources. Feel free to go beyond selected readings and become familiar with other parts of the books as well.

AUDIENCE

The primary audience for this book is K–12 science teachers, preservice teachers, and teacher leaders, science education specialists, instructional coaches, preservice instructors, and professional development providers. It is designed to be used individually or in a group setting, such as a professional learning community, professional development workshop or institute, or other venues for collaborative work.

ORGANIZATION

This book is divided into five chapters. Chapter 1 is an introduction to CTS. It describes what CTS is and what it is used for. It describes what a CTS guide is, what it is used for, and the resources it is used with.

Chapter 2 takes you on a tour of a CTS guide. It describes the different features of a CTS guide and provides descriptions of the resources used in the process.

Chapter 3 describes ways to get started with CTS. It provides guidance on selecting a guide and which resources to use, including how to access the resources that are available online for free. It describes how to gather and manage resources, especially when used in a group setting. It also provides a list and description of all the CTS guides. It provides guiding questions that can be used to study a topic and reflect on what was learned.

Chapter 4 addresses how to use CTS. It describes how CTS can be used for individual learning as well as suggestions for conducting CTS with groups. It provides practice in selecting a topic and section to address a question of practice. It describes different contexts CTS can be used in including building content understanding, curriculum, instruction, assessment, and professional learning.

Chapter 5 is the heart of the book. It includes the 103 CTS study guides, organized into categories. Category A includes the life science guides; category B, physical science guides; and category C, earth and space science guides. Each of these categories includes subcategories in which the guides are listed in alphabetical order. Category D includes scientific and engineering practices guides; category E includes crosscutting concept guides; and category F includes STEM connections, including the nature and enterprises of science, technology, and engineering.

There is also a CTS website at www.curriculumtopicstudy2.org where you can find additional information about CTS as well as crosswalks to new resources. Professional development opportunities are announced on the website.

Acknowledgments

This book began as a National Science Foundation–funded project at the Maine Mathematics and Science Alliance (MMSA). We would like to thank the National Science Foundation (NSF) for the original funding that supported the first edition of this book and the professional development for teachers and science leaders. We wish to thank our former MMSA colleagues for helping us shape and refine the CTS process. We are especially grateful to Project 2061 for helping us realize the critical importance of using a study procedure to understand the multifaceted components that support science literacy that led to our topic adaptation of the "study procedure." Thank you to our editor, Jessica Allan, for your patience and perseverance as we conceptualized and developed a second edition to revise CTS to fit the current landscape in science education. "To get it right" takes time! And thank you to NSTA Press for the collection of resources and the NGSS Hub available through NSTA that are utilized in this new edition.

About the Authors

Page Keeley and Joyce Tugel have been science education colleagues and professional development partners for two decades. As Page Keeley developed the first CTS process and guides and authored the first edition, Joyce Tugel was instrumental in designing and implementing the professional development. In this new, updated version of CTS, Joyce joins Page as coauthor. Like two sides of a coin, while Page and Joyce are practically inseparable when it comes to working together to support science teaching and learning, their separate bios are as follows:

PAGE KEELEY has been a leader in science education for over twenty years. She retired from the Maine Mathematics and Science Alliance (MMSA) in 2012 where she had been the senior science program director since 1996. Today she works as an independent consultant, speaker, and author providing professional development to school districts and organizations in the areas of formative assessment and teaching for conceptual understanding.

Page has been the principal investigator and project director on three National Science Foundation–funded (NSF) projects including the *Northern New England Co-Mentoring Network* (NNECN), *PRISMS: Phenomena and Representations for Instruction of Science in Middle School*, and *Curriculum Topic Study: A Systematic Approach to Utilizing National Standards and Cognitive Research*. In addition, she developed and directed state MSP projects including *Science Content, Conceptual Change, and Collaboration* (SC4) and *TIES K–12: Teachers Integrating Engineering Into Science K–12* and two National Semi-Conductor Foundation grants, *Linking Science, Inquiry, and Language Literacy (L-SILL)* and *Linking Science, Engineering, and Language Literacy* (L-SELL). She developed and directed the Maine Governor's Academy for Science and Mathematics Education Leadership, which completed its fourth cohort group of Maine teacher STEM leaders and is a replication of the National Academy for Science and Mathematics Education Leadership, of which she is a fellow.

Page is a prolific author of twenty-two national best-selling and award-winning books, including twelve books in the *Uncovering Student Ideas in Science* series, four books in the first edition *Curriculum Topic Study* series, and four books in the *Science and Mathematics Formative Assessment: Practical Strategies for Linking Assessment, Instruction, and Learning* series. Several of her books have received prestigious awards in educational publishing. She has authored over 45 journal articles and contributed to several book chapters. She is a frequent invited speaker at regional, national, and international conferences on the topic of formative assessment in science, understanding students' (and teachers') thinking, and teaching for conceptual understanding.

Prior to leaving the classroom to work at the Maine Mathematics and Science Alliance in 1996, Page taught middle and high school science for fifteen years. At that time she was an active teacher leader at the state and national level, serving two terms as president of the Maine Science Teachers Association and National Science Teachers Association (NSTA) District II director 1995–1998 and NSTA Executive Board member (prior to the board and council restructuring in 1997). She received the Presidential Award for Excellence in Secondary Science Teaching in 1992 and the Milken National Distinguished Educator Award in 1993.

Since leaving the classroom in 1996, her work in leadership and professional development has been nationally recognized. In 2008, she was elected the 63rd president of the NSTA, the world's largest organization of K–12, university, and informal science educators. In 2009, she received the National Staff Development Council's (now Learning Forward) Susan Loucks-Horsley Award for Leadership in Science and Mathematics Professional Development. In 2013, she received the Outstanding Leadership in Science Education award from the National Science Education Leadership Association (NSELA) and in 2018, the Distinguished Service to Science Education Award from NSTA. She has served as an adjunct instructor at the University of Maine, was a Cohort 1 fellow in the National Academy for Science and Mathematics Education Leadership, was a science literacy leader for the AAAS/Project 2061 Professional Development Program, and has served on several national advisory boards. She has a strong interest in global science education and has led science/STEM education delegations to South Africa (2009), China (2010), India (2012), Cuba (2014), Iceland (2017), Panama (2018), and Costa Rica (2019).

Prior to entering the teaching profession, Page was a research assistant for immunogeneticist Dr. Leonard Shultz at the Jackson Laboratory of Mammalian Genetics in Bar Harbor, Maine. She received her BS in life sciences/pre-veterinary studies from the University of New Hampshire and her master's degree in science education from the University of Maine. In her spare time she enjoys travel, reading, photography, and fiber art, and she dabbles in modernist cooking and culinary art. A Maine resident for almost forty years, Page and her husband currently reside in Fort Myers, FL, and Wickford, RI. Page can be contacted at pagekeeley@gmail.com or through her website at www.uncoveringstudentideas.org. You can follow Page on Twitter at @CTSKeeley.

JOYCE TUGEL has been a leader in science education since she left the classroom in 2001 to become the science professional development specialist at the TERC Eisenhower Regional Alliance. In 2005, she came to work with Page Keeley at the Maine Mathematics and Science Alliance, as a K–12 science specialist, where she managed and directed several projects and provided professional development support to schools and districts throughout New England and nationally. Recently retired from the Maine Mathematics and Science Alliance, today she works as an independent K–12 STEM/science education consultant. Her work primarily focuses on professional development in the areas of science curriculum, instruction and formative assessment, teacher leadership, service- and place-based learning, and implementation of the *Next Generation Science Standards*. She is the lead professional development consultant for Curriculum Topic Study.

Joyce is a contributing author of two books in the *Uncovering Student Ideas in Science* series (NSTA Press) and is a coauthor of *Science Curriculum Topic Study: Bridging the Gap Between Three-Dimensional Standards, Research, and Practice* (2nd edition, Corwin/NSTA Press). Joyce has also authored articles for the NSTA journals, *Science and Children* and *The Science Teacher*.

Prior to leaving the classroom, Joyce taught high school chemistry and physical science for ten years. As an active teacher leader, she collaborated with the Maine Department of Education and the Maine Mathematics and Science Alliance. During her teaching career, Joyce received the Presidential Award for Excellence in Secondary Science Teaching in 1998, the Milken Foundation National Distinguished Educator Award in 1999, and the New England Institute of Chemists Secondary Teaching Award in 1999. As a dedicated NSTA member, she was the NSTA District II director from 2000 to 2003 and the Professional Development Division director from 2003 to 2006. She has served on the board of directors of both the National Science Education Leadership Association and the National Science Teachers Association. She is also a Cohort 3 fellow of the National Academy for Science and Mathematics Education Leadership. Joyce also has a strong interest in global science education and has co-led science/STEM education delegations to Iceland (2017), Panama (2018), and Costa Rica (2019) with Page.

Prior to entering the teaching profession, Joyce was a researcher in environmental biogeochemistry at the University of New Hampshire and liked to "dabble in mud"! She received her BS and MS degrees in microbiology from the University of New Hampshire. In her spare time she enjoys photography, breadmaking, knitting, and exploring the world with her grandchildren. A New Hampshire resident for forty years, Joyce and her husband reside in Barrington, NH. Joyce can be contacted at jtugel@gmail.com.

CHAPTER 1

Introduction to Curriculum Topic Study (CTS)

- *Before I plan a lesson, I use Curriculum Topic Study to examine the research on students' commonly held ideas. I start with students' ideas in mind whenever I teach a new unit.*

- *Curriculum Topic Study has helped me think about new content I will be teaching in our middle school. I haven't used some of these ideas since I studied earth science in high school! Now I know where to turn to refresh my knowledge and gain a better understanding of the new content I will be teaching this year.*

- *Curriculum Topic Study helped our curriculum committee examine topics in our curriculum across grade bands to identify gaps, redundancy, relevant phenomena, and make sure we are teaching important ideas at a level developmentally appropriate for each grade. Without this tool, we would have been relying on our opinions instead of the evidence from the NGSS, Framework, and research.*

- *Our professional learning community uses Curriculum Topic Study to foster productive conversations about teaching and learning, especially as we implement our new standards that are based on the Framework for K–12 Science Education. It provides a filter for us to fine-tune our discussions and support our ideas with evidence from our readings.*

- *Our school has many of the books used for Curriculum Topic Study in our professional library. Until now, they just sat on the shelf. Now we have a purpose and a process for using them.*

- *I have been using Curriculum Topic Study with our preservice teachers. It has equipped them with the tools and processes they will need to develop and implement standards-based lessons that reflect the current vision their schools are moving toward in their science programs. As beginning teachers, they know they will have "expert resources" to turn to when they have a question about teaching or learning.*

- *Curriculum Topic Study has helped us understand how to weave the strands of disciplinary core ideas, practices, and crosscutting concepts together in a meaningful three-dimensional way rather than a force-fit. It has given us a common set of ideas and neutral, third-party perspective to work from at our science department meetings.*

- *I can't imagine designing a workshop for our elementary science teachers without studying the curricular topic first. Yesterday we examined our new instructional materials on matter and its properties. After the study I felt better prepared to explain how the lessons aligned with standards and how conceptual understanding develops from one lesson to the next.*
- *I used to use the first version of Curriculum Topic Study, and I'm so glad to have the new updated second edition. Even though our state has adopted the NGSS, we still find it tremendously helpful to turn to some of the descriptions in other resources to help us clarify our individual interpretations of the performance expectations and deepen our understanding of the three dimensions, STEM connections, and the nature of science.*

What Is Curriculum Topic Study (CTS)?

The above quotes are typical comments from science educators who use Curriculum Topic Study in their work as teachers, science coordinators, curriculum developers, preservice instructors, professional developers, science coaches, and leaders and members of collaborative professional learning communities. Curriculum Topic Study, referred to throughout this book by its acronym, *CTS*, is a methodical study process that uses a common set of resources to help science and STEM educators improve the teaching and learning of science- and STEM-related disciplines. CTS can be used with any set of state standards as well as with the *Next Generation Science Standards* (*NGSS*). By using the CTS process, science and STEM educators do the following:

- Improve their understanding of the science- or STEM-related content they teach.
- Identify the important K–12 ideas and discard irrelevant or unimportant content to maximize time for deeper learning.
- Clarify the learning goals in their curriculum and state standards so that interpretations are consistent with the intent of the learning goal and alignment is strengthened.
- Identify bundles or clusters of related learning goals within a curricular unit that form a network of ideas and practices.
- Improve the coherency of curriculum and instruction within a grade level as well as K–12 vertical articulation.
- Strengthen the alignment between assessment, instruction, and learning goals so that instructional strategies, formative assessment, contexts for learning, and relevant phenomena support effective teaching and deeper learning.
- Understand what makes the learning of some topics difficult for students, increase their awareness of preconceptions students bring to their learning, and become more aware of misconceptions that may develop during instruction.
- Rely less on their own opinions about teaching and learning and ground their decisions in evidence from learning research and a research-based framework for teaching and learning in science.
- Increase opportunities for all students to learn conceptually rather than merely memorizing facts, vocabulary, and disconnected ideas.

The CTS guides are used in a systematic process that engages individuals or groups in a scholarly study of a curricular topic using readings from a collective core set of professional science education resources. These readings have been vetted in advance and then organized into "Curriculum Topic Study Guides" (see example of a K–12 CTS guide for Chemical Reactions in Figure 1.1). The specific features of a CTS guide are described in Chapter 2.

FIGURE 1.1 Example of a K-12 CTS guide

Chemical Reactions
Grades K–12 Standards- and Research-Based Study of a Curricular Topic

Section and Outcome	Selected Sources and Readings for Study and Reflection Read and examine **related** parts of
I. **Content Knowledge**	**IA:** *Science for All Americans* • Ch. 4: Structure of Matter, pp. 46–48 • Ch. 10: Understanding Fire, pp. 153–155 **IB:** *Science Matters* (2009 edition) • Ch. 7: Chemical Reactions, pp. 121–124 **IC:** *Framework for K–12 Science Education:* Narrative Section • Ch. 5: PS1.B: Chemical Reactions, pp. 109–110 **ID:** *The NSTA Atlas of the Three Dimensions:* Narrative Page • 3.2: Chemical Reactions and Nuclear Processes (PS1.B)
II. **Concepts, Core Ideas, or Practices**	**IIA:** *Framework for K–12 Science Education:* Grade Band Endpoints • Ch. 5: PS1.A: PS1.B: Chemical Reactions, pp. 110–111 **IIB:** *Next Generation Science Standards:* Disciplinary Core Ideas Column • Grade 2: PS1.B: Chemical Reactions, p. 13 • Grade 5: PS1.B: Chemical Reactions, p. 28 • MS: PS1.B: Chemical Reactions, pp. 35, 37 • HS: PS1.B: Chemical Reactions, p. 70 **IIC:** *NSTA Quick Reference Guide to the NGSS K–12:* Disciplinary Core Ideas Column • Grade 2: PS1.B: Chemical Reactions, p. 97 • Grade 5: PS1.B: Chemical Reactions, pp. 112–113 • MS: PS1.B: Chemical Reactions, pp. 122–124 • HS: PS1.B: Chemical Reactions, pp. 147–148
III. **Curriculum, Instruction, and Formative Assessment**	**IIIA:** *Disciplinary Core Ideas: Reshaping Teaching and Learning* • K–2: Core Idea PS1, p. 17 • 3–5: Core Idea PS1, pp. 20–21 • MS: Core Idea PS1, pp. 23–24, 28 • HS: Core Idea PS1, pp. 26–27 **IIIB:** *Uncovering Student Ideas:* Assessment Probe and Suggestions for Instruction • USI.1 (2018 edition): Rusty Nails, pp. 93, 98 • USI.4: Burning Paper, pp. 23, 28; Nails in a Jar, pp. 31, 36 • USI.K-2: Back and Forth, pp. 63, 66 • USI.PS3: Will It Form a New Substance? pp. 131, 135–136; What Is the Result of a Chemical Change? pp. 137, 141; What Happens to Atoms During a Chemical Reaction? pp. 143, 146–147

(Continued)

(Continued)

Section and Outcome	Selected Sources and Readings for Study and Reflection Read and examine *related parts* of
IV. Research on Commonly Held Ideas	**IVA: *Benchmarks for Science Literacy:*** Chapter 15 Research • 4D: Chemical Changes, p. 337 **IVB: *Making Sense of Secondary Science: Research Into Children's Ideas*** • Ch. 10: Chemical Change, pp. 85–91 • Ch. 13: Composition of Air and Chemical Interactions of Air, p. 110 **IVC: *Uncovering Student Ideas:*** Related Research • USI.1 (2018 edition): Rusty Nails, pp. 96–97 • USI.4: Burning Paper, pp. 27–28; Nails in a Jar, pp. 35–36 • USI.K–2: Back and Forth, p. 65 • USI.PS3: Will It Form a New Substance? pp. 134–135; What Is the Result of a Chemical Change? pp. 140–141; What Happens to Atoms During a Chemical Reaction? p. 14
V. K–12 Articulation and Connections	**VA: *Next Generation Science Standards:*** Appendices: Progression • Appendix E: PS1.B: Chemical Reactions, p. 7 **VB: *NSTA Quick Reference Guide to the NGSS K–12:*** Progression • PS1.B: Chemical Reactions, p. 62 **VC: *The NSTA Atlas of the Three Dimensions:*** Map Page • 3.2: Chemical Reactions and Nuclear Processes (PS1.B & PS1.C)
VI. Assessment Expectation	**VIA: *State Standards*** • Examine your state's standards **VIB: *Next Generation Science Standards:*** Performance Expectations • Grade 2: 2-PS1-4, p. 13 • Grade 5: 5-PS1-2, 5-PS1-4, p. 28 • MS: MS-PS1-2, MS-PS1-5, MS-PS1-6, p. 37; MS-PS1-3, p. 35 • HS: HS-PS1-2, HS-PS1-4, HS-PS1-5, HS-PS1-6, HS-PS1-7, p. 70 **VIC: *NSTA Quick Reference Guide to the NGSS K–12:*** Performance Expectations Column • Grade 2: 2-PS1-4, p. 97 • Grade 5: 5-PS1-2, 5-PS1-4, pp. 112–113 • MS: MS-PS1-2, MS-PS1-3, MS-PS1-5, MS-PS1-6, pp. 122–124 • HS: HS-PS1-2, HS-PS1-4, HS-PS1-5, HS-PS1-6, HS-PS1-7, pp. 147–148

Review Chapter 2 instructions on how to use this guide.

Visit curriculumtopicstudy2.org for more information about CTS and additional resources.

Additional Readings:

There are 103 CTS Guides provided in this book. The CTS guides address disciplinary topics from life science, physical science, and earth and space science; scientific and engineering practices; crosscutting concepts; and STEM connections that include the nature and enterprises of science, technology, and engineering. Most guides address a topic across the K–12 grade span. Some topics are applicable to a lower grade span, such as Macroscopic Structure and Function of Organisms (Grades K–5) or a higher grade span such as Nuclear Processes (Grades 9–12). A listing of all the CTS topic guides and their descriptions can be found in Chapter 3.

Each CTS guide has six sections listed on the left column that describe the purpose of that section of the CTS study. The column on the right lists relevant professional readings from resources that are available in print and/or online. Many of these resources are often found in science educators' or schools' professional libraries. Several of the CTS resources are available free online. Multiple resources are listed as options for each section so that you may choose from resources that are available or that you are familiar with and prefer. You may decide to use one resource for each section of a guide or use multiple resources to get different perspectives. There is also space at the end of each guide to add your own additional supplementary resources. Figure 1.2 lists the collective resources used for CTS, their publisher, and whether they can be accessed free online.

FIGURE 1.2 CTS collective resources

Resource	Publisher	Free Online
Science for All Americans (AAAS, 1989)	American Association for the Advancement of Science (AAAS)	Yes
Benchmarks for Science Literacy—Chapter 15 Research Summaries (AAAS, 2009)	American Association for the Advancement of Science (AAAS)	Yes
Science Matters (Hazen & Trefil, 2009)	Anchor Books	No
A Framework for K–12 Science Education (NRC, 2012)	The National Academies Press	Yes
Next Generation Science Standards (NGSS Lead States, 2013)	The National Academies Press Achieve or NSTA NGSS Hub	Yes
Next Generation Science Standards: Appendices (NGSS Lead States, 2013)	The National Academies Press Achieve or NSTA NGSS Hub	Yes
Making Sense of Secondary Science: Research Into Children's Ideas (Driver, Squires, Rushworth, & Wood-Robinson, 1994)	Routledge Press	No
The NSTA Quick Reference Guide to the NGSS, K–12 (Willard, 2015)	NSTA Press	No
Uncovering Student Ideas in Science series (Keeley et al., 2007–2019)	NSTA Press	No
The NSTA Atlas of the Three Dimensions (Willard, 2019)	NSTA Press	No

(Continued)

(Continued)

Resource	Publisher	Free Online
Disciplinary Core Ideas: Reshaping Teaching and Learning (Duncan, Krajcik, & Rivet, 2016)	NSTA Press	No
Helping Students Make Sense of the World Using Next Generation Science and Engineering Practices (Schwarz, Passmore, & Reiser, 2017)	NSTA Press	No
State Standards	Available through each state	Yes

CTS provides a process for effective and efficient use of readings from science professional resources. While these readings are to some extent readily accessible, science educators would typically have to sift through a large amount of material to find the information they are looking for. Teachers who are unfamiliar with the professional resources currently used to support and inform science education may not even know where to look for the information they are seeking. The CTS guides identify a purpose for each reading and explicitly point out where to find the information related to that purpose. CTS guides do the vetting groundwork for busy educators.

For all the thought that went into the development of current science standards in each state—whether a state adopted NGSS, modified NGSS, or created their own standards—just having a set of standards is not enough. Standards need to be connected to the direction and shifts science education is currently taking from the classroom on up to the national level. Science educators need access to the research on how students learn science to better understand their own students' thinking related to the standards. Curricular, instructional, and assessment challenges can be better met by understanding the meaning and intent of standards. CTS helps science educators build a bridge between their standards, research, and effective curriculum, instruction, and assessment. It moves standards away from a checklist and into the hands and minds of teachers, leaders, preservice instructors, and professional development providers who now have a process for using them to guide their work in curriculum, instruction, or assessment.

Why Study a Curriculum Topic?

Recent shifts in national and state standards are driving changes in science curriculum, instruction, and assessment. While the content taught in schools has typically been organized in the curriculum by topics, it is essential for educators to make a bridge between these topics and the disciplinary core ideas, practices, and crosscutting concepts laid out in the *Framework for K–12 Science Education* (NRC, 2012) and used to develop the *Next Generation Science Standards* (NGSS Lead States, 2013). To do this, science educators need to understand the content, organization, and pedagogical implications of those specific ideas, practices, and crosscutting concepts. CTS provides a systematic and methodical process to accomplish this.

As districts strive to develop a coherent curriculum of core science concepts, ideas, and practices across multiple years of school, they are also challenged to figure out how to integrate STEM (science, technology, engineering, and mathematics) into their already overburdened science programs. How can schools authentically integrate STEM disciplines so teachers and students can "work smarter, not harder"? The CTS process can help educators intertwine developmentally appropriate and meaningful STEM-related practices and ideas with contemporary science topics.

As the following quote by a former director of the National Science Foundation's (NSF) Elementary, Secondary, and Informal Education Division indicates, teachers who are unfamiliar with the topics they teach tend to rely on textbooks, teach in a more didactic way, and often fail to make connections between important ideas in science:

> [When] teachers cover topics about which they are well-prepared, they encourage student questions and discussions, spend less time on unrelated topics, permit discussions to move in new directions based on student interest, and generally present topics in a more coherent way, all strategies described as standards-based teaching. However, when teachers teach topics about which they are less well-informed, they often discourage active participation by students, keep any discussion under tight rein, rely more on presentation than on student discussions, and spend time on tangential issues. (Kahle, 1999)

Although this quote preceded the release of the *Framework for K–12 Science Education* (NRC, 2012) and the *Next Generation Science Standards* (NGSS Lead States, 2013) two decades ago, it is still applicable today. Furthermore, the significant shifts made in current standards require a deeper level of engagement to translate standards into effective three-dimensional teaching and learning that incorporates relevant phenomena.

A plethora of general professional resources for teachers are available in schools, districts, and professional development settings. These "one-tool-fits-all disciplines" resources can be useful to teachers who know the content and structure of their disciplines well and are familiar with the cognitive research base on how students think and learn. But they fall short for novice teachers, generalists who teach all content areas, and others who may not have a sufficient knowledge base in the content or pedagogy of the science or STEM disciplines they teach.

For example, some general resources encourage teachers to use common misconceptions to construct questions or tasks for assessment purposes. What if teachers do not know what commonly held ideas students are likely to bring to their learning or develop during instruction? What if teachers hold similar misconceptions about the content? CTS provides content-specific tools to identify misconceptions that have been cited in research studies as well as examples of formative assessment probes designed to reveal the alternative ideas students may have about a science topic. CTS provides the science-specific information needed to make the general tools and processes utilized by school districts more effective.

In addition to adding science specificity to the general resources used in schools, teacher leaders and educators who work with in-service or preservice teachers benefit from using CTS by

- Having a common language and knowledge base about teaching and learning regardless of the school's demographics or science curriculum

- Making informed, evidence-based decisions as current programs are reshaped to reflect new standards and curriculum shifts
- Promoting collegiality among groups of colleagues engaged in intellectual discourse about science teaching and learning
- Providing a greater content focus to professional development activities

By taking the time to study a topic before planning a unit or lesson, teachers build a deeper understanding of the content they teach and effective ways to help students understand, apply, and transfer that content. Such a study process is right in line with commonly used curricular approaches such as "backwards design." Backwards design, popularized in Wiggins and McTighe's (2005) *Understanding by Design*, begins by identifying evidence of meeting desired standards before planning teaching and learning experiences. The backwards design model helps teachers determine what ideas and processes in a topic are worth teaching and learning and what it looks like when students demonstrate understanding. It emphasizes the importance of providing opportunities for students to actively construct meaning and to uncover abstract and counterintuitive ideas students may have related to the curricular topic.

Although processes such as backwards design are powerful when adopted by districts, it is often assumed that teachers using backwards design are knowledgeable about the science content, know what the most important ideas are, can make the connections that reflect the knowledge structure of the discipline and support student learning, and are aware of the commonly held ideas students are likely to bring to their learning. Unfortunately, this is not always the case. In reality, some teachers may not be aware of what the scientific community currently recognizes as the most important ideas in K–12 science. Some teachers fall back on what they were taught or what may be included in outdated or poorly developed instructional materials. Some teachers have had limited science coursework in their college education or few in-service professional development opportunities in science. Middle and high school science teachers who specialize in one discipline may not be as knowledgeable in other disciplines of science. Some teachers are not aware of research-identified alternative ideas students are likely to hold, and some teachers may harbor those very same misconceptions.

The critical, and often overlooked, first step in any effective curricular and instructional design involves having a clear understanding of the specific concepts, ideas, and practices that students need to learn, how they intertwine in three-dimensional learning, and the pedagogical implications for how students learn them. A careful study of a topic clarifies the "end in mind" and provides a guide for planning that combines recommendations from standards- and research-based publications with the wisdom of teacher practice. Without a process to compare their practice with current information from standards and research, teachers are likely to continue doing what they have always been doing. Recall the old adage, "If you always do what you've always done, you'll always get what you've always gotten."

All too often, science teachers strike out on their own, when a wealth of information and resources, carefully thought through by highly respected scientists and accomplished educators, sits at their fingertips. Even those teachers who understand the science behind the topics they teach can benefit from CTS. Knowing the content is distinct from knowing how to organize and represent it for student learning. Indeed, a deep familiarity with the content of a science topic can sometimes make it more difficult to identify difficulties a novice learner is likely to have. Designing learning experiences and facilitating learning requires a specialized kind of knowledge called *pedagogical content knowledge*. Shulman defined this knowledge back in 1986 as the special knowledge teachers have about teaching and content that makes the learning of specific topics easy or difficult for learners and that teachers use to develop strategies for representing and formulating content so it is accessible to learners. Through CTS, teachers with strong content backgrounds can gain new insights about specific concepts, ideas, or practices at a specific grade level that may have been overlooked; connections within and across topics; relevant phenomena for learning; developmental considerations based on learning progressions for introducing new ideas; and effective instructional strategies and assessment techniques. Thus CTS can help teachers at all levels, from preservice teachers to highly accomplished veteran teachers, continuously deepen and extend their pedagogical content knowledge.

WHY FOCUS ON TOPICS?

To understand why this book focuses on topics, it is important to clarify what we mean by *topic*. CTS regards topics as the broad organizers for a collective set of related concepts, ideas, and practices taught within a single lesson or a broader unit of instruction. Teachers often design instruction by organizing learning goals into unit topics. Topics provide an organizer for combining disciplinary core ideas, scientific and engineering practices, and crosscutting concepts. The *NGSS* also organizes the disciplinary core ideas and performance expectations by topic. One can access the *NGSS* either by topic or by disciplinary core idea. Figure 1.3 shows how the middle school *NGSS* can be accessed by topic on the NSTA NGSS Hub at http://ngss.nsta.org/AccessStandardsByTopic.aspx.

FIGURE 1.3 Middle school NGSS accessed by topic

Middle School		
Life Science	**Earth & Space Science**	**Physical Science**
Middle School Life Science Introduction	**Middle School Earth & Space Science Introduction**	**Middle School Physical Science Introduction**
MS. Structure, Function, and Information Processing	MS. Space Systems	MS. Structure and Properties of Matter
MS. Matter and Energy in Organisms and Ecosystems	MS. History of Earth	MS. Chemical Reactions
MS. Interdependent Relationships in Ecosystems	MS. Earth's Systems	MS. Forces and Interactions
MS. Natural Selection and Adaptations	MS. Weather and Climate	MS. Energy
MS. Growth, Development, and Reproduction of Organisms	MS. Human Impacts	MS. Waves and Electromagnetic Radiation
Middle School Engineering Design Introduction		
MS. Engineering Design		

NOTE: Graphic used with permission from NSTA. NGSS Hub at https://ngss.nsta.org/AccessStandardsByTopic.aspx.

Making this distinction of *topic* clear is important when using CTS. The intent of CTS is not to focus on teaching the topic per se but rather to focus on how curriculum, instruction, and assessment are organized around a coherent, related set of concepts, ideas, and practices. Some CTS topics have a fairly small grain size, such as Food Chains and Food Webs. There are also broad topics—such as Cycling of Matter and Flow of Energy in Ecosystems—of which the Food Chains and Food Webs CTS topic could be considered a subtopic. For a topic to be included as a CTS guide, it must include important content found in the *Framework for K–12 Science Education*. This content is usually found in the *NGSS* as well, although some ideas are not included or moved to a different grade span in the *NGSS*.

Several of the titles of the CTS guides match the components of the core ideas in the *Framework for K–12 Science Education* and the *NGSS*. Some examples are Chemical Reactions, Natural Selection, Relationship Between Energy and Forces, and Weather and Climate. There are CTS guides that match the scientific and engineering practices such as Developing and Using Models and guides that break the practices down even further, such as Defining Problems or Constructing Scientific Explanations. There are guides that match each of the seven crosscutting concepts in the *Framework for K–12 Science Education* and the *NGSS*, such as Cause and Effect.

There are also guides that can be used to make STEM connections in science, engineering, or technology, such as Engineering Design or Relationship Between Science, Engineering, and Technology. There are also guides that can be used to embed the nature of science into curriculum and instruction, such as Science as a Way of Knowing or Hypotheses, Theories and Laws.

CTS does not include contextual topics such as Butterflies, Dinosaurs, Balls and Ramps, and Rain Forests. These are considered themes. Unlike the CTS topics, by themselves, they are not considered critical content. When developing themes for curricular units, it is important to select a CTS guide that focuses on the core content that students would learn through that theme. For example, if an elementary theme is Balls and Ramps, a CTS guide that would help teachers focus on the critical content rather than the theme would be the CTS guide *Force and Motion*.

There are 103 topic study guides included in Chapter 5. But this does not imply that all 103 of these topics are addressed separately in the K–12 curriculum. Many of the CTS topics overlap, such as Weather and Climate and Global Climate Change. Some guides include traditional topic organizers (such as Organs and Systems of the Human Body); others reflect the organization of the *Framework for K–12 Science Education* and the *NGSS* (such as Group Behaviors and Social Interactions in Ecosystems). Some topics are subsumed by broader topics, such as Conservation of Matter in Properties of Matter. There are K–12 topic guides as well as guides that are organized for younger students, such as Characteristics of Living Things for Grades K–5 or Formation of the Earth, Solar System, and the Universe for Grades 6–12. This book provides enough examples of common curricular topics so that you can find a guide that best matches the topic you or the teachers you work with teach.

CTS Impacts Beliefs About Teaching and Learning

CTS draws out educators' knowledge and beliefs about a topic and how students learn it. The CTS process helps educators examine how new ideas and beliefs gained through CTS fit or do not fit with their previous ideas and beliefs. One way people change or reinforce their ideas or beliefs is through discourse in a social setting. CTS provides an opportunity for educators to engage in evidence-based discussions that may counter or reinforce one's beliefs. For example, elementary teachers who previously focused only on macroscopic properties in their matter unit may find through CTS that young children are capable of using abstract ideas about particles to explain phenomena. There is now strong evidence that young children are more capable of learning more abstract or challenging ideas than was once thought and that teaching some ideas can begin as early as kindergarten (NRC, 2007). Some teachers may find that "doing science" involves more than hands-on activities and labs—explanation construction, modeling, and argumentation are key practices in "doing science." High school teachers may find that they sometimes need to revisit an idea from an earlier grade, such as conservation of matter during a physical change, due to a strongly held naïve idea that may have never been surfaced and worked through. Many teachers find that some of their favorite activities they hold near and dear have little or no substantive connection to what is considered most important for students to learn or could contribute to or reinforce commonly held misconceptions.

Engaging in CTS discussions with colleagues is much like evidence-based argumentation in the science classroom. Supporting ways of thinking about curriculum instruction and assessment by citing evidence from CTS resources and resolving the dissonance between long held beliefs and new ways of thinking about teaching and learning is a powerful way to impact teacher practice.

Examining the research on learning as an integral part of the CTS process can have a strong impact on teachers' beliefs about teaching and learning. Significant shifts in beliefs and how middle and high school teachers saw themselves in their role as teachers were observed during the NSF-funded Curriculum Topic Study project that produced the first edition of this book. Teachers realized that just because they taught something did not mean their students learned it. They realized their students came to the classroom with preconceived ideas and that these ideas needed to be surfaced and worked through for students to learn. From their experiences with CTS, they moved away from the idea that some students were more capable of learning science than others to embracing the belief that all students can learn science if we start with where the student is, not where the teacher wants to go.

Teaching is a complex and demanding profession. It is enhanced when teachers operate from a body of specialized professional knowledge that is based on research on effective teaching and learning. CTS provides such a professional knowledge base. Teachers who use CTS can describe or defend their choices of curriculum materials and instructional strategies. One notable observation made when working with teachers who use CTS is hearing how their conversations become more grounded in evidence from research. For example, instead of sharing their own opinions, teachers are frequently heard to say, "According to the research, it is common for students to have trouble with this concept. What can we do differently to help them?" As they engage in conversations about what is and what is not important to teach, they would say, "Let's look at the core ideas and concepts."

To engage in these types of evidence-based professional conversations, teachers need access to professional resources. They not only need to know which professional resources will help them, but they also need to know where to look in those professional resources. CTS has done that work for the busy teacher. It has identified a core set of resources that teachers can select from and linked vetted readings from those resources to specific teaching and learning purposes, assembled into a CTS guide. In the next chapter you will learn more about the CTS guides and their components.

CHAPTER 2

The Curriculum Topic Study Guide and Its Components

THE CTS STUDY GUIDE

At the heart of the Curriculum Topic Study process is the CTS study guide as shown by Figure 2.2. There are 103 CTS guides in Chapter 5, representing a full range of specific to broad science topics. Figure 2.1 shows how the CTS guides are organized in categories and the number of guides for each category. CTS users select a topic guide and examine, study, and reflect on vetted readings to build their understanding of teaching and learning related to the topic. You may select any of the six sections (I–VI) in a guide, or them all, depending on what your purpose is for doing a CTS. Sections can be studied in any order or approached consecutively. All grade levels can be studied, or you can focus on a specific grade level. Teachers who have used some of the resources in the guides prior to CTS remark how easily and quickly they can now find information relevant to what they need when they use the study guide. CTS guides do the time-consuming work of identifying where to find the relevant information busy science educators need.

FIGURE 2.1 Organization of the CTS guides

Category	# of Guides
Category A: Life Science Guides	**26**
Biological Structure and Function	9
Ecosystems and Ecological Relationships	8
Life's Continuity, Change, and Diversity	9
Category B: Physical Science Guides	**27**
Matter	11
Force, Motion, and Energy	16

(Continued)

13

FIGURE 2.1 (Continued)

Category	# of Guides
Category C: Earth and Space Science Guides	**21**
Earth Structure, Materials, and Systems	8
Earth History and Processes That Change the Earth	6
Earth in Space, Solar System, and the Universe	7
Category D: Scientific and Engineering Practices	**10**
Category E: Crosscutting Concepts	**7**
Category F: STEM Connections	**12**
Engineering, Technology, and Applications of Science	6
Nature of Science	6

FIGURE 2.2 Inheritance of Traits CTS guide

Inheritance of Traits
Grades K–12 Standards- and Research-Based Study of a Curricular Topic

Section and Outcome	Selected Sources and Readings for Study and Reflection Read and examine **related** parts of
I. **Content** **Knowledge**	**IA:** *Science for All Americans* • Ch. 5: Heredity, pp. 61–62 **IB:** *Science Matters* (2009 edition) • Ch. 16: The Code of Life, pp. 273–291 **IC:** *Framework for K–12 Science Education:* Narrative Section • Ch. 6: LS3.A: Inheritance of Traits, p. 158 **ID:** *Disciplinary Core Ideas: Reshaping Teaching and Learning* • Ch. 8: LS3.A: Inheritance of Traits, pp. 146–151 **IE:** *The NSTA Atlas of the Three Dimensions:* Narrative Page • 4.3: Inheritance and Variation of Traits (LS3.A & LS3.B)
II. **Concepts,** **Core Ideas,** **or Practices**	**IIA:** *Framework for K–12 Science Education:* Grade Band Endpoints • Ch. 6: LS3.A: Inheritance of Traits, pp. 158–159 **IIB:** *Next Generation Science Standards:* Disciplinary Core Ideas Column • Grade 1: LS3.A: Inheritance of Traits, p. 10 • Grade 3: LS3.A: Inheritance of Traits, p. 20 • MS: LS3.A: Inheritance of Traits, p. 50 • HS: LS3.A: Inheritance of Traits, p. 85

Section and Outcome	Selected Sources and Readings for Study and Reflection Read and examine **related** parts of
	IIC: *NSTA Quick Reference Guide to the NGSS K–12:* Disciplinary Core Ideas • Grade 1: LS3.A: Inheritance of Traits, p. 94 • Grade 3: LS3.A: Inheritance of Traits, p. 104 • MS: LS3.A: Inheritance of Traits, p. 131 • HS: LS3.A: Inheritance of Traits, p. 158
III. **Curriculum, Instruction, and Formative Assessment**	**IIIA:** *Disciplinary Core Ideas: Reshaping Teaching and Learning* • Ch. 8: How Does Student Understanding of This Disciplinary Core Idea Develop Over Time? p. 155 • Ch. 8: Grades K–2, pp. 155–156 • Ch. 8: Grades 3–5, p. 156 • Ch. 8: Middle School Grades, pp. 156–157 • Ch. 8: High School Grades, p. 158 • What Approaches Can We Use to Teach About This Disciplinary Core Idea? pp. 160–161 **IIIB:** *Uncovering Student Ideas:* Assessment Probe and Suggestions for Instruction • USI.2: Baby Mice, pp. 129, 134–135 • USI.LS: DNA, Genes, and Chromosomes, pp. 129, 132; Eye Color, pp. 135, 138
IV. **Research on Commonly Held Ideas**	**IVA:** *Benchmarks for Science Literacy:* Chapter 15 Research • 5B: Heredity, p. 341 **IVB:** *Making Sense of Secondary Science: Research Into Children's Ideas* • Ch. 5: The Mechanism of Inheritance, pp. 51–52 **IVC:** *Uncovering Student Ideas:* Related Research • USI.2: Baby Mice, pp. 133–134 • USI.LS: DNA, Genes, and Chromosomes, p. 132; Eye Color, p. 138 **IVD:** *Disciplinary Core Ideas: Reshaping Teaching and Learning* • Ch. 8: Challenges to Student Understanding, pp. 158–159
V. **K–12 Articulation and Connections**	**VA:** *Next Generation Science Standards:* Appendices: Progression • Appendix E: LS3.A: Inheritance of Traits, p. 6 **VB:** *NSTA Quick Reference Guide to the NGSS K–12:* Progression • LS3.A: Inheritance of Traits, p. 71 **VC:** *The NSTA Atlas of the Three Dimensions:* Map Page • 4.3: Inheritance and Variation of Traits (LS3.A)

(Continued)

FIGURE 2.2 (Continued)

Section and Outcome	Selected Sources and Readings for Study and Reflection Read and examine **related** parts of
VI. Assessment Expectation	**VIA:** *State Standards* • Examine your state's standards **VIB:** *Next Generation Science Standards:* Performance Expectations • Grade 1: 1-LS3-1, p. 10 • Grade 3: 3-LS3-1, 3-LS3-2, p. 20 • MS: MS-LS3-1, MS-LS3-2, p. 50 • HS: HS-LS3-1, p. 85 **VIC:** *NSTA Quick Reference Guide to the NGSS K–12:* Performance Expectations • Grade 1: 1-LS3-1, p. 94 • Grade 3: 3-LS3-1, 3-LS3-2, p. 104 • MS: MS-LS3-1, MS-LS3-2, p. 131 • HS: HS-LS3-1, p. 158

Review Chapter 2 instructions on how to use this guide.
Visit curriculumtopicstudy2.org for more information about CTS and additional resources.

Additional Readings:

SECTIONS AND OUTCOMES COLUMN ON THE LEFT OF EACH GUIDE

Each CTS guide uses a standard template divided into two corresponding parts. The left-hand column, titled Section and Outcome, contains the Roman-numeral-numbered sections I–VI and is labeled with the purpose or outcome of the corresponding reading or readings. The following describes each of the CTS guide sections and their outcomes:

Section I. Content Knowledge: This section helps you improve or build on your science content knowledge. It identifies what all adults (including science teachers) should know and be able to do as a result of their K–12 science education, regardless of whether they go on to study science in post-secondary education. It also clarifies science concepts encountered in the media, public issues, and other everyday science venues.

Section II. Concepts, Core Ideas, or Practices: This section helps you identify the goals for K–12 learning. It includes the three dimensions: specific disciplinary concepts or ideas, scientific and engineering practices, and crosscutting concepts at the grade levels designated by the *NGSS*. It also includes the ideas about the nature and enterprises of science, engineering, and technology.

Section III. Curriculum, Instruction, and Formative Assessment: This section helps you examine considerations for curriculum, instructional suggestions, and formative assessment probes that can be used to understand what students are thinking before and at any point during instruction.

Section IV. Research on Commonly Held Ideas: This section identifies related research summaries that can be used to examine developmental considerations, common misconceptions, partial understandings, and common difficulties students may encounter when learning about the topic.

Section V. K–12 Articulation and Connections: This section helps you examine the K–12 growth in understanding as a coherent progression of ideas that build over time. It reveals important prerequisites for learning as well as connections within and between topics.

Section VI. Assessment Expectation: This section helps you link the previous sections (I–V) to your state or local assessment context. It includes the three-dimensional performance expectations from the *NGSS*. If your state did not adopt NGSS, it can be used to examine similar performance expectations in one's own state standards or it may reveal important assessment concepts, ideas, or practices that can be used to inform development of district or classroom assessments.

At the bottom of each guide is a link to the CTS website where you can find additional information. There is also a space at the end of each guide where you can list additional readings from other resources.

SELECTED READINGS COLUMN ON THE RIGHT OF EACH GUIDE

The right-hand column of a CTS guide, titled Selected Sources for Reading and Reflection, lists readings from the CTS common set of resources that can be used for conducting a study. There are thirteen collective resources in all, used in different CTS guides. Each section provides vetted selections from two or more resources (a few guides may contain only one or no resources for some sections). By providing multiple options from the collective set of resources (both free online or purchased), you will most likely be able to access resources needed for each section of a study guide. You may choose to read from just one of the resources listed in a section, or more than one. The choice depends on your personal preference, which resources you have available to use, and whether any of the resources repeat the same material.

Consider the following scenario: You decide to do a full study for your genetics unit using all six sections of the CTS Guide Inheritance of Traits, shown in Figure 2.2. You have a copy of the *NSTA Quick Reference Guide to the NGSS K–12*, so you decide you will use that for CTS section II, where you will start the study. You decide to focus only on grades 9–12. You read sections IIC to get a sense of what the important core ideas and concepts are for your grade level. In section II, the readings from IIA and IIB duplicate the same material as IIC so you do not need to read those.

Using the same resource, the *NSTA Quick Reference Guide to the NGSS K–12*, you now go to CTS section V and read VB to get a sense of the K–12 progression leading up to your grade level. For section V resources, VA repeats the same chart as VB so you can skip that reading. However, you decide to also use *The NSTA Atlas of the Three Dimensions* (VC) because unlike the progression chart for VB, the VC reading allows you to visually examine the progression of how one idea contributes to another and connections to other ideas.

Moving next to CTS section VI, you continue to use *NSTA Quick Reference Guide to the NGSS K–12* to examine the performance expectations in section VIC, noting how the three dimensions are used together for assessment.

You have an online version of the *Framework for K–12 Science Education* and really like how it describes the science content so you decide to use that resource for CTS section I by reading the narrative listed for section IC. That leaves CTS sections III and IV. You have a copy of the *Uncovering Student Ideas in Science* book listed as USI.2 in sections IIIB and IVC and decide you will use that resource, since it is available to examine a formative assessment probe you can use to uncover your students' ideas, and consider suggestions for instruction that can inform your teaching. You read the research on learning summaries to get a sense of the preconceptions some of your students are likely to bring to their learning.

Last, you access your state's high school biology standards online so you can connect the results of the study to your state context.

The above scenario shows that if you were to study all six sections of this guide, you would only need four of the CTS resources plus access to your state's standards. Two of the resources (*Framework for K–12 Science Education* and your state standards) are free online. You would use these four resources to conduct a study using eight readings from the CTS guide (IC, IIC, IIIB, IVC, VB and VC, VIA and VIC). Confused? As you become more familiar with the resources, you will know which ones are duplicative and which offer a different perspective when used together. This is explained further on in this chapter where the CTS resource books are described. For now, the takeaway is that you do not need all the books, nor do you need to read all the listed readings for each section.

Page numbers are provided for each reading. Heed the line under Selected Resources for Study and Reflection that alerts you to examine the **related** parts of the reading. This means that sometimes there is material included on the pages that is hard to separate out from the topic you are focused on. We tried to keep the readings as concise as possible. If you find useful information beyond the page numbers listed, you are certainly welcome to include that in your study as long as it is related to the topic and your purpose for the study.

Some sections use source material in only certain categories of CTS guides. For example, the resource *Disciplinary Core Ideas—Reshaping Teaching and Learning* is only used with guides for disciplinary content topics in the life, physical, earth, and space sciences topics and the Engineering, Science, and Technology guides. *Helping Students Make Sense of the World Using Scientific and Engineering Practices* is only used with scientific and engineering practices CTS topic study guides. The *Framework for K–12 Science Education* is only used in section III for guides on the scientific and engineering practices and the crosscutting concepts. It is not used in section III for the disciplinary content topic guides. Figure 2.3 lists all the source material for reading and reflection included in the entire collection of CTS study guides. The asterisk indicates duplicative material, meaning out of all the resources listed in that section, choose only one from the ones marked with an asterisk because they contain the same material. Parts of the *Framework* are the same as other section II resources but are not noted here as duplicative since some decisions were made to not include some *Framework* ideas in the *NGSS* or they were moved to a different grade level.

FIGURE 2.3 Links between outcomes and source material

If your outcome for studying a topic is to:	Then you would read related parts of:
Improve or enhance your science and engineering content knowledge (Section I)	*Science for All Americans*
	Science Matters
	Framework for K–12 Science Education* (for practices and crosscutting concepts)
	Disciplinary Core Ideas: Reshaping Teaching and Learning
	Helping Students Make Sense of the World Using Scientific and Engineering Practices
	NSTA Quick Reference Guide to the NGSS K–12*
	The NSTA Atlas of the Three Dimensions*

(Continued)

FIGURE 2.3 (Continued)

If your outcome for studying a topic is to:	Then you would read related parts of:
Identify the disciplinary core ideas, core concepts, scientific and engineering practices, and crosscutting concepts (Section II)	*Framework for K–12 Science Education* *Next Generation Science Standards** *NSTA Quick Reference Guide to the NGSS K–12**
Examine considerations for curriculum, instruction, and formative assessment (Section III)	*Disciplinary Core Ideas: Reshaping Teaching and Learning* *Helping Students Make Sense of the World Using* *Scientific and Engineering Practices* *Framework for K–12 Science Education** *Uncovering Student Ideas in Science* *NSTA Quick Reference Guide to the NGSS K–12**
Examine the research on commonly held preconceptions and difficulties students may have during the learning process (Section IV)	*Benchmarks for Science Literacy:* Chapter 15 *Making Sense of Secondary Science* *Uncovering Student Ideas in Science* *Disciplinary Core Ideas: Reshaping Teaching and Learning* *Helping Students Make Sense of the World Using Scientific and Engineering Practices*
Examine how ideas and practices progress across grade spans and the connections between ideas (Section V)	*Next Generation Science Standards** *NSTA Quick Reference Guide to the NGSS K–12** *The NSTA Atlas of the Three Dimensions*
Examine the performance expectations at the local, state, and national level (Section VI)	local assessments state standards *Next Generation Science Standards** *NSTA Quick Reference Guide to the NGSS K–12**

*Duplicative material

ADDING YOUR OWN SUPPLEMENTARY MATERIAL

Each guide lists suggested readings by page number. Some of the resources include several additional pages beyond what is listed that could make the reading too unwieldy. Thus when vetting the resources, we tried to pull out the most essential information. You should not feel limited by only the pages suggested in the guide. If you feel there is more information related to the topic that would be useful to you, feel free to read beyond the designated page numbers. If you are facilitating a CTS study, list the additional pages you feel would be useful to include for your participants. For example, in the CTS guide Constructing Scientific Explanations, there are several additional pages that could be read for section IIIC that are not listed on the guide.

Notice at the end of each guide there is a space where you can list additional resources you might come across to add to the study. For example, the *NGSS Bozeman Science* videos at http://www.bozemanscience.com/next-generation-science-standards/ can be used to supplement CTS section I. At the time of this CTS publication, the NSTA Press book *Crosscutting Concepts—Strengthening Science Learning*, edited by Jeffrey Nordine and Okhee Lee, was still in development and not published in time to include for the Crosscutting Concepts CTS guides. When this book is released, readings for sections I and III can be added. Articles appearing in NSTA's journals, *Science and Children*, *Science Scope*, and *The Science Teacher*, as well as other journal articles can be added as additional resources.

Also notice on some guides there is another topic study guide listed that could be combined with the guide you are using. For example, the Category A guide Cycling of Matter and Transfer of Energy in Ecosystems can be combined with the Category E guide Energy and Matter: Flows, Cycles, and Conservation. The first guide focuses on disciplinary core ideas. The second guide focuses on the crosscutting concepts. If readings from both guides were combined into one guide, it may make the process long and unwieldy. However, you are encouraged to look at the other guide because there may be some sections you could add.

CTS WEBSITE

At the end of each guide there is also a website, curriculumtopicstudy2.org, where you can access additional information about CTS. The website includes links to the resources and where you can purchase them or access them online. It also includes a downloadable CTS template so that you can customize your own CTS guide. The website includes ongoing information about CTS professional development and ways science educators have used CTS.

COLLECTIVE RESOURCES USED IN THE CTS GUIDES

Now that we have summarized the components of a CTS guide, it is time to become more familiar with each of the CTS resources. All professionals—lawyers, doctors, mechanics, accountants—have their own career-specific reference tools that inform their practice. Science educators have their own unique professional reference tools as well. The set of books used for CTS can be considered science teachers' "tools of the trade" and can often be found in teachers' or schools' professional libraries.

BUILDING A PROFESSIONAL COLLECTION: EXPERTS AT YOUR FINGERTIPS 24/7!

Several source materials used in CTS are developed, published by, and available to science educators through NSTA Press and the National Academies of Science. All the books listed in the CTS collection of resources are available through various booksellers. Some of the CTS resources precede the release of the *Framework for K–12 Science Education* and the *NGSS*, yet are still relevant and useful today. Some resources, such as *Disciplinary Core Ideas — Reshaping Teaching and Learning*, have been published to support the understanding and implementation of the current vision of K–12 science education laid out in the *Framework for K–12 Science Education* and the *NGSS*. Do not dismay if you are not familiar with all these resources. In our experience, many teachers do not even know about some of these resources or if they do know about them, they have never used or looked at them. Often, it is only this lack of awareness that prevents teachers from obtaining their own copies or copies for their school's professional library.

The source materials listed in Figure 2.3 make up a comprehensive "suite of standards and research-based resources" used by the science education community. For teachers who do not have their own copies and may not have funds available to purchase them for their classroom or school, many of these materials are available free online. See Chapter 3 for more information on accessing these resources online. Some teachers and schools have these resources on science standards, curriculum, instruction, and formative assessment but they "sit on the shelf" because they have never had a process for using them.

Teachers and school districts that have used CTS in the past have often purchased the selected resources, realizing how important it is for CTS to be used by teachers and curriculum specialists to support teaching and learning in science. CTS gets the standards- and research-based publications into the hands of science educators and into schools and classrooms, where they are deliberately and routinely used. The advantage of the CTS guide design is that teachers can select from multiple options for resources based on their needs, purposes, and availability of the source materials.

A good justification for having the resources used for CTS is that you will have experts at your fingertips at all times! Teachers who have used CTS refer to the use of the resources as "standing on the shoulders of giants." Imagine being able to vicariously turn to advice from leading experts whenever you need it! That is what CTS does: It provides users with an external perspective derived from some of the best thinking in the science education world.

Because resources such as the *Framework for K–12 Science Education* (NRC, 2012) were developed through peer review, discussion, and consensus by leading scientists and science educators from K–12 schools, universities, science organizations, research laboratories, and governmental agencies, they reflect the best thinking available about what is important to learn in science and how and when to effectively teach it. The research summaries have been conducted by science education researchers, who have dedicated their work to understanding how students learn. CTS shifts the burden from relying solely on your own intuition, opinion, and experience to providing sound recommendations and evidence-based information that can be combined with the wisdom of your own practice. Teachers we have worked with remark, "Now that I have these resources and the CTS guides to use them, I can't imagine being a professional science teacher without them!"

DESCRIPTIONS OF THE COMMON RESOURCES USED IN CTS

Some of these resources may be more familiar to science educators than others. A few have been around for the last two decades, others were published after 2012. Most of the resources used in CTS are available through NSTA Press at www.nsta.org/publications/press/, National Academy Press at nap.edu, amazon.com, or major booksellers. You can check the CTS website at curriculumtopicstudy2.org for links to the resources and suggestions for ordering them. Remember that a key feature of CTS is its flexibility. You can use as many or as few of the resources as your purpose and access demands. The following is a description of the collection of resources used in the CTS guides and the corresponding section(s) in which they are used. Note that the sections listed do not always apply to all the guides. For example, the *Framework for K–12 Science Education* is only used in CTS section III for the category D and category E guides, and not the other guides.

Science for All Americans (*SFAA*): CTS Section I

SFAA, authored by James Rutherford and Andrew Ahlgreen for the American Association for the Advancement of Science (AAAS), was first published in 1989 (AAAS, 1989). It represents the first phase of AAAS's science reform initiative, Project 2061. *SFAA* is a seminal document that remains relevant to this day. While there have been many recent advances in modern science, most of the fundamental big ideas in science remain the same. *SFAA* defines the enduring, interconnected knowledge all adults should have acquired after their K–12 education to ensure basic science literacy. Science literacy as described in *SFAA* includes science, technology, engineering, mathematics, and social sciences and the interconnections among them. It reflects the belief that

> [t]he scientifically literate person is one who is aware that science, mathematics, and technology are interdependent enterprises with strengths and limitations; understands key concepts and principles of science; is familiar with the natural world and recognizes both its diversity and unity; and uses scientific knowledge and scientific ways of thinking for individual and social purposes. (AAAS, 1989 p. 4)

Rutherford, F. J., & Ahlgren, A. (1990). *Science for all Americans*. New York: Oxford University Press. Used with permission from the American Association for the Advancement of Science.

SFAA softens the boundaries between traditional content domains and emphasizes the big ideas and connectedness of science topics, which come together to form a complete picture of science literacy. The scientific terminology used in *SFAA* is the terminology all adults are expected to be familiar with to understand science, mathematics, and technology in their everyday lives.

Teachers who have used SFAA praise its eloquent and easy-to-understand prose. A high school teacher, remarking on how beautifully written SFAA is, once told us, "*Science for All Americans* is about as close as you can get to poetry in science." Teachers who have never used it before quickly embrace it and have commented that it is a book you can open up and start reading anywhere, for pleasure as well as professional use. Many of the ideas in current standards documents can still be traced back to *SFAA*, which initiated the development of standards in science during the early and mid-1990s. Figure 2.4 illustrates the influential role of SFAA in the development of standards.

FIGURE 2.4 The Influence of SFAA

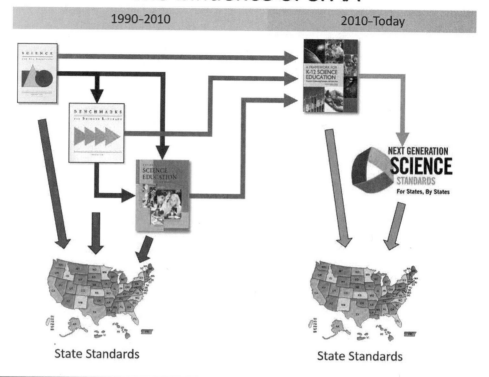

US Map Source: iStock.com/fourleaflover

Science Matters: Achieving Scientific Literacy, (2009), by Dr. Robert Hazen and Dr. James Trefil, Penguin/Random House. Used with permission.

Science Matters: Achieving Scientific Literacy: CTS Section I

Science Matters, first published in 1989, and updated in 2009 (version used in CTS), is authored by two renowned scientists, Dr. Robert Hazen and Dr. James Trefil. It is one of the few books that could be considered a compendium of the big ideas in science. *Science Matters* describes the knowledge needed to be an informed, decision-making citizen around issues related to science in the public arena and to cope with understanding concepts in the complex world of science and technology we live in. It differs from *SFAA* because it does not describe only the knowledge that culminates from a K–12 education. Some concepts and ideas described in *Science Matters* exceed basic science literacy but are included since they may appear in the media or other public venues (e.g., one may encounter the concept of dark matter when listening to a science podcast). The authors advocate for the importance of scientific literacy for all adults, not just scientists, using three lines of reasoning:

1. The argument from civics: Every citizen will be faced with public issues whose discussion requires some scientific background;

2. The argument from aesthetics: We live in a world that operates according to a few general laws of nature; and

3. The argument from intellectual connectedness: Scientific findings play a crucial role in setting the intellectual climate of an era (consider how Copernicus's discovery of the heliocentric universe played an important role in ushering in the Age of Enlightenment). (Hazen and Trefil, p. xvi)

The first edition of *Science Matters* was based on eighteen "great ideas" of science that Hazen and Trefil used to frame the natural world and advances in technology. When the second edition was updated in 2009 (the version used in CTS), the authors noted that the core concepts of science had not changed (Newton's Laws are still Newton's Laws) and that even with new advances in nanotechnology, archaea, LEDs, cloning, dark matter, extrasolar planets, and so on, these new findings fit into the *great ideas*. While the content of the chapters was updated, the authors decided only one new chapter needed to be added—a chapter on the advances in biotechnology, confirming the value of the *great ideas* approach to achieving scientific literacy.

The descriptions in *Science Matters* are interesting and comprehensible to those who have limited science backgrounds. It avoids the stilted, fragmented descriptions typical in textbooks and provides examples of phenomena, analogies and metaphors, and other clever ways to explain a core idea. Several of the teachers we have worked with have commented that they have used some of the descriptions and explanations in the book in their lesson designs.

A Framework for K-12 Science Education: Practices, Crosscutting Concepts, and Core Ideas (Framework): CTS Sections I, II, III

The *Framework* was published in 2012 by the National Research Council (NRC). The *Framework* can be thought of as the foundation for the construction of new science standards, including the *NGSS*. The *Framework* highlights the power of integrating a three-dimensional understanding of the core ideas of science with engagement in the practices of science and engineering and use of crosscutting concepts. The intent of the *Framework* is to promote a coherent, cohesive approach to teaching and learning, with a de-emphasis on facts and terminology, and a focus on depth over breadth. The *Framework* is based on a growing body of research on teaching and learning in science, articulates the foundational knowledge and skills for K–12 science and engineering, and organizes goals for learning using the three dimensions: Scientific and Engineering Practices, Crosscutting Concepts, and Disciplinary Core Ideas.

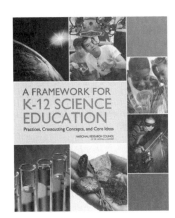

National Research Council. 2012. *Framework for K-12 Science Education: Practices, Crosscutting Concepts, and Core Ideas*. Washington, DC: The National Academies Press. https://doi.org/10.17226/13165. Used with permission.

The *Framework* is the published result of the NRC's Committee on a Conceptual Framework for New K–12 Science Education Standards findings, which concluded:

> K–12 science and engineering education should focus on a limited number of disciplinary core ideas and crosscutting concepts, be designed so that students continually build on and revise their knowledge and abilities over multiple years, and support the integration of such knowledge and abilities with the practices needed to engage in scientific inquiry and engineering design. (NRC, 2012, p. 2)

Teachers and districts that are implementing new standards are finding that reading the *Framework* provides invaluable insight into the intent of standards statements. The chapters on disciplinary core ideas include framing questions for the core ideas and narratives that clarify the content. It also includes grade band endpoints that describe the specific content students should learn by the end of grades 2, 5, 8, and 12. The chapter on scientific and engineering practices includes narratives that describe how each practice is used in science. It lists the goals for student learning that should be achieved by the end of grade 12 and the types of opportunities students should have to learn about and utilize the practices as they progress from elementary grades to middle school and on to high school. The chapter on crosscutting concepts includes narratives that describe each crosscutting concept, how it bridges disciplinary boundaries, and how each crosscutting concept is fundamental to an understanding of science and engineering. It also includes descriptions of the ways students use the crosscutting concepts as they progress from one grade span to the next.

Next Generation Science Standards (NGSS): CTS Sections II, V, and VI

NGSS Lead States, 2013. *Next Generation Science Standards For States, By States.* Washington, DC: The National Academies Press. https://doi.org/10.17226/13165 NGSS is a registered trademark of Achieve. Neither Achieve nor the lead states and partners that developed the Next Generation Science Standards were involved in the production of this product, and do not endorse it. Image reprinted with permission from NGSS, nextgenscience.org.

The *NGSS* were released in 2013, authored by forty-one writers through a consortium of twenty-six states. The three-year development process was carried out in collaboration with the National Research Council, National Science Teachers Association, American Association for the Advancement of Science, and Achieve, Inc.

The *Framework for K–12 Science Education* served as the foundation for the *NGSS*, and the standards are arranged across the life, physical, and earth and space science disciplines and engineering design, with the intent of providing all students an opportunity to experience science concepts and ideas through the practices of science. The *NGSS* are arranged in two ways: by topic or by disciplinary core idea. The *NGSS* include connections to engineering and technology and nature of science and are aligned by grade level and cognitive demand with the English language arts and mathematics Common Core State Standards.

The *NGSS* performance expectations incorporate three dimensions that provide a context for how the science idea is acquired and understood and how it is connected to a more universal understanding of science. The three dimensions are consistent with the *Framework*, including the Disciplinary Core Ideas (DCIs), Science and Engineering Practices (SEPs), and Crosscutting Concepts (CCCs).

When used with CTS sections II and VI, the *NGSS* details the grades K–12 performance expectations for life, physical, earth and space science, and engineering design and identifies in columns beneath the performance expectations the three-dimensional components that make up the performance expectation. Clarifications and boundaries for the performance expectations are also provided. Grades K–5 performance expectations and each of the three dimensions are articulated by individual year, while the middle and high school components are defined by grade spans (6–8 and 9–12). The summaries preceding each set of performance expectations provide useful glimpses of what an *NGSS* classroom might look like.

Included with the *NGSS* and used in CTS section V are appendices that further clarify the intent of the *NGSS*. Whether working as an individual, small group, or district team, it is beneficial to examine and discuss the conceptual shifts in the *NGSS* (Appendix A); disciplinary core ideas progression (Appendix E); student capabilities for engaging in the practices of science for grade bands K–2, 3–5, 6–8, and 9–12 (Appendix F); understanding of the nature of science in the *NGSS* (Appendix H); and considerations for engineering design in the *NGSS* (Appendix I).

NSTA Quick Reference Guide to the NGSS, K-12: CTS Sections I, II, III, V, VI

Since the 2013 release of the *NGSS*, the National Science Teachers Association (NSTA) has played a leading role in helping educators become familiar with the standards. The NGSS@NSTA Hub (http://ngss.nsta.org/) provides links to the performance expectations, the three dimensions that make up the performance expectations, the *NGSS* appendices, overviews of the standards, and classroom and professional resources, and serves as an organized go-to page for all things *NGSS*.

Cover used with permission from NSTA.

While the NGSS@NSTA Hub is a very useful online resource, many teachers, curriculum and professional developers, and administrators were looking for a publication that compiled the primary materials in hard copy in an easily accessible way. In response to this need, NSTA Press published *The NSTA Quick-Reference Guide to the NGSS, K–12* in 2015. Edited by Ted Willard, the guide compiles the same information included in the *Framework* and *NGSS* that describe and list the science and engineering practices, disciplinary core ideas, and crosscutting concepts; NGSS performance expectations; and progressions across grade spans, and compares the *NGSS* practices to the Common Core State Standards in mathematics and English language arts. It is the one resource that provides the same information one would get from the *NGSS* when using CTS so it is not necessary to have and use both books. Either one can be used for CTS. It also includes narrative sections from the *Framework* that are used in CTS sections I and III for the category 4 and category 5 CTS guides and a progression that can be used for CTS section V. If one were to choose the resource that is most versatile and could be used for several sections of a CTS guide, it would be the *NSTA Quick Reference Guide to the NGSS, K–12*.

Disciplinary Core Ideas: Reshaping Teaching and Learning: CTS Sections I, III, and IV

Disciplinary Core Ideas: Reshaping Teaching and Learning, edited by Ravit Colan Duncan, Joseph Krajcik, and Ann Rivet and published by NSTA Press in 2017, provides insight into how the teaching of science can be organized around disciplinary core ideas as described in the *Framework* and outlined in the *NGSS*. Helen Quinn, chair of the NRC committee that developed the *Framework*, notes that

Cover used with permission from NSTA.

> [s]tudents must build new knowledge by refining and revising prior knowledge (or their preconceptions, if the topic is new to

them). Teaching that ignores what has come before and does not capitalize on research into what makes a topic difficult to learn is at best inefficient and at worst ineffective. Hence, it is important not just to have a science curriculum for the current year but to have one that is designed to build knowledge and deepen understanding progressively across multiple years. (p. vii)

Disciplinary Core Ideas: Reshaping Teaching and Learning is a useful resource for educators who are examining subject matter through the lens of new standards, wanting to answer questions such as: What are the concepts behind the disciplinary core ideas? Why are they important? How does student understanding of this idea develop over time? What approaches can we use to teach about this core idea at the elementary, middle, and/or high school level? Chapters are organized by the core ideas of physical, life, earth and space science, and engineering, technology, and applications of science.

Disciplinary Core Ideas: Reshaping Teaching and Learning provides important background information, allowing teachers to deepen their understanding of science concepts and clarify the intent of the core ideas for each grade span. When coupled with other resources used in CTS, the conceptual shifts in teaching and learning can be more fully realized.

The drawback to this book is that the chapters are not organized by a common format. Therefore, sometimes the information useful for a CTS topic is scattered across different sections of a chapter. Also, some guides include readings in three sections, and other guides may include readings in only one or two sections. For example, not all chapters included a section on commonly held ideas that can be used with CTS section IV. Also, there is more to read for a topic than just the page numbers designated on the guide. If the guide were to include all the relevant pages, the reading would be quite lengthy, especially if used in a group jigsaw format; therefore, selected pages were included but you are encouraged to read beyond those pages if time and interest allow.

Helping Students Make Sense of the World Using Next Generation Science and Engineering Practices: CTS Sections I and III

Helping Students Make Sense of the World Using Next Generation Science and Engineering Practices, edited by Christina Schwarz, Cynthia Passmore, and Brian Reiser and published by NSTA Press in 2017, offers a variety of real-world examples that show what practice-centered teaching and learning look like at all grade levels. The authors describe the rationale for shifting from scientific inquiry to practices:

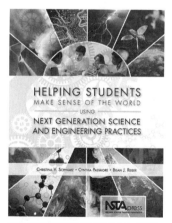

Cover used with permission from NSTA.

> The emphasis on science and engineering practices attempts to build on prior reforms and take advantage of what research has revealed about the successes and limitations of inquiry classrooms. We like to think of the focus on practices as a kind of Inquiry 2.0—not a replacement for inquiry but rather a second wave that articulates more clearly what successful inquiry looks like when it results in building scientific knowledge. The configuration of inquiry classrooms typically allows students to

explore the relationship between two variables ...; but, often this empirical exploration is not taking place in an ongoing process of questioning, developing, and refining explanatory knowledge about the world.... [Taking] our ideas of inquiry in science beyond designing investigations and testing hypotheses has led to the fuller articulation of inquiry as the scientific and engineering practices that enable us to investigate and make sense of phenomena in the world.... (pp. 5–6)

The book addresses three overarching questions: How will engaging students in science and engineering practices help improve science education? What do the eight practices look like in the classroom? How can educators engage students in practices? *Helping Students Make Sense of the World Using Next Generation Science and Engineering Practices* is a useful resource for K–12 science teachers, curriculum developers, and administrators as they strive to help students generate and revise knowledge through the lens of the practices.

Used with CTS section I, this resource helps you understand what the practice is and how it is used in science and/or engineering. Used with section III, it provides extensive information about using the practice in the classroom, accompanied by vignettes or examples from instructional materials, and descriptions of what the practice is **not** about. It also describes how one practice connects to other practices.

Similar to the preceding resource, the drawback to this book is that the chapters are not organized by a common format. Therefore, sometimes the information useful for a CTS topic is scattered across different sections of a chapter. Also, there is more to read for a topic than just the page numbers designated on the guide. If the guide were to include all the relevant pages, the reading would be quite lengthy, especially if used in a group jigsaw format; therefore, selected pages were included but you are encouraged to read beyond those pages if time and interest allow.

Uncovering Student Ideas series (USI): CTS Sections III and IV

Uncovering Student Ideas in Science (*USI*) is a series of books (eleven books as of 2019) published by NSTA Press that includes diagnostic questions designed to uncover and probe what students really think about core disciplinary ideas in science. When these diagnostic questions are used to make informed decisions about curriculum and instruction, they become formative assessment probes. The books also include teacher notes that describe implications for curriculum and instruction as well as specific suggestions for instruction that address commonly held ideas. One valuable use of the teacher notes is the Related Research section. This section summarizes and cites the research on the commonly held ideas students have related to the concept addressed by the probe. Since many teachers do not have access to the research journals cited in the *USI* books, this series of books makes the research accessible to teachers. As more books are published and second editions are released, crosswalks to the CTS topic guides are posted on the CTS website at www.curriculumtopicstudy2.org. The following codes are used to identify which books in the series are used in a CTS guide:

USI.1: *Uncovering Student Ideas in Science: 25 Formative Assessment Probes*, 2nd edition (Keeley, 2018)

Cover used with permission from NSTA.

USI.2: *Uncovering Student Ideas in Science: 25 More Formative Assessment Probes* (Keeley, Eberle, & Tugel, 2007)

Cover used with permission from NSTA.

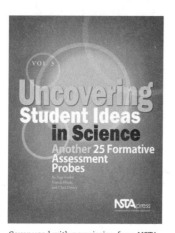

USI.3: *Uncovering Student Ideas in Science: Another 25 Formative Assessment Probes* (Keeley, Eberle, & Dorsey, 2008)

Cover used with permission from NSTA.

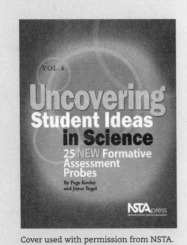
Cover used with permission from NSTA.

USI.4: *Uncovering Student Ideas in Science: 25 New Formative Assessment Probes* (Keeley & Tugel, 2009)

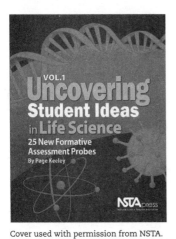
Cover used with permission from NSTA.

USI.LS: *Uncovering Student Ideas in Life Science: 25 New Formative Assessment Probes* (Keeley, 2011)

Cover used with permission from NSTA.

USI.PS1: *Uncovering Student Ideas in Physical Science: 45 New Force and Motion Formative Assessment Probes* (Keeley & Harrington, 2010)

(Continued)

(Continued)

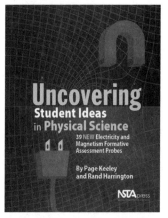

USI.PS2: *Uncovering Student Ideas in Physical Science: 39 New Electricity and Magnetism Formative Assessment Probes* (Keeley & Harrington, 2014)

Cover used with permission from NSTA.

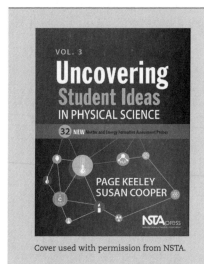

USI.PS3: *Uncovering Student Ideas in Physical Science: 32 New Matter and Energy Formative Assessment Probes* (Keeley & Cooper, 2019)

Cover used with permission from NSTA.

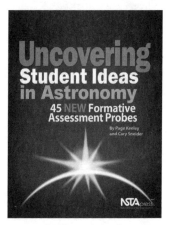

USI.A: *Uncovering Student Ideas in Astronomy: 45 New Formative Assessment Probes* (Keeley & Sneider, 2012)

Cover used with permission from NSTA.

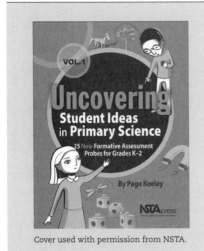

USI.K2: *Uncovering Student Ideas in Primary Science: 25 New Formative Assessment Probes for Grades K–2* (Keeley, 2013)

Cover used with permission from NSTA.

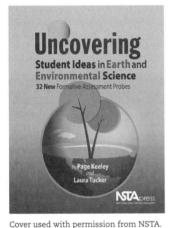

USI.E&ES: *Uncovering Student Ideas in Earth and Environmental Science: 45 New Formative Assessment Probes* (Keeley & Tucker, 2015)

Cover used with permission from NSTA.

Benchmarks for Science Literacy (Benchmarks):
Chapter 15 The Research Base: CTS Section IV

Benchmarks was released in 1993 (updated online in 2009), and chapters correspond to the chapters in *SFAA*. *Benchmarks* describes the specific steps along the way to achieving the science literacy described in *SFAA* by listing specific goals for student learning at K–2, 3–5, 6–8, and 9–12. *Benchmarks* was developed with input from hundreds of educators, scientists, university faculty, and science education specialists. They represented our first coherent set of standards, and many states developed their initial science standards based on the *Benchmarks*. Today the *Benchmarks* have been replaced by the *Framework* and the *NGSS* (*Benchmarks* contributors were involved in the development of

Project 2061 (American Association for the Advancement of Science). (1993). Benchmarks for science literacy. New York: Oxford University Press.

the *Framework* and much of the disciplinary core ideas are based on the *Benchmarks*). Chapter 15 contains the research summaries that informed development and placement of the learning goals. Many of these research summaries are still useful today and are included in CTS section IV.

Making Sense of Secondary Science: Research Into Children's Ideas: CTS Section IV

Driver, Rosalind, Squires, Ann, Rushworth, Peter, and Wood-Robinson, Valerie. 2013. *Making Sense of Secondary Science*. Taylor & Francis. Milton, United Kingdom. Used with permission.

Making Sense of Secondary Science, by Rosalind Driver, Ann Squires, Peter Rushworth, and Valerie Wood-Robinson (1994), is a comprehensive summary of research into students' ideas in life and physical sciences, with some earth and space science. While the title implies upper grades, the book addresses students' ideas from grades K–12. The summaries, which were first published in 1994 by Driver's research group at the University of Leeds, England, followed the release of *Benchmarks* and extended the research findings in *Benchmarks'* Chapter 15. References to the original studies are provided. The book is arranged by topics and corresponds to several of the topics in the CTS guides. This book is a useful resource that helps science teachers deepen their understanding of how students think about major ideas in science and how these ideas affect their learning. The authors explain:

> In planning teaching it is useful for teachers to think in terms of helping pupils to make a number of "small steps" towards the big ideas. The sequencing of these "small steps" can be informed by what is known about the progression of children's understanding. However, it is important to bear in mind that some of these "small steps" may, in themselves, present learners with difficulty. (Driver et al., 1994, p. 13)

Teachers who are interested in and are aware of the importance of identifying students' potential misconceptions as a way to bridge students' ideas with scientific ones have found this book to be a valuable resource. Unfortunately it is difficult to find this book in the United States today. If your district or organization used the first edition of CTS, they may have a copy.

Cover used with permission from NSTA.

The NSTA Atlas of the Three Dimensions: CTS Sections I, III, and V

The *NSTA Atlas*, by Ted Willard (NSTA Press, 2019), is a collection of visual maps, which show how expectations for students build over their K–12 education. The maps, used for section V, graphically depict the Disciplinary Core Ideas, Science and Engineering Practices, Crosscutting Concepts, and other elements of the *NGSS*, including the Nature of Science. Maps take the elements from the *NGSS* on a particular topic (such as models, patterns, or definition of energy) and organize them on a single page. The elements from grades K–2 are at the bottom of the page, and elements from grades 9–12 are at the top.

Elements are connected with arrows to indicate how achievement of one element can be useful in learning another. Individual elements can appear on multiple maps and connections between maps are illustrated to indicate connections between topics.

The NSTA Atlas also includes a narrative page preceding each map. The narrative is the same as the one in the *Framework* and is used with CTS section I to explain the topic addressed by the map. The narrative is also used in section III with the category 4 and category 5 CTS guides to describe how students can use the practice or crosscutting concept. Since the *Framework* did not include a narrative for the Nature of Science, leading nature of science researchers were commissioned by NSTA to write the narratives preceding the nature of science maps.

The *NSTA Atlas* is a useful tool to help you clarify the intended meaning of individual elements and to plan curriculum, instruction, and assessment for something as small as single unit of instruction to something as expansive as a full K–12 curriculum. Users of the *NSTA Atlas* maps are encouraged to check the NSTA NGSS Hub at ngss.nsta.org for professional development opportunities to learn how to use the maps.

State Standards: CTS Section VI

Because we recognize the importance of being able to link back to your state context and align your findings from CTS with state accountability factors, educators in each state and across districts combine findings from the CTS common resources and link them to the standards that are specific to their states, districts, or school settings. From Hawaii to Maine, each state has its own set of standards. Some are adopted from the *NGSS*, some are adapted from the *NGSS*, and some are based on the *Framework* without using the performance expectations in the *NGSS*.

Now that you know the background of the common set of resources and have a description of the CTS guides and the sections used for study, the next chapter provides information on how to get started with a curriculum topic study.

CHAPTER 3

Using Curriculum Topic Study

GUIDING THE CTS PROCESS

While the CTS guides are at the heart of the CTS process, understanding how to use the guides is important. How do you get started using CTS? If you are already familiar with CTS, you can go directly to Chapter 5, select a CTS guide, and get started with minimal direction. If you are new to CTS, we suggest starting with a few introductory steps to become acquainted with the materials and the process.

DECIDING WHICH RESOURCES TO USE

As described in the previous chapter, there are thirteen collective resources used with the CTS guides. However, not all these resources are used with every guide. The resources you choose to use depend on the purpose of your study, your personal preference, whether any of the resources repeat the same material, and which resources you have access to.

Some science educators prefer to use the *Framework* and *NGSS* as their primary resources for CTS sections I, II, III, V, and VI. Others prefer to use the *NSTA Quick Reference Guide to the NGSS K–12*, because it incorporates key elements of the *Framework* and *NGSS* into one concise resource that can be used with CTS sections I (for scientific and engineering practices and crosscutting concept guides), II, III (for scientific and engineering practices and crosscutting concept guides), V, and VI. For example, since the crosscutting concepts, core ideas, and practices are available in the *NGSS* as well as the *NSTA Quick Reference Guide to the NGSS K–12*, it would not be necessary to utilize both resources when studying sections that include both of those resources. The *NSTA Quick Reference Guide to the NGSS K–12* is the most versatile of all the CTS resources.

Accessing the Resources That Are Free Online

If availability of print resources is limited or cost of purchasing resources is a barrier, you may opt to utilize those resources that are available for free online. The following describes how to access the resources that are online and find the sections for reading that are designated on the guide. Keep in mind that sometimes web pages are changed or updated. As of the date of this publication (2019), these are the links that will take you to the resources.

Science for All Americans at www.project2061.org/publications/sfaa/online/sfaatoc.htm.

This link will take you to the Table of Contents. Click on the chapter listed on the CTS guide. There are no page numbers on the online version of *SFAA*. Scroll through the chapter until you find the name of the chapter reading listed on the CTS guide. Scan that section for material that is related to the topic. This resource is only used in section I.

Framework for K–12 Science Education at the National Academies Press website at nap.edu. Enter *Framework for K–12 Science Education* in the search bar. This and several other books will pop up. Click on the title of this book, which will take you to the *Framework* page. Click the Read Online button. Enter the page number listed on the CTS guide in the top page bar if you are using this online. This resource can also be downloaded for free to use offline, after registering with the National Academies Press, and used as a PDF book. This resource is used in sections I and II in all CTS guides and additionally in section III in the category D and category E CTS guides.

Next Generation Science Standards at nextgenscience.org. This link will take you to the Next Generation Science Standards website. Place your cursor on the Standards tab at the top of the page and scroll down to and click on Read the Standards. The NGSS are arranged by topic in the CTS guides. In the boxes on the right, Choose Standards by Topic if you are accessing them online or download the PDF of standards by topic by selecting Download Topic Arrangements. For the Appendices, click on To learn more about the NGSS Appendices at the top of the page. This will take you to the list of Appendices. Click on the appendix listed on the CTS guide.

Another way to access this resource is through the NSTA NGSS Hub. Go to ngss.nsta.org. This site contains "everything NGSS," and you may want to explore it further. For CTS, click on View the NGSS or go to https://ngss.nsta.org/AccessStandardsByTopic.aspx to download the PDF version with page numbers listed, click on the blue download button beneath the horizontal tabs bar. For the Appendices, find the list of appendices on the right side of the page. Click on the appendix listed on the CTS guide. The *NGSS* resource is used with sections II, V, and VI.

Benchmarks for Science Literacy: Chapter 15 Research Base at www.project2061.org/publications/bsl/. There is a link in the box on the right to read this book online. Click on that link and it will take you to the *Benchmarks* Table of Contents. Under Background, click on Chapter 15: The Research Base. Scroll down to C. Research Findings by Chapter and Section. Click on the chapter listed on the guide. There are no page numbers on the online version. Scroll until you find the name of the reading listed on the guide.

GATHERING AND MANAGING CTS RESOURCES

Figure 1.2 on page 5 lists the complete set of collective resources used in CTS. If you do not have your own personal copies of these resources, check with your colleagues, science department, or school or district professional library to see if they have copies and whether you can borrow them. In our work with new users of CTS, many have been surprised to find that several of these resources—*Framework for K–12 Science Education*, *Next Generation Science Standards*, *Science for All Americans (SFAA)*, and *Benchmarks for Science Literacy* in particular—are gathering dust somewhere on shelves in their schools or districts.

In many schools we have worked with, administrators have recognized the power of these resources and purchased sets for their schools, committees, and even individual teachers. In combination, these resources comprise a good investment in ongoing professional development in science teaching that supports both individual and collaborative learning.

To conduct a full CTS, with the fewest number of purchased resources, we suggest using the free online versions of *SFAA*, the *Framework*, *NGSS*, and *Benchmarks Ch. 15*. Once you have identified the free online resources you will use, you might consider purchasing one or more of the other resources.

If you prefer print materials, the most versatile, comprehensive resource used for CTS is the *NSTA Quick Reference Guide to the NGSS K–12*. This resource can be used with CTS sections II (Concepts, Core Ideas, or Practices), IV (K–12 Articulation and Connections), and VI (Assessment Expectations). In addition, if you are studying a scientific and engineering practice (category D guides) or crosscutting concept topic (category E guides), the *Quick Reference Guide* can be used for CTS sections I (Content Knowledge) and III (Curriculum, Instruction, and Formative Assessment)—eliminating the need to have access to both the *Framework* and the *NGSS*.

Managing CTS Materials When Working With a Group

Managing CTS materials is a critical element to success when working with a group. If you are designing professional development that incorporates CTS, facilitators should first acquire their own copies of the CTS parent book (this book) and all the CTS resources that they plan to use to conduct the topic study. Second, facilitators must decide how to provide access to the CTS resources for participants. If the purpose of the professional learning is to introduce teachers to the CTS process with the expectation that they will continue to use CTS on their own after professional development, CTS resources need to be available for them to use when needed. There is a variety of options for acquiring and managing materials. Choose an option or combination of options that best fit your audience; professional development context; and which resources—print or free online—you wish to use:

1. ***Provide a set of resources to each participant.*** This is obviously the most expensive and resource intensive approach. Groups who choose this option often have grant funds available for participant materials.

2. ***Provide a set of resources to each team.*** If participants are working as teams and your funding source can support it, consider providing a set of materials to each team of up to five teachers from the same school. These materials can then continue to be used by the school team and their colleagues for their CTS-related work.

3. ***Obtain and provide a session set of resources.*** If you regularly use CTS in your work, consider acquiring your own set of materials to provide for use during your CTS professional development. For each group of six teachers, provide one copy of each of the resources (minimum of four resources to conduct a full CTS) that you plan to use with the teachers.

4. ***Obtain a partial session set and borrow or have participants bring resources they may have.*** You may wish to purchase a few sets of resources and find an organization or school that can lend you additional copies. You might also contact teachers in advance who may already have copies of some of the resources and ask them to bring them.

5. ***Use paired readings.*** If you have session copies but not enough for everyone in the group, consider having teachers read the CTS sections in pairs.

6. ***Provide handouts of the readings.*** If resource book sets are not available, consider making copies of each of the readings and distributing the copies to groups (check permissions by the publisher first). If you do this, be sure to put a full set of the resources on a "library display" so that teachers can look at the actual books during the breaks.

7. ***Combine books with electronic access.*** If you have a wireless internet connection at your setting, provide access to the online versions to the teachers with digital devices and provide the books to those who do not have online access.

8. ***Notify participants in advance to download the online free materials.***

One of the laments leaders frequently hear after introducing CTS to teachers is that they embrace the process but won't be able to use it after the session because their schools do not have the funds to buy the materials. From our experience, we have found that teachers who are excited about using CTS go back to their schools and share its value with their administrators. As a result, most of the teachers we have worked with have convinced their administrators to provide the resources that are not available for free online, either as a set for school-based PLCs or grade-level teams, or as a school set to keep in the school's professional library. What they originally perceived as a roadblock to using CTS was overcome when their administrators saw the value of investing in these resources to build the capacity of their science teachers to grow and collaborate as professionals for the purpose of improving student learning. One of the teachers we worked with remarked how her principal shifted from bringing in the "one-shot wonder" or motivational speaker that cost the school a lot of money for one day to using those same funds to provide eighteen sets of CTS materials for teachers to use. Additionally, many of the teachers we have worked with have gone on to buy their own sets of materials to be part of their professional library that they will own and keep throughout their professional career. In summary, don't let remarks about not having the resources discourage you, especially when several are free online!

SELECTING A GUIDE, DEFINING YOUR PURPOSE, AND CHOOSING YOUR OUTCOMES

The first step in using CTS is to select the topic you want to study. Figure 3.1 lists all the CTS guides arranged by category and subcategory, grade levels, a description of the study, and concepts addressed through the study. Guides are listed in alphabetical order for each category or subcategory, A–C, and listed in the order they appear in the *NGSS* or *Framework* for the other categories.

FIGURE 3.1 Descriptions of the 103 CTS guides

Category A: Life Science Disciplinary Content Guides
Life Science Subcategory: Biological Structure and Function—Nine guides

CTS guide	Grade levels	Description of the CTS guide	Concepts within the CTS guide
Behavior, Senses, Feedback, and Response	K–12	A comprehensive study of an organism's behavior, regulation, and the nervous system; includes plant tropisms. Study is at the level of an individual organism.	Behavior, senses, nervous system, feedback, response, homeostasis, regulation, and control
Brain and the Nervous System	3–12	A study of the brain and how parts of the nervous system work together.	Nervous system, brain
Cell Division and Differentiation	9–12	A study of mitosis, growth of an organism, and gene expression that results in different types of cells.	Mitosis, growth and development, gene expression
Cells and Biomolecules: Structure and Function	6–12	A study of basic microscopic structure and function including cells, cell organelles, and molecules within cells.	Cells, cells as systems, chemistry of life, levels of organization
Characteristics of Living Things	K–5	A comprehensive study of what distinguishes living things from nonliving, including their characteristics and needs.	Concept of living, needs of plants and animals, plant and animal characteristics, life processes
Energy Extraction, Food, and Nutrition	K–12	A comprehensive study of how organisms need food for energy, growth, and repair of body structures.	Energy for living things, respiration, food, nutrition
Macroscopic Structure and Function of Organisms	K–5	A study of plant and animal observable (macroscopic) body parts and processes that support life.	Parts and wholes of living things, macroscopic structure and function, organ systems

(Continued)

FIGURE 3.1 (Continued)

CTS guide	Grade levels	Description of the CTS guide	Concepts within the CTS guide
Organs and Systems of the Human Body	K–12	A study of organs and how they work together in human body systems.	Organs, organ systems, human body systems
Photosynthesis and Respiration at the Organism Level	K–12	A comprehensive study of organisms' requirements for matter and energy, how they make or acquire food, and what they do with it. Emphasis is at the level of the organism.	Photosynthesis, respiration, food

Life Science Subcategory: Ecosystems and Ecological Relationships—Eight guides

CTS guide	Grade levels	Description of the CTS guide	Concepts within the CTS guide
Cycling of Matter and Flow of Energy in Ecosystems	3–12	A comprehensive study of how matter cycles and energy are transferred through ecosystems. May be combined with the crosscutting CTS guide Matter and Energy.	Matter cycles in ecosystems, energy transfer in ecosystems
Cycling of Matter in Ecosystems	3–12	A study of how matter is cycled between organisms and their environment. May be combined with the crosscutting CTS guide Matter and Energy.	Matter cycling between organisms, matter cycling between organisms and the environment, decomposers, decay, food chains, food webs, carbon cycle
Decomposition and Decay	3–12	A study of decomposition of living matter including the role of microorganisms in ecosystems.	Decomposers, decay, microorganisms
Ecosystem Stability, Disruptions, and Change	3–12	A study of how ecosystems maintain stability and can change over time. May be combined with Biodiversity and Human Impact.	Sustainability, stability and change, populations, communities
Food Chains and Food Webs	K–8	A study of how matter and energy are transferred from one organism to another and the interconnections between organisms.	Food chains, food webs, transfer of matter and energy
Group Behaviors and Social Interactions in Ecosystems	3–12	A study of how social interactions and group behaviors support the survival of individuals and groups.	Social interaction, group behavior

CTS guide	Grade levels	Description of the CTS guide	Concepts within the CTS guide
Interdependency in Ecosystems	K–12	A study of relationships between organisms, populations, communities, and their environment.	Needs of organisms and populations, interactions between living and nonliving resources, interdependency
Transfer of Energy in Ecosystems	6–12	A study of how energy is transferred between organisms and flows through ecosystems. May be combined with the crosscutting CTS guide Matter and Energy.	Energy transfer between organisms, energy transfer between organisms and the environment, conservation of energy

Life Science Subcategory: Life's Continuity, Change, and Diversity—Nine guides

CTS guide	Grade levels	Description of the CTS guide	Concepts within the CTS guide
Adaptation	3–12	A study of how organisms are adapted to their environment and the effect on populations of changes to their environment.	Adaptation, populations, natural selection, changes to the environment
Biodiversity and Human Impact	3–12	A study of speciation, the diversity of organisms in an environment, and human impact on species, including extinction.	Biodiversity, speciation, human impact on species, extinction
Biological Evolution	3–12	A comprehensive study of biological evolution that includes evidence of common ancestry, natural selection, adaptation, and biodiversity.	Biological evolution, natural selection, adaptation
Diversity of Species and Evidence of Common Ancestry	3–12	A study of the similarity and differences among present-day and past organisms that provide evidence of biological evolution.	Diversity, biological classification, relatedness among organisms, fossil evidence
DNA, Genes, and Proteins	6–12	A study of the structure of DNA and proteins and the role of DNA, genes, chromosomes, and proteins in genetics.	DNA, genes, chromosomes, proteins
Inheritance of Traits	K–12	A study of how genetic information is passed on from parent to offspring and the influence of the environment on traits.	Chromosomes, genes, DNA, traits, genetics
Natural Selection	3–12	A study of one of the mechanisms of biological evolution.	Natural selection, species, change in populations

(Continued)

FIGURE 3.1 (Continued)

CTS guide	Grade levels	Description of the CTS guide	Concepts within the CTS guide
Reproduction, Growth, and Development	K–12	A comprehensive study of reproduction and how organisms grow and change throughout their life cycle.	Asexual and sexual reproduction, mitosis, meiosis, growth and development, life cycles
Variation of Traits	K–12	A study of the structures and processes that contribute to differences between parents and offspring and variation within a population.	Variation, mutation, influence of environment, meiosis, traits

Category B: Physical Science Disciplinary Content Guides

Physical Science Subcategory: Matter—Eleven guides

CTS guide	Grade levels	Description of the CTS guide	Concepts within the CTS guide
Atoms and Molecules	6–12	A study of the structure of atoms and molecules and their interactions.	Atoms, molecules, parts of atoms, structure of matter, electrical forces
Behavior and Characteristics of Gases	3–12	A study of the properties and behavior of gases.	Properties of gases, particulate nature of matter, kinetic molecular theory
Chemical Bonding	6–12	A study of the types of chemical bonds, how they form, and how they affect the properties of materials.	Structure of matter, ionic and covalent bonds, electrical forces
Chemical Reactions	K–12	A study of chemical changes in matter.	Chemical reactions, chemical changes, changes in properties
Conservation of Matter	3–12	A study of how matter is conserved ranging from the level of objects to materials, to substances, to atoms and molecules.	Parts and wholes, conservation of matter in physical changes, conservation of matter in chemical changes
Elements, Compounds, and the Periodic Table	3–12	A study of pure substances, how they interact, and how patterns of the periodic table describe characteristics of and interactions between elements.	Trends in the periodic table, patterns of chemical properties, relative properties of elements, compounds
Mixtures and Solutions	K–12	A study of mixtures, solutions, and the process of dissolving.	Mixtures, solutions, dissolving, physical processes
Nuclear Processes	9–12	A study of phenomena and processes that involve the nucleus of an atom.	Nuclear processes, fusion, fission, radioactive decay, radiation, energy

CTS guide	Grade levels	Description of the CTS guide	Concepts within the CTS guide
Particulate Nature of Matter	K–12	A comprehensive study of matter and its parts from observable matter and parts of matter to a basic understanding of invisible particles, to atoms and parts of atoms.	Parts and wholes, microscopic and macroscopic properties, particles, atoms, molecules
Properties of Matter	K–12	A comprehensive study of physical and chemical properties of matter.	Matter, physical properties, chemical properties, characteristic properties
States of Matter	K–8	A study of solids, liquids, and gases and transitions from one state to another.	Macroscopic and microscopic properties of solids, liquids, and gases, changes of state, kinetic molecular theory

Physical Science Subcategory: Force, Motion, and Energy—Sixteen guides

CTS guide	Grade levels	Description of the CTS guide	Concepts within the CTS guide
Concept of Energy	3–12	A study of how energy is defined, described, and conceptualized.	Energy, forms of energy, heat, temperature, thermal energy
Conservation of Energy	6–12	A study of the how energy is conserved as it moves from one place or form to another.	Conservation of energy, transfer of energy
Electric Charge and Current	3–12	A study of the forces and energy involved in electrical interactions.	Electricity, electric charge, current, transfer of energy, electric circuits, fields
Energy in Chemical Processes and Everyday Life	3–12	A comprehensive study of energy produced from life processes and natural resources.	Production of energy, energy from natural resources, energy conversion
Force and Motion	K–12	A comprehensive study of forces and how forces change motion.	Force, motion, pushes and pulls, Newton's Laws
Forces Between Objects	K–12	A study of types of forces and their interactions.	Collisions, gravity, electrical and magnetic interactions
Gravitational Force	3–12	A study of gravitational force and interactions with objects.	Gravity, gravitational field, Newton's Law of Universal Gravitation
Kinetic and Potential Energy	6–12	A study of kinetic and potential energy and the connection to motion of objects.	Kinetic energy, potential energy
Magnetism	3–12	A study of magnetism and magnetic interactions.	Magnets, magnetic force, electromagnetism, magnetic fields

(Continued)

FIGURE 3.1 (Continued)

CTS guide	Grade levels	Description of the CTS guide	Concepts within the CTS guide
Nuclear Energy	9–12	A study of nuclear energy.	Atomic nuclei, fission, fusion
Relationship Between Energy and Forces	K–12	A study of how energy is transferred when forces are exerted.	Forces, transfer of energy, force fields
Sound	K–12	A study of how sound is made and how it travels.	Sound, vibration, waves
Transfer of Energy	3–12	A study of how energy is transferred from one place to another and transformed from one form to another.	Energy transfer, transformation, heat
Visible Light and Electromagnetic Radiation	K–12	A study of light and other forms of electromagnetic radiation and how it interacts with matter.	Light, electromagnetic radiation, waves
Waves and Information Technologies	K–12	A study of how waves are used to develop technologies that detect and transmit information.	Waves, information technologies, instrumentation
Waves and Wave Properties	K–12	A study of the characteristics and properties of waves.	Waves, wave properties, transmission of waves

Category C: Earth and Space Science Disciplinary Content Guides

Earth Science Subcategory: Earth Structure, Materials, and Systems—Eight guides

CTS guide	Grade levels	Description of the CTS guide	Concepts within the CTS guide
Earth's Materials and Systems	K–12	A comprehensive study of Earth's interacting systems and the materials that interact within those systems.	Systems, system interactions, geosphere, atmosphere, hydrosphere, biosphere, water, air, soil, rock, living things
Earth's Natural Resources	K–12	A study of Earth's water, mineral, energy, and biological resources used by humans.	Natural resources, renewable and non-renewable resources
Global Climate Change	6–12	A study of the Earth's changing climate and relationship to human activity.	Climate change, climate, global warming, climate models, human activity
Human Impact on Earth Systems	K–12	A study of the impact of human activities on Earth's land, water, air, and natural resources.	Environmental impact, natural resource depletion, human population growth, sustainability
Structure of the Earth	6–12	A study of the internal structure of Earth and the solid materials that make up the lithosphere. Emphasis is on composition of the geosphere.	Internal structure of Earth, lithosphere, soil, rocks, minerals, core, mantle, crust

CTS guide	Grade levels	Description of the CTS guide	Concepts within the CTS guide
Water Cycle and Distribution	K–8	A study of where water is found on Earth and how it cycles between land, bodies of water, and the atmosphere.	Water cycle, water distribution, evaporation and condensation, precipitation, groundwater
Water in the Earth System	K–12	A comprehensive study of the hydrosphere.	Hydrosphere, water cycle, water distribution, currents, water's ability to dissolve and move materials
Weather and Climate	K–12	A study of weather and climate and the conditions that affect weather and climate.	Weather, climate, greenhouse effect, atmosphere

Earth Science Subcategory: Earth History and Processes That Change the Earth—Six guides

CTS guide	Grade levels	Description of the CTS guide	Concepts within the CTS guide
Biogeology	K–12	A study of how living organisms impact Earth's processes and structures.	Evolution, atmosphere, cycles, weathering, reservoirs, role of microbes, feedback mechanisms
Earthquakes and Volcanoes	K–8	A study of the patterns, formation, and effects of earthquakes and volcanoes. Can be combined with the study of natural hazards.	Earthquakes, volcanoes, plate tectonics, mapping
Earth's History	K–12	A study of how the Earth has changed over time, including how the rock and fossil record is used to determine ages and events.	Geologic time, rock record, rock cycle, fossils, tectonic events
Natural Hazards	K–12	A study of catastrophic events that affect Earth and humans.	Hurricanes, tornadoes, earthquakes, volcanoes, floods, tsunamis, droughts, weather disasters
Plate Tectonics	K–12	A study of Earth's plate movements and their patterns and resulting landforms.	Earth's interior, plates, plate tectonics, mountain formation, mapping Earth's features
Weathering, Erosion, and Deposition	K–12	A study of processes that wear down and transport Earth materials and change landforms.	Weathering, erosion, deposition

(Continued)

FIGURE 3.1 (Continued)

Space Science Subcategory: Earth in Space, Solar System, and the Universe—Seven guides

CTS guide	Grade levels	Description of the CTS guide	Concepts within the CTS guide
Earth and Our Solar System	K–12	A comprehensive study of Earth's place in our solar system, solar system objects, motions, and phenomena.	Earth, planets, moons, solar system objects, orbits, scale, cycles
Earth, Moon, Sun System	K–12	A comprehensive study of the relationship between the Earth, sun, and moon.	Phases of the moon, eclipses, tides, motions of moon and Earth
Earth-Sun System	K–8	A comprehensive study of the relationship between Earth and the sun.	Day-night cycle, seasons, shadows, sunrise, sunset, Earth's orbit
Formation of the Earth, Solar System, and the Universe	6–12	A study of how Earth, the solar system, and the universe were formed.	Big bang theory, formation of Earth and solar system, gravity
Phases of the Moon	K–8	A study of observations of and explanations for the phases of the moon.	Observing moon phases, explaining moon phases
Seasons and Seasonal Patterns in the Sky	K–8	A study of the cause of seasons and seasonal patterns in the sky.	Seasons, star patterns, shadows
Stars and Galaxies	K–12	A comprehensive study of our sun, stars, galaxies, and their place in the universe.	Sun, stars, galaxies, stellar evolution

Category D: Scientific and Engineering Practices—Ten guides

CTS guide	Grade levels	Description of the CTS guide	Skills and processes within the CTS guide
Asking Questions in Science	K–12	A study of the practice of questioning in science and what makes a good scientific question.	Ask investigable questions, identify testable and non-testable questions, evaluate questions
Defining Problems in Engineering	K–12	A study of the practice of defining a problem in engineering.	Define design problems that can be solved through development of a tool, object, or process
Developing and Using Models	K–12	A study of the concept of a model and the practice of developing and using a model in science and engineering.	Develop and use models to explain phenomena or predict or test ideas

CTS guide	Grade levels	Description of the CTS guide	Skills and processes within the CTS guide
Planning and Carrying Out Investigations	K–12	A study of how investigations are planned and carried out in science and in engineering.	Plan and conduct investigations, make observations and measurements, use tools to collect data
Analyzing and Interpreting Data	K–12	A study of how data are used in both science and engineering.	Construct, analyze, and interpret graphical displays of data to provide evidence for phenomena, apply concepts of statistics, analyze data to refine a design
Using Mathematics and Computational Thinking	K–12	A study of how mathematics is used in both science and engineering.	Organize data sets to reveal patterns and relationships, use mathematical representations to describe or support scientific conclusions and design solutions
Constructing Scientific Explanations	K–12	A study of the concept of a scientific explanation and a theory and how explanations are constructed in science.	Construct explanations based on evidence, apply scientific principles to explain real-world phenomena or events, relate theories to how the past, present, and future world works
Designing Solutions	K–12	A study of the central role of solving problems in engineering.	Design, construct, and implement a solution that meets specific design criteria and constraints
Argumentation	K–12	A study of the role of evidence, reasoning, and argument in science and engineering.	Construct arguments from evidence that support or refute claims for explanations or solutions about the natural or designed world
Obtaining, Evaluating, and Communicating Information	K–12	A study of the ways ideas and results are communicated in science and engineering and the validity of information.	Gather, read, and synthesize information, evaluate the merit and validity of ideas and methods, communicate scientific and technical information

(Continued)

FIGURE 3.1 (Continued)

Category E: Crosscutting Concepts—Seven guides

CTS guide	Grade levels	Description of the CTS guide	Concepts within the CTS guide
Patterns	K–12	A study of how patterns in nature guide organization, and classification and explain relationships and causes underlying them.	Organization, classification, and patterns in natural and designed systems
Cause and Effect	K–12	A study of causal relationships and the mechanisms by which they are mediated.	Causal and correlational relationships, predictions in natural and designed systems
Scale, Proportion, and Quantity	K–12	A study of size, time, and energy scales and the proportional relationships between quantities as scales change.	Time, space, and energy scales, proportional relationships, orders of magnitude
Systems and Systems Models	K–12	A study of how a group of related objects or components interact and influence each other and how models can be used to understand and/or predict the behavior of systems.	Parts and wholes, interactions in systems and subsystems, feedback loops, value and limitations of models
Energy and Matter: Flows, Cycles, and Conservation	K–12	A study of how matter cycles and energy are transferred through systems. May be combined with the CTS guide, Cycling of Matter and Flow of Energy in Ecosystems or Cycling of Matter in Ecosystems.	Cycling of matter in systems, flow of energy in systems, conservation of matter
Structure and Function	K–12	A study of how the structure or shape of an object determines many of its properties and functions.	Parts and wholes (macroscopic and microscopic), structure and function of living organisms and designed objects
Stability and Change	K–12	A study of conditions that affect stability and factors that control rates of change.	Stability and instability in natural and designed systems, changes over time (gradual and sudden), dynamic equilibrium, feedback mechanisms

Category F: STEM Connections

STEM Connections Subcategory: Engineering, Technology, and Applications of Science—Six guides

CTS guide	Grade levels	Description of the CTS guide	Concepts within the CTS guide
Defining and Delimiting an Engineering Problem	K–12	A study of identifying an engineering problem, the goals or criteria it must meet, and limitations or constraints.	Problem identification, criteria, limitations, constraints
Developing Possible Design Solutions	K–12	A study of the process engineers use to design potential solutions, including use of models.	Models, mathematical models, data analysis, brainstorming
Engineering Design	K–12	A comprehensive study of engineering design that is about the engineering design process (as opposed to doing engineering that is in the practices study).	Engineering, problem definition, models, criteria, constraints, optimization, defining and delimiting problems, evaluating solutions, risks, benefits, trade-offs
Improving Proposed Designs	K–12	A study of how engineers use optimization to compare and choose the best design.	Optimization, criteria, trade-offs
Interdependence of Science, Engineering, and Technology	K–12	A study of how the fields of science, engineering, and technology are mutually supportive of each other.	Interdependence of science, engineering, and technology
Influence of Science, Technology, and Engineering on Society and the Natural World	K–12	A study of how science, engineering and the technologies developed through science and engineering impact peoples' lives and the natural world.	Societal impact, environmental impact, science and society, benefits and risks of technology

STEM Connections Subcategory: Nature of Science—Six guides

CTS guide	Grade levels	Description of the CTS guide	Concepts within the CTS guide
Methods of Scientific Investigations	K–12	A study of the various ways science investigations are guided and conducted.	Scientific methods, scientific values, tools and techniques
Scientific Knowledge Demands Evidence	K–12	A study of how science relies on empirical evidence to construct explanations and that explanations are subject to change when new evidence is available.	Patterns, empirical evidence, certainty and durability of knowledge, change in light of new evidence, importance of argument

(Continued)

FIGURE 3.1 (Continued)

CTS guide	Grade levels	Description of the CTS guide	Concepts within the CTS guide
Hypotheses, Theories and Laws	K–12	A study of what hypotheses, theories, and laws are and how they are used by scientists. Also includes models and mechanisms.	Hypothesis, theory, law, explanatory models, mechanisms, explanations
Science as a Way of Knowing	K–12	A study of science as a unique way of knowing about the natural world that encompasses both a body of knowledge and processes.	Scientific knowledge, processes and practices
Science as a Human Endeavor	K–12	A study of how people of diverse backgrounds work as scientists and engineers and the human qualities that influence their work.	Diversity, careers, creativity and imagination, societal influences
Science Addresses Questions	K–12	A study of how scientists question natural systems and the limitations to scientific questions.	Questions about natural systems, ethical and societal limitations

Once you have selected a topic that is relevant to your work or what you teach, decide on your purpose or desired outcome. What do you want to learn from this study?

- Do you want to improve your own understanding of the content (section I)?

- Do you want to examine the specific concepts and ideas that make up the three dimensions of teaching and learning (section II)?

- Do you want to examine curricular considerations and suggestions for effective teaching, including formative assessment (section III)?

- Do you want to learn more about the commonly held ideas students bring to their learning or difficulties that may arise during instruction (section IV)?

- Do you want to examine the progression of K–12 learning, see how one idea builds from another, or examine connections within and across topics (section V)?

- Do you want to examine the performance expectations and how the three dimensions are intertwined, clarify their meaning and intent, identify boundaries, and link back to your own local or state assessments (section VI)?

You might choose just one of the outcomes and study only one section of a CTS guide or you might choose several or all the outcomes. The outcomes you select will determine the resources you will need and the sections you will read, study, and reflect on. It is also important to keep the grade or grade span relevant to your work and purpose of the study in mind. If your work and outcome require a K–12 perspective, then you should examine all the grade level readings. If your focus is on a particular grade or grade span, you might examine only the readings that include that grade or grade span.

GETTING STARTED

If you are doing a CTS for the first time or you are studying an unfamiliar topic, we recommend you study all six sections. Before you begin, activate your own prior knowledge about the topic. Brainstorm a list of what you already know about the topic and generate a list of questions you have about teaching and learning related to the topic and grade span you selected. What do you want to find out as you study the topic? As you proceed through the study, note the things you are learning that may differ from what you thought you knew about the topic. Also note where the study answers the questions you started with what you wanted to find out.

Reading the CTS Sections

CTS is meant to be a flexible process. You do not have to use *all* the resources or study *all* the sections or *all* the grade levels. CTS is not a linear process. You can start anywhere with sections I–VI. If a topic is unfamiliar, start with section I. Section VI is often studied last so that you can look at it through the lens of the other sections. Conversely, you can study it first and work backward to understand what is needed to achieve the performance expectation. You can use just one of the resources in each section or you can combine them (the ones that do not duplicate the exact same information) to get different perspectives or additional information that you might not get from using only one resource. Remember if you are using the *NSTA Quick Reference Guide to the NGSS K–12*, you won't need the *Framework* (except for CTS section I for the disciplinary content guides in categories A–C) or the *NGSS* as this handy reference guide includes the same information used in CTS from both of those resources. While the resource *The NSTA Atlas of the Three Dimensions* is used primarily for CTS section V, the page preceding the map provides the same information that is in the *Framework* that can be used with section I or section III for category D and category E guides.

Don't forget to note that critical point in Figure 3.2 that is highlighted on each guide! As you read and study each section, filter out the information that does not apply to the topic or purpose of your study. Focus on what is most related to the topic and outcome of the section you are reading. Sometimes the related information is only a paragraph or one or two lines on the included page. On some guides, the resource may list where to start or stop your reading.

FIGURE 3.2 Important consideration for reading!

Developing and Using Models
Grades K–12 Standards- and Research-Based Study of a Curricular Topic

Section and Outcome	Selected Sources and Readings for Study and Reflection
	Read and examine related parts of

GUIDING QUESTIONS FOR GETTING STARTED WITH CTS

When starting out with CTS, we suggest you use guiding questions to frame your study. If you are a facilitator of CTS, we suggest you choose the questions you want your participants to focus on during the study. Sample questions are provided below, or you can develop your own questions. General questions that can be used with any of the CTS guides, any of the CTS sections I–VI, and any of the selected resources include

1. What new knowledge and/or insights did you gain from this study?
2. How are you thinking differently about teaching or learning after studying this topic?
3. If you used more than one resource during the study, what similarities and differences did you find between the different resources? How did using the resources collectively provide insight into the topic?
4. What were the most important takeaways from this study?
5. What other questions do you have about this topic that were not answered from your study?

There are also questions that are specific to each of the three dimensions: disciplinary core idea topics (CTS guide categories A, B, C, and F), scientific and engineering practices topics (CTS guide category D), and crosscutting concepts topics (CTS guide category E). Figure 3.3 lists guiding questions that can be used with these three dimensions of CTS guides.

FIGURE 3.3 Guiding questions for study and reflection to use with the three dimensions of CTS

CTS section	Disciplinary content CTS guides (Categories A, B, C, and F)	Scientific and engineering practices CTS guides (Category D)	Crosscutting concepts CTS guides (Category E)
I. Content Knowledge	• What are the core concepts and ideas that make up this topic? • Why is knowledge of this topic important? • What contexts, examples, or phenomena were used to explain this topic? • If you used more than one resource, how did the combined resources contribute to your understanding?	• How would you generally describe this practice? • How do scientists and/or engineers use this practice in their work? • Why is this practice important to science and/or engineering? • What examples were helpful in describing this practice?	• How does this crosscutting concept bridge disciplinary boundaries? • How do scientists and/or engineers use this crosscutting concept? • What are some of the "big ideas" related to this crosscutting concept?
	Reflect: What new knowledge or insights did you gain from this section?	**Reflect:** What new knowledge or insights did you gain about this practice?	**Reflect:** What new knowledge or insights did you gain about this crosscutting concept?

CTS section	Disciplinary content CTS guides (Categories A, B, C, and F)	Scientific and engineering practices CTS guides (Category D)	Crosscutting concepts CTS guides (Category E)
II. Concepts, Core Ideas, and Practices	• What concepts and specific ideas make up the goals for learning in this topic? • How do these concepts and ideas help clarify what is important to teach and learn? • How does the language used in these goals for learning help you identify terminology students should know and use?	• What elements of this practice make up the goals for learning? • How does this section help you clarify how the practices are used at different grade levels?	• What specific ideas related to this crosscutting concept make up the goals for learning? • How does this section help you clarify how the crosscutting concepts are used at different grade levels? • Which crosscutting concepts are emphasized in which grade(s)?
	Reflect: How do the learning goals in this section align with your district or classroom learning goals? What new knowledge or insights did you gain from this section?	**Reflect:** What ways of using this practice are missing from or need more emphasis at your grade level(s)? What new knowledge or insights did you gain from this section?	**Reflect:** What ways of including crosscutting concepts are missing from or need more emphasis at your grade level(s)? What new knowledge or insights did you gain from this section?
III. Curriculum, Instruction, and Formative Assessment	• What are some considerations for designing curriculum? • What are some suggestions for effective instruction? • What contexts, phenomena, representations, or experiences can help students learn the ideas in this topic? • How can a formative assessment probe (if available) give you insight into how students think about the topic?	• What are some suggestions for effectively teaching students to use this practice? • How do curricular emphases and/or instructional opportunities or strategies differ between grade spans?	• What are some suggestions for effectively teaching students to use this crosscutting concept? • How do curricular emphases and/or instructional opportunities differ between grade spans?
	Reflect: What are the implications for your curriculum, instruction, or assessment? What changes are you thinking of making?	**Reflect:** What are the implications for your curriculum, instruction, or assessment? What changes are you thinking of making?	**Reflect:** What are the implications for your curriculum, instruction, or assessment? What changes are you thinking of making?

(Continued)

FIGURE 3.3 (Continued)

CTS section	Disciplinary content CTS guides (Categories A, B, C, and F)	Scientific and engineering practices CTS guides (Category D)	Crosscutting concepts CTS guides (Category E)
IV. Research on Commonly Held Ideas	• What are some common misconceptions, alternative ideas, or difficulties students have? • Is there information about where or how those commonly held ideas develop? • Are some commonly held ideas more prevalent or resistant at certain ages? • Are suggestions given on how to address some of the commonly held ideas?	• What does research (if any) say about how students think about this practice and difficulties they may have? • How can the research help you address ways your students might think about or use this practice?	• What does research (if any) say about how students think about this crosscutting concept? • How can the research help you address ways your students might think about or use this crosscutting concept?
	Reflect: What insights did you gain about how your students are likely to think about this topic? Are there changes you will make based on this section?	**Reflect:** What new insights did you gain about how students think about or use this practice?	**Reflect:** What new insights did you gain about how students think about or use this crosscutting concept?
V. K–12 Articulation and Connections	• How would you summarize the progression of ideas from K to 12? • Which prerequisites need to be considered for your grade level? • Do you notice any "storylines" or threads you can follow from one idea to the next in the *Atlas*? • Are there connections to ideas on other maps shown on the *Atlas*?	• How would you summarize how use of this practice progresses from K to 12? • Which prerequisites need to be considered for your grade level? • Are there connections to other practices shown on the *Atlas*?	• How would you summarize how this crosscutting concept progresses from K to 12? • Which prerequisites need to be considered for your grade level? • Are there connections to other crosscutting concepts and practices shown on the *Atlas*?
	Reflect: What insights did you gain that will help you organize the content in your standards or curriculum coherently?	**Reflect:** What new insights did you gain that will help you organize this practice coherently?	**Reflect:** What new insights did you gain that will help you organize this crosscutting concept coherently?

CTS section	Disciplinary content CTS guides (Categories A, B, C, and F)	Scientific and engineering practices CTS guides (Category D)	Crosscutting concepts CTS guides (Category E)
VI. Assessment Expectation	• How did the readings from sections I–V help you better understand the intent of similar expectations in your state's standards? • How do your standards match up with the performance expectations in this section? • What practices and crosscutting concepts combine to support the disciplinary content in the *NGSS*? • How do the *NGSS* clarification statements help you know what kinds of experiences and curricular emphases will prepare students for assessment? • How do the *NGSS* boundary statements help you know where to draw the line in what should be included in assessment?	• How did the readings from sections I-V help you better understand the intent of similar processes or practices in your state's standards? • How do your state's standards include this practice? • Which elements of a practice need more emphasis in your district or classroom assessment? • Which disciplinary core ideas are assessed using this practice?	• How did the readings from sections I-V help you better understand the intent of similar concepts in your state's standards? • How do your state standards include this concept? • Which concepts need more emphasis in your district or classroom assessments? • Which disciplinary core ideas does this concept support?
	Reflect: How can you use this section to inform your classroom and district assessments?	**Reflect:** How can you use this section to inform your classroom and district assessments?	**Reflect:** How can you use this section to inform your classroom and district assessments?
Other Resources	• If an additional resource was provided, what did you learn from the resource?	• If an additional resource was provided, what did you learn from the resource?	• If an additional resource was provided, what did you learn from the resource?

The *Framework for K–12 Science Education*, *NGSS*, and the *NSTA Quick Reference Guide to the NGSS K–12* (which combines and organizes the *Framework* and the *NGSS*) are considered the central resources for CTS. Development of the *NGSS* was based on the *Framework*. However, not all the *Framework* ideas are included in the *NGSS*. Some of the CTS resources offer additional information that may not be included in the other resources or include information that is organized differently. The following resource-specific questions can be added to the questions in Figure 3.3:

SCIENCE FOR ALL AMERICANS: ADDITIONAL QUESTIONS

Section I: Content Knowledge

- What should all science-literate adults, regardless of whether they go on to study science, know about this topic after their K-12 education?
- How does the clarity of the prose help you understand the topic?
- What connections across other areas of science are described?
- To what extent is technical terminology used? What terminology should all adults be familiar with related to this topic?

SCIENCE MATTERS: ADDITIONAL QUESTIONS

Section I: Content Knowledge

- What knowledge did you gain from this reading that exceeds K-12 science literacy yet was helpful in understanding the topic?
- What examples, analogies, phenomena, or other descriptions used by the authors were useful in understanding the ideas in the topic?
- How do the authors' explanations contribute to public understanding of science in the media or societal issues?

FRAMEWORK FOR K-12 SCIENCE EDUCATION: ADDITIONAL QUESTIONS

Section I: Content Knowledge

- What is the framing question (for categories A-C topics)? How do the concepts and ideas described in the narrative address the framing question?
- (Additional reading) The chart on pages 50-53 of the *Framework* (Box 3-2) shows how the practice is used by scientists or in science and with engineers or in engineering. Summarize how the practice you studied is used in one or both fields and by the people who work in those fields.
- What are some specific examples of ways this crosscutting concept is used?

Section II: Concepts, Core Ideas, and Practices

- What is the core idea and component idea(s) this topic relates to?
- The *Framework* grade band endpoints were used by the *NGSS* to inform placement of ideas by grade. In some cases the placement in the *NGSS* differs from the *Framework*. In other cases some ideas in the *Framework* were not included in the *NGSS*. Do you see any examples of that for this topic?

DISCIPLINARY CORE IDEAS: RESHAPING TEACHING AND LEARNING: ADDITIONAL QUESTIONS

Section I: Content Knowledge

- Why is this disciplinary core idea important for understanding science?

Section III: Curriculum, Instruction, and Formative Assessment

- What does a progression of instruction and learning look like from one grade span to the next?
- What kinds of instructional approaches or opportunities (e.g., tasks, curricular examples, investigations, phenomena, representations, etc.) support learning this topic?

Section IV: Research on Commonly Held Ideas

- What are some challenges that students face in learning these ideas?

HELPING STUDENTS MAKE SENSE OF THE WORLD: ADDITIONAL QUESTIONS

Section I: Content Knowledge

- What does it mean to participate in this practice?
- How does this practice relate to other practices?

Section III: Curriculum, Instruction, and Formative Assessment

- What does this practice look like in the classroom?
- What is not included in or intended by this practice?
- How can you support equity when using this practice?
- What are some considerations for assessing this practice?
- What are some suggestions for getting started with this practice?

THE NSTA ATLAS OF THE THREE DIMENSIONS: ADDITIONAL QUESTIONS

Section I: Content Knowledge

- Nature of Science is an area that was added to the *NGSS* after the *Framework* and is therefore not described in the *Framework*. Leading researchers in the Nature of Science developed the narrative preceding a map to describe the nature of science ideas. How does the narrative help you understand that aspect of the nature of science?

Section V: K-12 Articulation and Connections

- How does the visual layout of the map help depict how the elements of the disciplinary core ideas, practices, or crosscutting concepts progress from one grade span to the next?
- Are there any threads you can follow to establish a storyline from one lesson to the next?

(Continued)

(Continued)

- Can you find an example of how one element on the map contributes to learning another element? Are some ideas necessary prerequisites?

- Are there ideas on a map that you can bundle for a lesson or sequence of lessons?

- What are some connections to other disciplinary core ideas, practices, or crosscutting concepts on other maps?

Now that you have learned about how to organize and manage resources and have examined questions to guide your study, the next chapter addresses ways to use CTS to answer questions of practice and apply CTS to different contexts.

CHAPTER 4

Using and Applying Curriculum Topic Study

WHAT CTS IS NOT

The previous chapters helped you understand what CTS is and the resources used for CTS. In this chapter you will examine different ways CTS is used in both individual professional learning situations as well as in various group professional learning contexts. Before you consider ways you might use CTS in your work, keep in mind that CTS helps you seek answers to questions of practice related to content, curriculum, instruction, or assessment. CTS provides the analytic lens and process to help you think through questions and issues related to your work in science teaching and learning and analyze the information you gather through the process. It does not provide the specific answers or solutions you may be seeking. The answers or solutions come from you as you synthesize the information you gather through the CTS process and relate it to the purpose of your study. The following describes what CTS is **not** intended to do:

- CTS is **not** a remedy for weak science content knowledge. It can be used to enhance and support content knowledge but it is not a replacement for a serious lack of content knowledge.

- CTS is **not** a curriculum or collection of lesson plans and activities. It will help you make better decisions about curriculum and instructional materials, including teacher-designed lessons.

- CTS is **not** an instructional how to. It does not prescribe a particular instructional model or set of instructional strategies. It can help you think through ways to make your instruction more coherent and effective, including embedding formative assessment to help you understand students' thinking prior to or throughout an instructional sequence.

- CTS is **not** a source of performance assessments. It does identify performance expectations that can be used to inform assessment development.

- CTS is **not** a quick fix. It takes serious, dedicated time to read, analyze, and reflect on the findings from CTS.

- CTS is **not** a stand-alone or end-all for professional or preservice learning. It is used with other types of professional or preservice learning, such as backwards planning for curriculum. Additionally, CTS often reveals additional learning needs.

- CTS is **not** solely a *Next Generation Science Standards* (*NGSS*) resource nor does it replace other *NGSS* resources. CTS can be used with any set of standards. Furthermore, when used to support *NGSS*, CTS is just one of a set of processes and tools developed to support *NGSS*. Besides CTS, there are many other resources developed by Achieve and others that are not included in this book but are valuable tools to consider.

INDIVIDUAL USE OF CTS

Individual teachers can use CTS on their own to strengthen their content knowledge, clarify their learning goals, inform curriculum and assessment planning, design or modify lessons, examine ideas students are likely to bring to their learning, anticipate difficulties students may have, understand what comes before and after their grade level, and answer specific questions related to their practice. CTS increases teachers' knowledge of what is important to teach and how to effectively organize and teach their curriculum (pedagogical content knowledge). Perhaps you are switching to a new grade level, teaching new subject matter, or shifting from a focus on teaching to a focus on learning. CTS can help you make that transition by grounding your practice in the important ideas from the standards, deepening your knowledge about coherent and effective curriculum and instruction, utilizing research on learning, and understanding how the three dimensions intertwine and support each other.

When using CTS as an individual, remember you can start with any of the sections I–VI on a CTS guide. You might use only one section to ask a specific question about your practice or you might do a full topic study to have a deep understanding of the topic you teach. Keep notes as you study each section, recording the information from the resources that will be used as "evidence" when you make curricular or instructional decisions. Figure 4.1 provides short snapshots that you can use to practice first steps of CTS by identifying a topic study guide and the section you can use to answer a specific question of practice. Try out two or three of these snapshots if you are new to CTS. Figure 4.2 provides an answer key to the guide and section(s) that can be used for the snapshots.

FIGURE 4.1 Snapshots for practicing CTS Guide and sections (I–VI) selection

A. I am planning a lesson in which students will investigate how matter is conserved during a chemical reaction. What alternative ideas and potential difficulties should I be aware of?	B. Our seventh graders will be developing models to explain what causes the phases of the moon. What important aspects of this practice should I focus on and how can it support learning about moon phases?	C. I am not sure about the depth or breadth I should go into when including ideas about atoms and molecules in my lessons. What is the current thinking about what is important at my grade level?
D. My students are designing and testing different types of parachutes to safely land a fragile object. How can I ensure this activity is addressing the engineering practice of designing solutions?	E. Our elementary students have a difficult time understanding what happens to matter when one organism eats another. What instructional considerations should I take into account as I plan my lessons?	F. The concept of waves is embedded in several of our secondary physical science units. How do ideas about wave properties develop from the middle grades through high school?
G. I notice my students have problems interpreting graphs. Are there certain types of graphs students have problems interpreting? What common difficulties related to analyzing graphs should I be aware of?	H. The concept of systems cuts across many of our instructional units. How can I build my understanding of what systems are and how they can help us understand phenomena?	I. We are writing assessments for our geology unit. I'm wondering how a performance expectation about movement of earth's plates could inform our learning targets for a three-dimensional performance task?
J. Our elementary teachers are planning a lesson on the water cycle. How can I use a formative assessment probe to elicit their initial ideas about the water cycle and find instructional suggestions to inform the planning of our lessons?	K. I'm teaching a course on astronomy for the first time. Our standards ask students to construct an explanation of the Big Bang Theory. I need to refresh my knowledge about the Big Bang. What basic introduction to this theory could help me?	L. When we ask students to construct explanations, do they know what that means? How can I help them distinguish everyday explanations from scientific explanations?
M. We are working on revising our approach to energy in our K–12 curriculum. When do students encounter the concept of energy, and how does this concept progress from one grade span to another?	N. Our district is revising our K–12 curriculum to include crosscutting concepts. How are concepts and procedures in mathematics used with the crosscutting concept of scale, proportion, and quantity?	O. How some organisms in a population may have traits that allow them to survive changes in their environment is a major idea in our life science curriculum. What commonly held ideas should I anticipate that students may have about this idea?

FIGURE 4.2 Snapshots answer key

Snapshot	Guide(s)	Sections(s)
A	Conservation of Matter, or Chemical Reactions	IV
B	Developing and Using Models and Phases of the Moon	II and III
C	Atoms and Molecules	II or V
D	Designing Solutions	II and III
E	Cycling of Matter in Ecosystems, or Food Chains and Food Webs	III
F	Waves and Wave Properties	II or V
G	Analyzing and Interpreting Data	IV
H	Systems and System Models	I
I	Plate Tectonics	VI
J	Water Cycle and Distribution	III
K	Formation of the Earth, Solar System, and the Universe	I
L	Constructing Scientific Explanations	III
M	Concept of Energy	V
N	Scale, Proportion, and Quantity	II or V
O	Adaptation	IV

GROUP USE OF CTS

Group use of CTS can be as simple as two colleagues studying a topic together, an instructional coach and a new teacher clarifying learning goals, several people on a committee using CTS to inform their curriculum work, an after-school professional learning community (PLC) deepening their understanding of crosscutting concepts, a preservice methods course using CTS before planning a lesson, or a large workshop setting in which groups combine their studies to examine three-dimensional learning. Whatever the format, topic, or audience is, professional learning and collaboration is enhanced when colleagues have opportunities work together to analyze, discuss, and apply findings from a curriculum topic study.

In addition to managing the resources used for CTS, as described in Chapter 2, leaders of small- or large-group use of CTS must consider ways to organize, discuss, and share results of CTS, especially when a group is doing a full CTS of all the sections (I–VI). A commonly used method for dividing a task among a group is the jigsaw strategy. A jigsaw builds on the idea that we learn best when we have to teach others. It is also a way to reduce the amount of reading an individual would have to do by distributing the readings among a group, with each reading summarized and shared by the group member assigned to that CTS section. Below are some suggested ways to jigsaw a CTS:

Large Group Jigsaw with Expert Groups. In this option, CTS readings are assigned to "table groups." Each table group becomes the expert for their assigned reading. After the participants in each expert group have read their assigned section, they discuss the reading within their group. For example, one table group might be assigned to read CTS sections IA and IC for the CTS study of Natural Selection. After participants have discussed that reading, they prepare a summary of sections IA and IC to share with the other groups that were assigned other readings from the Natural Selection CTS guide. When all groups have finished their summaries, a member of each group meets with others to form a group that includes all the readings. In this new group, participants share out each of their summaries, completing a full topic study. For example, a member of the group that summarized sections IA and IC meets with a member of the group that summarized the section II reading, a member of the group that summarized the section III reading and so on forming a new group of six, with each person sharing their summary of the reading that was assigned to them.

Small Group Jigsaw. In this option, each person in a small group is assigned to be the "expert" for a specific reading or section of a CTS guide. Everyone in the group has a different assignment in which they read their section and prepare a summary for their group. Each person takes a turn sharing the summary of their section with their whole small group.

Assigning Jigsaw Readings

There are a variety of ways to assign jigsaw readings. The breakdown you choose depends on the resources you have available, the grade spans you want to focus on, and the sections of the CTS guide that fit the purpose of your study in case you choose not to do all six sections. Individuals can be assigned a reading or it can be done in pairs. The following are options for assigning readings:

Assign by CTS sections. I, II, III, IV, V, VI, deciding within each section which subsections to include, for example: IA, IC, IIC, IIIA, IVC, VC, VIA and VIC. If you are the facilitator, review the readings first and combine short ones so that everyone is reading for about the same amount of time.

Assign by Book. Assign readings from different sections using the same book or a combination of books. For example, one person might do all the readings from the *Framework*, whereas another does all the readings from the *Atlas*.

Assign by Grade Span. If you are doing a full K–12 topic study, consider dividing the readings for sections II, III, and VI by grade span (as seen in Figures 4.3 and 4.5).

FIGURE 4.3 Teachers engaged in CTS

CHAPTER 4 • USING AND APPLYING CURRICULUM TOPIC STUDY

Other Group Strategies

Some professional learning groups prefer jigsaws and others prefer not to be restricted in their readings and discussions. Whether you choose to jigsaw or not depends on your audience, the topic(s) chosen, and the time frame available for doing the study. Some other group strategies are

> ***Large Group Discussion, All Readings.*** This option works best when each participant has access to at least one CTS resource for each section and readings can be done in advance or in a longer time frame. Each participant reads all the sections selected by the facilitator, makes notes, and shares findings and insights during a whole group discussion. While some participants may not have had time to read all sections, it is likely that someone has read one of the sections assigned and can share with the whole group.
>
> ***Small Group Discussion, Assigned Expert.*** This option is the same as the small group jigsaw but instead of reading only the section that is assigned to you, you may read any or all the sections. However, you are assigned one reading you have the responsibility to be the expert for, in case others don't get to that reading. Once you complete that section, you can read any of the other sections, as well as participate in the discussion of any sections you read. This option ensures that all the readings are covered but provides flexibility for everyone to read sections they are most interested in.

These are just some of the ways you can organize and structure a group CTS. As you become familiar with CTS, you will find ways to facilitate the process that works best in your professional setting.

UTILIZING CURRICULUM TOPIC STUDY FOR DIFFERENT PURPOSES

CTS is a versatile process that can be tailored to fit your purpose and the context you are working in. Think of it as the Swiss Army knife of content, curriculum, instruction, and assessment. Like a Swiss Army knife, it is a versatile tool that serves several different purposes. You can use a single part of the tool or several in combination.

CTS can be used by preservice and classroom teachers, curriculum and assessment developers, science specialists, instructional coaches, professional developers, preservice instructors, informal educators, and anyone interested in improving science teaching and learning. It can be used to build understanding of science content; clarify learning goals; inform curricular, instructional, and assessment decisions; and understand how students think about science concepts, ideas, and practices. It can be used to examine a single grade level, a grade span (e.g., grades 3–5), or a vertical progression K–12. The remainder of this chapter presents examples of how CTS is used in different ways educators encounter in their work. As you become familiar with CTS, you may find other ways to utilize the process and resources that are not listed here.

Using CTS to Enhance Content Knowledge

A science-literate adult is one who has a basic understanding of the science needed to be a productive and informed participant in today's world. A science-literate person is not necessarily one who has majored in one of the sciences in college or works in a science-related career. Science literacy applies to all adults, regardless of their post-secondary education or career choice. It applies to all teachers of science including those who have a college degree in science as well as education generalists who teach all subject areas, including science. When using CTS with teachers, it is important to explain what science literacy means in the context of CTS.

A strategy teachers frequently use to gain or refresh their own adult content knowledge before teaching a new topic is to look in their textbooks, teacher guides, or the internet to get the information they need. While instructional materials may include the same content encountered by students, they often only superficially cover the content with an emphasis on terminology and facts. Furthermore, they often do not provide rich explanations, relevant phenomena, interesting examples, or explanatory models that support adult understanding of the content.

CTS provides a different alternative to developing or enhancing one's science (and engineering) content knowledge. CTS does not replace formal content coursework or content-focused professional development. Instead it provides a systematic way to engage with the content a science-literate adult should understand and be able to use, through selected readings that can be used on a "need-to-know" basis or embedded within a variety of professional learning formats.

In addition to disciplinary content, section I of CTS also clarifies the content that makes up the understandings for engaging in scientific or engineering practices and using crosscutting concepts. For example, crosscutting concepts are new to some teachers. Section I readings can build teachers' own understanding of a crosscutting concept, such as Patterns, in which the reading from the *Framework* explains the relationship between the way data are represented, pattern recognition, and the development of mathematical expressions.

Most new standards and curricula include scientific and engineering practices, such as developing and using models to understand and explain phenomena or make predictions. A teacher might decide to learn more about models before designing learning experiences in which students develop or use models. Section I of the *Developing and Using Models* CTS guide describes what models are and how they are used in science or engineering. An excerpt from IA *Science for All Americans* can be used to understand what a model is and the different types of models. The reading goes on to describe each of the different types of models and considerations for selecting a model.

> A model of something is a simplified imitation of it that we hope can help us understand it better. A model may be a device, a plan, a drawing, an equation, a computer program, or even just a mental image. Whether models are physical, mathematical, or conceptual, their value lies in suggesting how things either do work or might work. For example, once the heart has been likened to a pump to explain what it does, the inference may be made that the engineering principles used in designing pumps could be helpful in understanding heart disease. When a model does not mimic the phenomenon well, the nature of the discrepancy is a clue to how the model can be improved. Models may also mislead, however, suggesting characteristics that are not really shared with what is being modeled. Fire was long taken as a model of energy transformation in the sun, for example,

but nothing in the sun turned out to be burning. (American Association for the Advancement of Science, 1989, p. 168)

The section IA reading from *Science for All Americans* can be combined with the section IB narrative in the *Framework for Science Education*, which focuses more on conceptual models. Additional descriptions of ways models are used in science and engineering are provided in the *Framework* reading. Taken together, the two readings provide background information about models that can help teachers build their own understanding of models to effectively incorporate the practice of developing and using models in their curriculum, instruction, and assessment.

When teachers have a strong understanding of the content they teach, they are able to be more versatile in quickly and effectively responding to students' questions, ideas, and learning needs. A teacher with a working knowledge of the content they teach knows the best question to ask to push student thinking and is better able to steer and guide students' learning down the most appropriate path.

Whether you are using CTS individually to examine what every science-literate adult should know about a topic or you are using it in a professional learning format such as a workshop, science methods class, summer institute, or an online professional learning community, the following are suggestions for using CTS to build content knowledge:

- Identify and discuss the big ideas or components of practices that are the culmination of a K–12 science education.

- Identify and discuss examples that illustrate and explain key ideas or practices.

- Look for relevant terminology all adults should be familiar with and clarify definitions.

- Find examples of ways the content integrates across the sciences, mathematics, technology, or engineering.

- Look for vivid descriptions, real-life examples, phenomena, and analogies that help make the content comprehensible.

- Look for and explain theories, laws, scientific principles, or generalizations.

- Make connections between the content from the CTS reading and what you are doing in a professional development setting (e.g., content immersion with a scientist).

Using CTS to Inform Curriculum

Curriculum is the way content is organized and generally consists of a scope and sequence of learning goals and activities for learning. It refers to the knowledge and practices in topic areas that teachers teach and that students are expected to learn (NRC, 2012). Standards by themselves are not a curriculum. They outline the learning goals, which inform the selection of materials, activities, tasks, discussions, and so on, that make up the curriculum.

Backwards Design. Backwards design, a process and framework developed by Grant Wiggins and Jay McTighe (2005), is a method used to develop curriculum by first establishing the goals for learning and being clear about what an understanding of those goals looks like, before selecting instructional methods, activities, and assessments. It is a focus on the output before selecting the input. CTS can be thought of as the "upfront part of backwards design." By first doing CTS, learning goals that support the performance expectation can be clarified, boundaries for assessment can be established, relevant phenomena can be selected, and suggestions for effective instruction can inform the development of activities and tasks.

Unpacking Learning Goals or Standards into Subcomponents. Unpacking refers to the practice of breaking down a broad learning goal or standard into its component parts to more precisely identify what students should know and be able to do. It involves a process of clarifying what the true meaning and intent of a learning goal or standard is. Section VI of CTS describes the performance expectation(s) for a particular topic. It is the expectation of what students should be able to do by the end of a given grade, not by the end of the unit. It is used for assessment purposes but curriculum and instruction involve more than the performance expectation. To achieve the performance expectation, unpacking the disciplinary core idea(s), practice, and crosscutting concept that contribute to a performance expectation using CTS section II or section V is crucial. In designing curriculum and instruction, the elements that make up the disciplinary core ideas, practices, and crosscutting concepts are unpacked. Examining the elements of the three dimensions in section II (for example, the bullets in the foundation boxes of the *NGSS*) and discussing them with colleagues provide insight and clarification into what is to be learned by students and are used to inform the curriculum. The *Atlas* in section V helps educators visualize how one idea builds off another and the connections between ideas. The progressions of elements in section V are also used to unpack learning goals. Examining the clarification statements and boundaries for the performance expectation that includes the dimension(s) being unpacked helps pinpoint what is and what is not important to know. Furthermore, unpacking the practices allows educators to make decisions about multiple contexts in the curriculum in which a practice is used to figure out phenomena or solve a problem.

A single dimension, whether it is a disciplinary core idea, a scientific or engineering practice, or a crosscutting concept can be unpacked into the specific subideas or subpractices that make up that dimension. For example, after a middle school study of plate tectonics, the disciplinary core idea from NGSS ESS2.B, "Maps of ancient land and water patterns, based on investigations of rocks and fossils, make clear how Earth's plates have moved great distances, collided, and spread apart," can be broken down into the following subideas using CTS sections I, II, V, and VI:

- Rock formations and the fossil record help scientists reconstruct where land and oceans once were located.

- The rock layer beneath earth's surface is made up of huge sections of thick, solid rock called plates.

- Continents and ocean basins are part of these plates.

- These plates have moved great distances, collided, and spread apart.

- The movement of plates results in most continental (continental shelves, mountains) and ocean floor features (ridges, fracture zones, trenches).

These subideas come from the readings of the narrative sections as well as the ideas described in the *Framework's* grade band endpoints, NGSS disciplinary core ideas, and elements in the *Atlas*. For unpacking the three dimensions in more detail, including identifying student learning challenges (CTS section IV), brainstorming phenomena, and looking at intersections of crosscutting concepts with the other two dimensions, NSTA provides some very good tools on their NGSS Hub website at https://ngss.nsta.org/ngss-tools.aspx. Figure 4.4 summarizes a three-dimensional unpacking process.

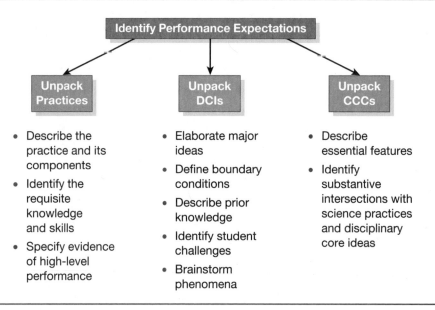

FIGURE 4.4 Unpacking the three dimensions of a performance expectation

SOURCE: Graphic by Ted Willard based on procedures described in Creating and Using Instructionally Supportive Assessments in NGSS classrooms (NSTA Press, in press) by Harris, C. J., Krajcik, J. S., & Pellegrino, J. W.

Three-Dimensional Curriculum. While earlier science standards and frameworks communicated the importance of inquiry, most publications placed these expectations at the front of the standards document—as a separate section. When teachers were running short on time, they typically focused on what they perceived to be the highest priority—the content, attempting to efficiently "cover it all," often with an emphasis on facts and terminology.

The *Framework for K–12 Science Education* articulates a vision where students, over multiple years of school, actively engage in scientific and engineering practices and apply crosscutting concepts to deepen their understanding of the core ideas in science and engineering. The *NGSS* and most state standards now expect students to be learning science by "being scientists." It is no longer sufficient to know facts and terminology; students are expected to make sense of the world around them through developing explanations of concepts and phenomena and designing solutions supported by evidence-based arguments and reasoning.

The term *three-dimensional learning* refers to the three pillars that support each standard in the *NGSS* or other similar standards. These three dimensions are the Science and Engineering Practices, Crosscutting Concepts, and Disciplinary Core Ideas. As teachers shift toward a three-dimensional learning approach, students are making sense of phenomena by building models, designing investigations, sharing ideas, analyzing data, making claims based on evidence, engaging in scientific argumentation, and applying new knowledge to other situations. They use crosscutting concepts such as cause and effect or systems to develop a coherent and usable understanding of science. The nature of science and engineering is also naturally embedded in the classroom environment, strengthening important 21st-century skills such as communication, collaboration, and critical thinking.

This three-dimensional approach to teaching and learning is exciting but complex. CTS guides can help teachers clarify the intent of each of the three dimensions, as well as what is intended by "Nature of Science." We recommend that before a teacher, school, or district interweave these dimensions into a unit of study, a curriculum, or an instructional approach, they take the time to examine each of the specific dimensions of interest using a CTS guide.

For example, a PLC group decides to select a topic from the life science CTS guides, a scientific practice guide, and a crosscutting concept guide. They decide to focus on CTS sections II, III, V, and VI for their grade level. They also study section I to ground their own knowledge of the topic and section IV to be aware of difficulties students might have in understanding the key ideas in their unit of instruction. They conduct a study of each of the three guides and merge the results to discuss the implications for organizing their curricular unit topic using a three-dimensional approach.

Another approach to developing curricular units is to use *bundles*. Bundling is when related performance expectations are combined to create the endpoints for a unit of study. Bundles allow for more efficient use of instructional time by teaching ideas that were traditionally taught in separate units, to be taught together as a coherent unit of interconnected ideas. Sample bundles and tools for bundling can be viewed on the NGSS website at https://www.nextgenscience.org/resources/bundling-ngss. After conducting a CTS, the results from the study can be used to create your own bundles or examine existing bundles through the lens of CTS.

Curriculum Coherence and Articulation Across Multiple Grades. Putting together a multigrade science curriculum (e.g., K–5; 6–12, K–12) aligned with standards is not an easy task. It is even more difficult when committee members lack the necessary tools and resources to undertake this arduous work. A curriculum scope and sequence can be compared to a jigsaw puzzle:

> Imagine that we are faced with a pile of jigsaw puzzle pieces and told to put them together. Our first reaction might be to ask for the picture. When we put together a jigsaw puzzle, we usually have a picture to guide us. None of the pieces means anything taken alone; only when the pieces are put together do they mean something. (Beane, 1995, p. 1)

CTS provides the picture needed to put the necessary pieces together in a way that they make sense for students (coherence). Examining a curricular topic (section II) is like holding a jigsaw puzzle piece up to see roughly what area of the puzzle you should put it in. After getting some of the initial pieces laid out, the interconnections between concepts, ideas, and practices in CTS section V help connect groups of puzzle pieces and the progression links groups of pieces together. The CTS results are the *picture* you keep looking at to make sure the pieces fit together and are not disconnected, such as a K–12 or other grade span curriculum. Figure 4.6 shows elementary teachers sharing results of a CTS as a lens to examine their curriculum.

A coherent curriculum is one that holds together, that makes sense as a whole; and whether you use two dimensions or three dimensions, the dimensions are unified and connected by that sense of the whole. This involves thinking through the flow of ideas and practices in the three dimensions and across grades to determine

- The important set of concepts, ideas, and practices students should learn
- Which crosscutting concepts and practices will support a disciplinary core idea, and across the curriculum, how to ensure they all receive sufficient attention
- The connections among the concepts, ideas, and practices that support three-dimensional learning
- Which concepts, ideas, and practices need to recur frequently and in varied contexts across disciplines and grades
- Important prerequisites leading to increasing sophistication of concepts, ideas, and use of practices
- Connections to the nature and enterprises of science, engineering, and technology

FIGURES 4.5 (TOP) AND 4.6 (BOTTOM)   Elementary teachers using CTS for curriculum decisions

Selecting Curriculum. The choice of instructional materials can have as much of an impact on student learning as improvements in pedagogy (Chingos & Whitehurst, 2012). CTS is helpful for reviewing and selecting curriculum materials. Beware of instructional materials that claim to be standards or *NGSS* "aligned." Regardless of whether you adopt the *NGSS* or other standards, CTS is used as a lens to examine instructional materials with an eye for determining whether they are informed by standards and research. CTS is not a replacement for a rigorous and thorough curriculum analysis procedure. There are several very good processes for reviewing curriculum materials. CTS can help you use these processes with increased validity and reliability by first studying the curricular topic. For example, the EQuIP rubric (Educators Evaluating the Quality of Instructional Products) provides criteria used to evaluate the extent to which lessons and units are designed to meet the *NGSS*. The *NGSS* Lesson Screener includes fewer criteria and is less rigorous than the EQuIP rubric yet provides a quick screen to see if a sequence of lessons is on track. These resources can be used with CTS and are found at https://www.nextgenscience.org/resources/equip-rubric-lessons-units-science.

Implementing Curriculum. Curriculum implementation involves the classroom use of new instructional materials. As schools and districts adopt new instructional materials that reflect the current vision of standards, teachers will need to understand their part in a multiyear scope and sequence and how to support students in building on their prior knowledge and use of the three dimensions (NRC, 2015). They may have to learn the major concepts and ideas of new disciplinary content they have not taught before or even deepen their understanding of familiar disciplinary content and how it is interwoven with practices and crosscutting concepts. To do this teachers need opportunities to communicate and collaborate with other teachers teaching with the same materials as well as across grade levels. CTS provides a process to help teachers do this. It helps them improve their understanding of the content of the curricular topic(s) they are teaching (CTS section I); understand the meaning and intent of the curricular objectives (CTS section II); be aware of the research on learning that may have informed the development of the materials or that they should be aware of as they use the materials (CTS section IV); and understand how one concept, idea, or practice contributes to or connects with another (CTS section V); as well as examine the assessment expectation, and link back to section III by examining the curricular unit's lessons and other support material.

If you are leading curriculum implementation for a grade level, consider creating a customized CTS guide to match the curricular unit teachers will be implementing. There is a template for creating a customized guide on the CTS website at www.curriculumtopicstudy2.org. Include only readings that relate to the curriculum unit being implemented. For example, fifth-grade teachers using the FOSS Mixtures and Solutions kit may combine the CTS guide Mixtures and Solutions with the Developing and Using Models CTS guide, focusing on readings for grade 5 using the CTS resources available to teachers. For section III, they look at the module after completing the other sections of CTS and discuss how the lessons support what they learned through the CTS, including supplemental suggestions that might strengthen the unit. Figure 4.7 shows an example of a customized guide for curriculum implementation. For this guide, the teachers are using the online resources and the *NSTA Quick Reference Guide to the NGSS*. Since there are not enough copies of the Section IV resource, the facilitator shares results from that section using her copy of *Making Sense of Secondary Science*. After completing the CTS, teachers have a lens through which to view their materials and focus their instruction.

FIGURE 4.7 Grade 4 FOSS Mixtures and Solutions Module

FOSS Mixtures and Solutions Module
Grade 4 Standards- and Research-Based Study of a Curricular Topic

Section and Outcome	Selected Sources and Readings for Study and Reflection Read and examine **related** parts of
I. Content Knowledge	**IA:** *Science for All Americans* • Ch. 11: Models, pp. 168–172 **IB:** *Framework for K–12 Science Education:* Narrative Section • Ch. 5; PS1.A: Structure and Properties of Matter, pp. 106–107
II. Concepts, Core Ideas, or Practices	**IIA:** *Framework for K–12 Science Education:* Grade Band Endpoints • Ch. 5: PS1.A: Structure and Properties of Matter, p. 108 (focus on grades 3–5) • Ch. 3: Developing and Using Models, p. 58 **IIB:** *NSTA Quick Reference Guide to the NGSS K–12:* Disciplinary Core Ideas Column • Grade 5: PS1.A: Structure and Properties of Matter, pp. 112–113; PS1.B: Chemical Reactions, pp. 112–113 • 3–5: Developing and Using Models, p. 100
III. Curriculum, Instruction, and Formative Assessment	**IIIA:** *FOSS Mixtures and Solutions Module* • Examine the lessons and the teacher support material after completing the other sections of CTS
IV. Research on Commonly Held Ideas	**IVA:** *Benchmarks for Science Literacy:* Chapter 15 Research • 11B: Models, p. 357 **IVB:** *Making Sense of Secondary Science: Research Into Children's Ideas* • Ch. 8: Mixtures and Substances, pp. 74–75 • Ch. 9: Dissolving, pp. 83–84 • Ch. 10: Mixtures of Substances, p. 85 • Ch. 11: Particle Ideas About Solutions, p. 95 • Ch. 12: Dissolving Substances in Water, pp. 100–101
V. K–12 Articulation and Connections	**VB:** *NSTA Quick Reference Guide to the NGSS K–12:* Progression • PS1.A: Structure and Properties of Matter, p. 61 • PS1.B: Chemical Reactions, p. 62 • Developing and Using Models, Condensed Practices, p. 51
VI. Assessment Expectation	**VIA:** *State Standards* • Examine your state's standards **VIB:** *NSTA Quick Reference Guide to the NGSS K–12:* Performance Expectations • Grade 5: 5-PS1-1, 5-PS1-2, 5-PS1-3, 5-PS1-4, pp. 112–113

Integration and Interdisciplinary Connections. Many schools, districts, and afterschool programs are striving to bring "the real world" into their curriculum through interdisciplinary learning. We applaud these efforts and appreciate their value in supporting the 21st century skills of collaboration, creativity, critical thinking, and problem solving. The challenge lies in ensuring that the interdisciplinary connection goes beyond providing an engaging, fun activity to one that possesses academic rigor and content integrity. We have worked with several projects and initiatives where CTS has been used to ensure that students learn science by applying important ideas through use of scientific and engineering practices.

Project Based Learning (PBL). PBL is a teaching method in which students gain knowledge and skills by investigating an authentic question, problem, or challenge. In-school and after-school educators who embark on a PBL approach benefit from conducting a CTS study of the science topic that will be investigated, asking: What should we as the facilitators of learning know? What are the concepts, ideas, and practices we can incorporate as we design and facilitate our PBL curriculum? What are appropriate expectations for our age group of students? "Gold standard" project-based learning has design elements that include focusing on key knowledge, understanding, and success skills. CTS can clarify these elements so that students' opportunity to learn the science is not lost in the project.

Service Learning. The National Youth Leadership Council defines *service learning* as an approach to teaching and learning in which students use academic knowledge and skills to address genuine community needs. We have collaborated extensively with in-school and after-school programs that are working with the KIDS (Kids Involved Doing Service learning) model, which is based on three key principles: Academic Integrity, Apprentice Citizenship, and Student Ownership. When our service-learning partner first experienced a CTS study of Ecosystems, she exclaimed, "This is the best thing since sliced bread!" She finds that CTS helps her participants, who often do not have a science background, understand how science learning goals can be supported through service-learning projects. For an example of a free downloadable resource that uses CTS in service learning projects, go to http://harkinsconsultingllc.com/products/integrating-scientific-practices-and-service-learning-engaging-students-in-stem/. (Note: This resource uses the first edition of Curriculum Topic Study.)

Stem Integration. Engineering has become a component of many science programs, as well as a stand-alone course, with the expectation that all students have an opportunity to experience the engineering design process across the K–12 continuum. Many schools struggle with how they can fit engineering into an already overburdened curriculum. Groups who conduct a CTS study of the Science and Engineering Practices or Engineering Design see the relatedness of engineering with science concepts they teach and are able to combine the two—either applying their understanding of a science concept to generate creative solutions to a problem, or arriving at a "need to know" moment during their engineering design lesson, where a science concept is then explored. For section III, a useful resource can be added that will be available in 2020: *Uncovering Student Ideas About Engineering and Technology* (Keeley, Sneider, & Ravel, in press). This resource can be used with the category D and F guides.

The Committee on STEM Integration (National Academy of Engineering and National Research Council, 2014) advocates a more integrated approach to K–12 STEM education, particularly in the context of real-world issues. While many teachers, schools, or districts want to bring STEM into the classroom, it is frequently implemented in a disjointed manner, resulting in S, T, E, and M as separate disciplines. CTS can help STEM educators understand what is meant by science, technology, engineering, and mathematics integration, so groups can move beyond the superficial STEM integration to a deeper, cohesive instructional design that interweaves but does not force-fit the disciplinary fields. The category D guides can be used to study the engineering practices as well as the science and engineering practice of using mathematics and computational thinking. The category F contains the STEM Connections guides. These guides are divided into two sections. The engineering, science, and technology guides focus on understanding the engineering design process, including how science is utilized in the process and how new technologies result. It includes both the distinctions of and the relationships between engineering, science, and technology and their impacts on society and the natural world we live in. The Nature of Science guides focus on understanding the nature of scientific knowledge. It is distinct from engaging in the scientific practices because the focus is more on understanding the enterprise of science. Together all twelve guides in category F can strengthen teachers' understanding of how STEM is represented in the classroom.

Using CTS to Inform Instruction

Instruction refers to methods of teaching and the sequencing of learning activities used to help students achieve the learning objectives specified in their curriculum (NRC, 2012). It includes both the active involvement of the teacher and the student in carrying out activities, investigations, discussions, and other learning opportunities that help them learn and use science concepts, ideas, and scientific and engineering practices. For example, a group of elementary teachers summarize their findings from section III of the Developing and Using Models CTS guide (Figure 4.8). They will use the summary to inform their instruction when they have their students develop models.

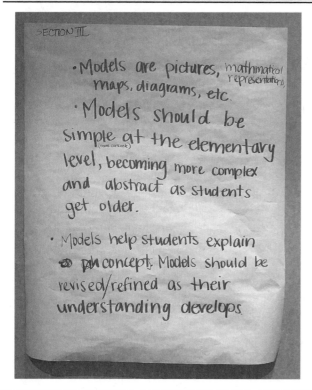

FIGURE 4.8 Charting CTS results from a study of Developing and Using Models

Modifying Lessons. The vision of the *Framework for K–12 Science Education* calls for coherent investigation of core ideas across multiple years of school, with a seamless blending of the Science and Engineering Practices, Crosscutting Concepts, and Disciplinary Core Ideas. It is imperative that educators acknowledge the implications this shift has for planning lessons.

It is unrealistic to expect that districts will start from scratch, and experienced teachers often have lessons that have been shown to engage students and stimulate their curiosity. CTS can be used to examine such lessons to determine how they might fit into the current vision and goals for science teaching and learning. By studying a curricular topic, teachers can examine their lessons through the lens of CTS, asking

- Does my lesson provide an opportunity for students to use a scientific or engineering practice and crosscutting concept?

- Does my lesson uncover preconceptions and connect to prior experiences my students have had?

- Is my lesson based on an anchoring phenomenon that will generate interest, stimulate curiosity, and raise questions related to a scientific concept? Or is my lesson based on an interesting problem to solve?

- Does my lesson take the time to examine, investigate, and puzzle through ideas and phenomena and construct initial concepts and explanations or solve a problem using the engineering design process?

- How will students develop conceptual and procedural understanding? How will I introduce formal concepts and vocabulary linked to students' experiences so they can engage in sense making and construction of scientific explanations or solutions to problems?

- How will students apply concepts, skills, and explanations to new contexts, and transfer learning to new, related situations?

- If I am going to deepen understanding by incorporating a 5E approach of Engage, Explore, Explain, Elaborate, Elaborate, then this will take more time. How can my lesson fit within this instructional model?

- Does my lesson provide an opportunity for students to learn the concepts and ideas and use the practices that will later be assessed?

Teachers are encouraged to make these shifts one step at a time. Take one lesson or unit of study that relates to an important topic in science such as Cycling of Matter in Ecosystems, Properties of Matter, or Forces Between Objects. Combine the topic with the study of a scientific or engineering practice and a crosscutting concept that can be part of a lesson. Examine your existing lesson, and use the results from CTS sections II, III, IV, V, and VI to modify your lesson based on your study's findings. After you implement the modified lesson in the classroom, take a moment to reflect on how your lesson has changed. How did these shifts impact your students? How can you incorporate these insights into other lessons?

From Inquiry-Based Teaching to Scientific and Engineering Practices. Teachers have worked hard to make inquiry the centerpiece of their science teaching for several decades. While inquiry is still very much a part of doing science, the focus has shifted to scientific and engineering practices where the role of inquiry has been enhanced and involves simultaneously building new knowledge while using the range of intellectual and social processes required to participate in science and engineering. There are ten category D CTS guides that address a scientific practice, an engineering practice, or a practice that is used in both science and engineering. To design instruction that combines learning content with use of a practice, a CTS guide from categories A, B, or C can be combined with a category D guide. Each topic is studied together, and the results are merged to understand how students will use the practice to learn important concepts and ideas.

For example, a study of the Earth, Moon, and Sun System, focusing on middle school, is combined with a study of Developing and Using Models. Questions that guide the study might include

- **CTS section II:** What are the specific ideas students learn about the Earth, Moon, and Sun System? What are the ways students are expected to develop and use models? How can I merge these ideas with these elements of the practice?

- **CTS section III:** What are some effective ways students learn about the Earth, Moon, and Sun System? What are some ways students are supported in using this practice? What instructional experiences can students have that combine learning about the Earth, Moon, and Sun system with developing and using a model to explain phenomena associated with this system? Is there a formative assessment probe I could use to uncover their initial ideas about Earth, Moon, and Sun phenomena that could also reveal how students use a model?

- **CTS section IV:** What does the research say about commonly held ideas and difficulties I need to be aware of when students are trying to make sense of Earth, moon, and sun phenomena? What can the research tell me about difficulties students might have using models that I should be aware of when using Earth, moon, and sun models?

- **CTS section V:** In looking at the progression and prerequisite ideas for both topics, how will I be building on prior ideas and use of the practice? Are there prerequisites I should first check on to make sure my students understand and can use prior ideas and elements of the practice?

- **CTS section VI:** How will my instruction prepare students for the assessment? Will they be able to demonstrate how the practice is used with Earth, Moon, Sun system phenomena? What boundaries for the assessment should I be aware of?

Selecting Phenomena. One of the major shifts in instructional practices today involves phenomenon-based teaching and learning. Natural phenomena are observable events or processes that occur in the natural world that are explanatory or predictive. To develop knowledge of the core concepts and ideas in science, students build ideas by gathering and using evidence that can be used to explain or predict phenomena. Phenomena are also used in engineering. Problems may arise from phenomena (such as how to prevent erosion of a section of a riverbank) that can be solved using the engineering design process. By centering instruction on phenomena, the teaching and learning phase shifts from learning about a topic to figuring out why or how something happens related to the topic. A good resource for learning more about phenomena is on the Next Generation Science Standards website at https://www.nextgenscience.org/resources/phenomena.

The challenge for educators is to select phenomena that are relevant and can be used to drive instruction. CTS can help with this process. CTS section I provides an overview of the content to be learned and may give insight into contexts that include phenomena. Sometimes a phenomenon is described in CTS section I. For example, in section IA of the CTS guide Visible Light and Electromagnetic Radiation there is an explanation for why the sky looks blue. This phenomenon could be used with middle or high school students to explain how the scattering of different wavelengths of light affects the color we see. CTS section III describes instructional contexts and may offer suggestions for relevant phenomena, including how the phenomenon is used to drive instruction. For example, in section IIIA of the Inherited Traits CTS guide, a third-grade teacher might be looking for phenomena that can be used to explain how some traits are inherited, some are influenced by the environment, and some are both. Examples of phenomena such as how a flamingo's pink color is influenced by its diet and how burrowing and nesting behaviors can be both instinctive (inherited) and learned provide explanatory evidence that traits can be inherited and/or influenced by the environment.

A collection of phenomena is archived at https://www.ngssphenomena.com. After studying a disciplinary topic, phenomena on this site can be analyzed through the lens of the CTS study to determine its usefulness in a phenomenon-based lesson. Questions to ask about a phenomenon after doing CTS include

1. What is the phenomenon and how do you explain it?
2. What key idea(s) can be used to explain the phenomenon?
3. Where, when, and how can the phenomenon be used in the lesson?
4. What are some difficulties to anticipate or commonly held ideas students might have about the phenomenon?
5. What prerequisite ideas are needed for students to use the phenomenon?
6. How can the phenomenon be used for assessment?
7. Overall reflection: How does the phenomenon support students' learning?

Using CTS to Inform Classroom Assessment

The new vision of teaching and learning presents not only considerable challenges but also a unique and valuable opportunity for assessment (NRC, 2014). *Classroom assessment* refers to assessments designed or selected by the teacher that can be used to inform instruction, provide feedback to the learner, monitor changes in students' thinking and learning, or measure and document the extent to which students' have achieved learning goals after they have had an opportunity to learn following a lesson, activity, or curriculum unit. The first step in using CTS to inform assessment is to define the purpose and stage in the assessment process:

Diagnostic Assessment. What preconceptions and existing ideas do my students have about the scientific concepts and ideas in the topic? CTS sections II (or V) and IV can be used to examine the specific concepts and ideas in the topic that students will need to know and understand as well as the research on how students think about concepts and ideas. Section III provides examples of questions that reveal students' thinking. All this information can be used before and during classroom instruction to elicit ideas and ways of thinking that students bring to their learning.

Formative Assessment. When teachers use diagnostic assessment data to inform their instruction, it becomes formative assessment. Formative assessment also includes a feedback loop between student(s) and the teacher and the teacher and student(s). During formative assessment the teacher gathers evidence of how students are building their understanding of concepts and ideas as well as how they use scientific and engineering practices. It is used as a checkpoint at any point during an instructional cycle. The information feeds back into the teachers' instructional plan to make modifications that will move students toward the learning goal(s). CTS section II helps teachers identify clear goals students are moving toward. CTS section IV helps teachers anticipate the commonly held ideas students may bring to or develop during their learning. Together these two sections can be used to develop formative assessment probes. Section III helps inform the questions, representations, contexts, phenomena, and other instructional strategies and experiences that may be effective in building a bridge between students' existing ideas and use of practices to the scientific or engineering understandings and use of practices that make up the goals for learning. Section V indicates possible gaps or steps along the way that may need to be revisited or assessed.

Summative Assessment. Summative assessments are given after students have had the opportunity to learn and use the concepts, ideas, and practices. CTS sections II and V clarify the specific learning goals that make up a disciplinary core idea, practice, or crosscutting concept. Section VI clarifies how these three elements come together in the form of a performance expectation. CTS can be used as a lens to determine if the assessment fairly targets the learning goals. It can also be used to determine other contexts that may be used to gather evidence of whether students can transfer their learning to other situations, phenomena, or problems.

CTS provides information teachers can use to help teachers develop classroom assessments or be better consumers of assessment and assessment data. Some specific assessment applications of CTS include the following.

Developing Learning Intentions and Success Criteria. Many districts mandate that for each lesson or period of instruction, the teacher must post the learning objective and share it with students. Often this becomes what has been referred to as the "wallpaper objective" (Wiliam, 2011). In other words, it is posted and shared with the students at the beginning of the lesson but then ignored for the rest of the lesson. This token approach is not what is meant by sharing, clarifying, and using learning goals so that students can monitor their own progress toward meeting a learning goal, a key component of formative assessment that provides feedback to the learner.

Learning intentions (sometimes called learning targets) are used to determine the goal for a lesson and make the purpose of the lesson explicit to students so they know what they are expected to learn and do during the lesson (Keeley, 2015). Learning intentions are accompanied by success indicators that gauge the extent to which a learning goal has been met. They are sometimes referred to as "I can" statements. Taken together they do the following:

- Give students a clear idea of what will be learned and why
- Transfer the responsibility for learning to the student (no teacher can do the learning for the student)
- Provide students with a way to monitor their learning
- Help students focus on the purpose of the lesson and what they should be learning rather than merely on the completion of the activity
- Help teachers review progress and provide a clearer focus for instructional next steps
- Help break down broad standards or goals

To develop learning intentions and success criteria, teachers can use CTS section VI to first determine what students will be expected to know and perform on a final assessment. CTS sections II and V describe the specific elements and the progression or "steps along the way." Teachers examine their lessons and develop learning intentions and success criteria that are specific to a lesson or lesson sequence.

For example, in a fourth-grade unit on light, the teacher used the CTS guide Visible Light and Electromagnetic Radiation. Using CTS section VI, the teacher examines the performance expectation that will be required as part of their district end-of-year testing program. The performance expectation states: "Develop a model to describe that light reflecting from objects and entering the eye allows objects to be seen" (NGSS Lead States, 2013). As she unpacks this performance indicator she notes that one of the things students will need to know to achieve this performance expectation is how light reflects off objects. She then uses CTS section II to examine the disciplinary core idea that states, "An object can be seen when light reflected off its surface enters the eye." Using CTS section V, she looks at what students learned about light and reflection prior to grade 4. In grades K–2 they learned that some objects give off their own light and that mirrors can redirect beams of light. Examining CTS section IV she finds that students have a commonly held idea that only shiny things like mirrors reflect light.

She uses this information to develop a lesson that all objects we see can reflect light and that they do not have to be shiny like mirrors to reflect light. She has students design an investigation to collect evidence that ordinary objects, both dull and shiny, can reflect light. She develops her learning intention and success criteria for the lesson, which she will use to make the purpose of their learning explicit and allow students to self-monitor how well they are meeting the goal of the lesson. Her learning intention and success criteria, informed by CTS sections VI, II, IV, and IV, are:

Learning Intention #1: Understand how light is reflected from objects.

- I can design an investigation to test whether an object reflects light.
- I can show what happens when light strikes a mirror or other type of smooth, shiny object.
- I can show what happens when light strikes an object that is not smooth or shiny.

Following the lesson on reflection, she designs a lesson that will involve students in developing models to explain how they see the light from a reflected object. Using the same sections of CTS, she develops another learning intention:

Learning Intention #2: Understand the role of light in how we see objects.

- I can draw a diagram that shows the path of light to and from an object.
- I can use my model (the diagram) to explain how we see objects.
- I can use my model to describe what happens when you look at an object in a totally dark room.

Because learning is not in the materials or tasks themselves, CTS helps teachers develop learning intentions and success criteria that help students make conceptual connections and use the practices of science and engineering as they manipulate materials and complete activities and investigations. Thus they become active participants in their learning rather than passive recipients (Heritage, 2010).

Formative Assessment Probes. Assessment probes are used diagnostically to elicit students' ideas. When the data are used to inform instruction, and monitor changes in students' thinking, they become formative assessment probes. Formative assessment probes are two-tiered. The first tier includes a prompt with selected answer choices, and the second tier has students construct an explanation to support their answer choice. Formative assessment probes are designed to uncover how students are thinking about a phenomenon or concept. The use of formative assessment probes helps make students' thinking visible to themselves, their peers, and the teacher. Figure 4.9 is an example of a formative assessment probe that utilized the CTS guide Weathering and Erosion.

FIGURE 4.9 Example of a formative assessment probe: Grand Canyon

Six friends were standing along the rim of the Grand Canyon. Looking down, they could see layers of rock and the Colorado River at the bottom. They wondered how the Grand Canyon formed. They each had a different idea. This is what they said:

Natara: I think the Grand Canyon was formed when Earth formed. It has just gotten bigger over time.

Cecil: I think the Grand Canyon formed from earthquakes that cracked open the land and pulled it apart.

Garth: I think the Colorado River and streams slowly carved out the Grand Canyon.

Robert: I think a huge flood rushed through the land and formed the Grand Canyon.

Kumiyo: I think the river got so heavy that it sunk down through the rock and formed the walls of the Grand Canyon.

Luna: I don't agree with any of your ideas. I think the Grand Canyon was formed in some other way.

Who do you think has the best idea? _____ Explain your thinking.

CTS has been used to develop assessment probes for the *Uncovering Student Ideas in Science* series as well as with teachers to develop their own formative assessment probes. Steps in using CTS to design formative assessment probes are:

1. Identify the CTS guide for the unit topic you are teaching. For the example in Figure 4.9, the guide Weathering and Erosion was selected.

2. Examine section II or V to identify the specific concepts and ideas in your curricular unit. For the example in Figure 4.9, the disciplinary core idea, the role of water in Earth's surface processes, was identified. One of the elements that make up this disciplinary core idea describes how water's movement causes weathering and erosion and can change land's surface features. This idea will be the focus of the formative assessment probe.

3. Examine CTS section IV to learn more about commonly held ideas students might have related to weathering and erosion.

4. Select a phenomenon that can be the focus of the prompt that will elicit students' ideas about the role of water in the weathering and erosion of surface features. For the example in Figure 4.9, the formation of the Grand Canyon was selected as the phenomenon.

5. Develop the prompt and answer choices. Distracters should mirror the commonly held ideas from the study of the research in CTS section IV. Include a best answer, which for the example in Figure 4.9 is answer choice Garth.

6. Add a second part to the probe in which the student provides an explanation for the answer choice.

For the example in Figure 4.9, the teacher uses the formative assessment as an initial elicitation to uncover ideas students bring to their learning that they use to explain a long-term weathering and erosion phenomenon such as the Grand Canyon. The teacher then uses the information about students' thinking to design instructional experiences. During instruction students will revisit their initial ideas, modifying them as they gather evidence from investigations and other instructional opportunities. The formative assessment probe can be used again after students have had the opportunity to figure out the phenomenon and use their ideas about how moving water, which carried small pieces of rock over a very long period of time, carved out the Grand Canyon. The assessment probe can now be used to provide evidence of the extent to which students are able to use the specific ideas about weathering and erosion.

The sections in teacher notes that accompany each of the probes in the *Uncovering Student Ideas in Science* series mirror the results of a CTS. For example, CTS section I informs the explanation. CTS section II and VI informs the section that lists related disciplinary core ideas and performance expectations. The summaries of related research in the teacher notes are informed by CTS section IV. The suggestions for further assessment and instruction are informed by CTS section III. These teacher notes are a good example of how CTS can be used in a variety of ways, including developing support materials for teachers.

Summative Three-Dimensional Classroom Assessment Tasks. To measure the three-dimensional science learning described in the *Framework* and the *NGSS* requires assessment tasks that examine students' performance of scientific or engineering practices in the context of crosscutting concepts and disciplinary core ideas (NRC, 2014). To develop such assessments that are aligned with three-dimensional performance expectations requires careful thought and development. CTS can be used to develop rich, culminating performance tasks. After selecting a performance expectation or bundle of performance expectations from CTS section VI, the appropriate CTS topic guides are selected for each dimension. CTS section II is used to unpack the ideas and practices that will be assessed in a multipart performance task. CTS section V is examined to determine if students have the necessary prerequisites to complete the task and if multiple related and connected ideas can be combined in a task. CTS section III can be used to determine if students had appropriate instructional opportunities to learn the ideas and practices leading up to the task. All this information feeds into the development of a complex task and is further supported by using assessment tools and resources available on the *NGSS* website at https://www.nextgenscience.org/assessment-resources/assessment-resources and guidance from the resource, *Developing Assessments for the Next Generation Science Standards* (NRC, 2014).

OTHER APPLICATIONS OF CTS

There are more applications of CTS other than the ones described above. There are also other existing tools and resources that can be used with CTS such as the *NGSS* Storylines, the *NGSS* course descriptions, curated *NGSS* lessons, and more. Many of these tools and resources that complement CTS can be accessed on the NSTA *NGSS* Hub at ngss.nsta.org. As you become familiar with CTS, you may find there are additional tools and resources that can be used with CTS or you might develop your own tools and templates to use for various CTS applications. Consider sharing tools you have developed that can be used with CTS. With your permission and citation, these tools can be shared on our *Curriculum Topic Study Second Edition* website. Contact either of the authors of this book (see bios for contact information) if you would like to share your CTS tools and applications.

CHAPTER 5

The Curriculum Topic Study Guides

This chapter contains the complete set of Curriculum Topic Study Guides. The organization and use of these guides is described in Chapters 2 and 3. There is also a template on the CTS website www.curriculumtopicstudy2.org for customizing a guide for use with groups that have a specific purpose or to develop your own CTS guide for a topic not included in this book. Since curriculum topics often overlap and intertwine, two or more guides can be combined and intertwined. A three-dimensional approach involves combining a disciplinary content guide with a scientific and engineering practice guide and a crosscutting concept guide.

The disciplinary content guides (categories A, B, C) and the STEM Connections and Nature of Science guides (category F) are arranged in alphabetical order within each subcategory. The scientific and engineering practices guides (category D) and the crosscutting concept guides are arranged in order of their appearance in the Framework.

CATEGORY A: LIFE SCIENCE GUIDES

The twenty-six CTS guides in this section are divided into three subsections. The guides focus on structures, processes, and relationships of living organisms. They range in scale from biomolecules, to cells, to organisms, populations of organisms, and ecosystems. Guides focus on processes that happen in the blink of an eye to processes that occur over billions of years.

Biological Structure and Function Guides focus on structure and function from the level of molecules to whole organisms. This section addresses the parts of organisms and how they function to support life. The alphabetically arranged guides in this section include

- Behavior, Senses, Feedback, and Response
- Brain and the Nervous System
- Cell Division and Differentiation
- Cells and Biomolecules: Structure and Function
- Characteristics of Living Things
- Energy Extraction, Food, and Nutrition
- Macroscopic Structure and Function of Organisms
- Organs and Systems of the Human Body
- Photosynthesis and Respiration at the Organism Level

Ecosystems and Ecological Relationships Guides focus on organisms' interactions with each other and with their physical environment, including obtaining and using resources, changing the environment and its effects on organisms, and social interactions and group behavior. This section ranges in scale from interactions between individual organisms, to populations, to entire ecosystems. The alphabetically arranged guides in this section include

- Cycling of Matter and Flow of Energy in Ecosystems
- Cycling of Matter in Ecosystems
- Decomposition and Decay
- Ecosystem Stability, Disruption, and Change
- Food Chains and Food Webs
- Group Behaviors and Social Interactions in Ecosystems
- Interdependency in Ecosystems
- Transfer of Energy in Ecosystems

Life's Continuity, Change, and Diversity Guides focus on reproduction, heredity, and the diversity of life over time. It focuses on the flow of information between parent and offspring, and between generations of organisms over both short and very long periods of time. This section addresses changes in populations and the factors that support both unity of life and life's diversity. The alphabetically arranged guides in this section include

- Adaptation
- Biodiversity and Human Impact
- Biological Evolution
- Diversity of Species and Evidence of Common Ancestry
- DNA, Genes, and Proteins
- Inheritance of Traits
- Natural Selection
- Reproduction, Growth, and Development
- Variation of Traits

NOTES FOR USING CATEGORY A GUIDES

Overall

- *Atlas* page numbers have not been provided because *The NSTA Atlas of the Three Dimensions* was produced concurrently with this edition. Titles and map codes are accurate.
- The same resources are not always included for each section. For example, some guides include readings from *Science Matters* for section I, others do not.
- Eliminate redundancy. Some readings include the exact same information. However, even when the information is the same, there may be an advantage in how the information is presented. Select the reading based on the resources you have available to use for CTS and/or the advantage of using one over the other.

Section I

- Readings from the *Framework* and *Atlas* narrative are exactly the same. Choose one of these resources for this section.
- When reading the *Framework*, stop at Grade Band Endpoints.
- Some *Atlas* maps combine topics. When using the *Atlas* narrative for this section, if there is more than one core idea included in the narrative, focus on the one that is listed on your CTS guide.
- When reading *Disciplinary Core Ideas* for this section, focus on the content that helps you understand this topic. There are also suggestions for instruction embedded in this reading that can be added to CTS section III.

Section II

- Readings from the *Framework*, *NGSS*, and *NSTA Quick Reference Guide* are practically the same. In a few cases, a *Framework*, idea was not included in the *NGSS* or was moved to a different grade span. In the *Framework*, goals are described in grade bands K–2, 3–5, 6–8, and 9–12. In the *NGSS* and *NSTA Quick Reference Guide*, they are phrased as disciplinary core ideas and designate a specific grade.

- The readings from the *NGSS* and the *NSTA Quick Reference Guide* are exactly the same. Choose one of these resources. An advantage to using the *NSTA Quick Reference Guide* is that the disciplinary core idea is matched to the performance expectation (section VI) on the chart.

Section III

- Readings listed for *Disciplinary Core Ideas* are the longest readings. In some guides these have been broken down into subsections. You can read all the designated pages or focus on the subsections that are of interest to you. If doing CTS with a group, you may consider assigning subsections so one person does not have to read all the designated pages.

- Sometimes readings about instructional implications from *Disciplinary Core Ideas* may mention and cite a research study on students' commonly held ideas or difficulties. This can be combined with section IV.

- Several readings may be listed from the *Uncovering Student Ideas in Science* series. This does not mean you need to read them all. Choose ones from the books you have available. The page numbers list the probe that can be used to elicit students' ideas, and the teacher notes contain suggestions for instruction that are designed to address the ideas elicited by the probe. Some of these books also have a section on curricular considerations. This can be added to your reading.

- There are always new books released in the *Uncovering Student Ideas in Science* series that may include probes that are not listed on the topic guides. Check the CTS website for a list of new probes published after 2019 that can be used with CTS.

Section IV

- If you are using the research summaries from the *Uncovering Student Ideas* series, they usually include references to the commonly held ideas identified in *Benchmarks* Chapter 15 Research and *Making Sense of Secondary Science* so there is no need to use all three of the resources listed. In addition, the *Uncovering Student Ideas* series includes research published after *Benchmarks* and *Making Sense of Secondary science*.

Section V

- Readings from the *NGSS* Appendix E and the *NSTA Quick Reference Guide* both describe a progression. The difference is the progression is summarized for each grade span in the *NGSS* Appendix E. In the *NSTA Quick Reference Guide*, they are listed by the disciplinary core ideas along with the code for the performance expectation. You can choose one or both of these resources, depending on the level of specificity you desire.

- The *Atlas* includes the same information as the *NSTA Quick Reference Guide*, but the visual mapping of the ideas allows you to see precursor ideas and connections, as well as connections to other maps. Be sure to read the front matter of the *Atlas* before using the maps. The information will help you use the maps effectively.

- The *Atlas* map often combines two or more topics. Follow the connections that match the topic you are studying.

Section VI

- The *NGSS* and the *NSTA Quick Reference Guide* provide the exact same information. Clarifications and assessment boundaries are also included in both. Choose one of these resources based on the advantages listed below or the resources you have available.

- An advantage to using the *NSTA Quick Reference Guide* is that the performance expectation is listed next to the disciplinary core idea included in that performance expectation.

- An advantage to using the *NGSS* that is not evident in the *NSTA Quick Reference Guide* is that the scientific and engineering practice and crosscutting concept chart, included below the performance expectations, shows the other two dimensions that are part of the performance expectation.

Behavior, Senses, Feedback, and Response
K–12 Standards- and Research-Based Study of a Curricular Topic

Section and Outcome	Selected Sources and Readings for Study and Reflection Read and examine **related** parts of
I. **Content Knowledge**	**IA:** *Science for All Americans*<ul><li>Ch. 6: Basic Functions, pp. 76–78</li><li>Ch. 6: Learning, pp. 78–80</li></ul>**IB:** *Framework for K–12 Science Education:* Narrative Section<ul><li>Ch. 6: LS1.A: Structure and Function, pp. 143–144</li><li>Ch. 6: LS1.D: Information Processing, p. 149</li></ul>**IC:** *Disciplinary Core Ideas: Reshaping Teaching and Learning*<ul><li>Ch. 6: LS1.D: Information Processing, pp. 105–106</li></ul>**ID:** *The NSTA Atlas of the Three Dimensions:* Narrative Page<ul><li>4.1: The Structure and Function of Organisms (LS1.A & LS1.D)</li></ul>
II. **Concepts, Core Ideas, or Practices**	**IIA:** *Framework for K–12 Science Education:* Grade Band Endpoints<ul><li>Ch. 6: LS1.A: Structure and Function, pp. 144–145</li><li>Ch. 6: LS1.D: Information Processing, pp. 149–150</li></ul>**IIB:** *Next Generation Science Standards:* Disciplinary Core Ideas Column<ul><li>Grade 1: LS1.B: Growth and Development of Organisms; LS1.D: Information Processing, p. 10</li><li>Grade 4: LS1.A: Structure and Function; LS1.D: Information Processing, p. 25</li><li>MS: LS1.D: Information Processing, p. 45</li><li>HS: LS1.A: Structure and Function, p. 80</li></ul>**IIC:** *NSTA Quick Reference Guide to the NGSS K–12:* Disciplinary Core Ideas Column<ul><li>Grade 1: LS1.B: Growth and Development of Organisms; LS1.D: Information Processing, p. 94</li><li>Grade 4: LS1.A: Structure and Function; LS1.D: Information Processing, p. 107</li><li>MS: LS1.D: Information Processing, p. 129</li><li>HS: LS1.A: Structure and Function, p. 154</li></ul>
III. **Curriculum, Instruction, and Formative Assessment**	**IIIA:** *Disciplinary Core Ideas: Reshaping Teaching and Learning*<ul><li>Ch. 6: Expectations for Grades K–2: LS1.A: Structure and Function; LS1.D: Information Processing, p. 110</li><li>Ch. 6: Expectations for Grades 3–5: LS1.A: Structure and Function; LS1.D: Information Processing, pp. 110–112</li><li>Ch. 6: Expectations for Middle School, LS1.D: Information Processing, p. 114</li><li>Ch. 6: Expectations for High School, LS1.D: Information Processing, p. 115</li></ul>**IIIB:** *Uncovering Student Ideas:* Assessment Probe and Suggestions for Instruction<ul><li>USI.2: Plants in the Dark and Light, pp. 107, 110–111</li><li>USI.LS: Rocky Soil, pp. 79, 81–82</li><li>USI.K-2: Senses, pp. 35, 37–38</li></ul>
IV. **Research on Commonly Held Ideas**	**IVA:** *Benchmarks for Science Literacy:* Chapter 15 Research<ul><li>6C: Nervous System, p. 345</li></ul>**IVB:** *Making Sense of Secondary Science: Research Into Children's Ideas*<ul><li>Ch. The Nervous System, pp. 46–47; The Responses of Plants, p. 47</li></ul>

Section and Outcome	Selected Sources and Readings for Study and Reflection Read and examine **related** parts of
	IVC: *Uncovering Student Ideas:* Related Research • USI.2: Plants in the Dark and Light, p. 110 • USI.LS: Rocky Soil, p. 81 • USI.K-2: Senses, p. 37
V. K–12 Articulation and Connections	**VA: *Next Generation Science Standards:*** Appendices: Progression • Appendix E: LS1.A: Structure and Function; LS1.D: Information Processing, p. 4 **VB: *NSTA Quick Reference Guide to the NGSS K–12:*** Progression • LS1.A: Structure and Function, p. 68 • LS1.D: Information Processing, p. 69 **VC: *The NSTA Atlas of the Three Dimensions:*** Map Page • 4.1 The Structure and Function of Organisms (LS1.A, LS1.B, & LS1.D)
VI. Assessment Expectation	**VIA: *State Standards*** • Examine your state's standards **VIB: *Next Generation Science Standards:*** Performance Expectations • Grade 1: 1-LS1-1, 1-LS1-2, p. 10 • Grade 4: 4-LS1-1, 4-LS1-2, p. 25 • MS: 4-LS1-8, p. 45 • HS: HS-LS1-3, p. 80 **VIC: *NSTA Quick Reference Guide to the NGSS K–12:*** Performance Expectations Column • Grade 1: 1-LS1-1, 1-LS1-2, p. 94 • Grade 4: 4-LS1-1; 4-LS1-2, p. 107 • MS: MS-LS1-8, p. 129 • HS: HS-LS1-3, p. 154

Review Chapter 2 instructions on how to use this guide.

Visit curriculumtopicstudy2.org for more information about CTS and additional resources.

NOTE: *Atlas* page numbers have not been provided because *The NSTA Atlas of the Three Dimensions* was produced concurrently with this edition. Titles and map codes are accurate.

Additional Readings:

Available for download at www.curriculumtopicstudy2.org

Copyright © 2020 by Corwin Press, Inc. All rights reserved. Reprinted from *Science Curriculum Topic Study: Bridging the Gap Between Three-Dimensional Standards, Research, and Practice* (2nd ed.) by Page Keeley and Joyce Tugel. Thousand Oaks, CA: Corwin, www.corwin.com. Reproduction authorized for educational use by educators, local school sites, and/or noncommercial or nonprofit entities that have purchased the book.

Brain and the Nervous System
Grades 3–12 Standards- and Research-Based Study of a Curricular Topic

Section and Outcome	Selected Sources and Readings for Study and Reflection Read and examine **related** parts of
I. Content Knowledge	**IA: *Science for All Americans*** • Ch. 6: Basic Functions, pp. 76–78 **IB: *Framework for K–12 Science Education:*** Narrative Section • Ch. 6: LS1.D: Information Processing, p. 149 **IC: *Disciplinary Core Ideas: Reshaping Teaching and Learning*** • Overview Ch. 6: LS1.D: Information Processing, pp. 105–106 **ID: *The NSTA Atlas of the Three Dimensions:*** Narrative Page • 4.1: The Structure and Function of Organisms (LS1.D)
II. Concepts, Core Ideas, or Practices	**IIA: *Framework for K–12 Science Education:*** Grade Band Endpoints • Ch. 6: LS1.D: Information Processing, pp. 149–150 **IIB: *Next Generation Science Standards:*** Disciplinary Core Ideas Column • Grade 4: LS1.D: Information Processing, p. 25 • MS: LS1.A: LS1.D: Information Processing, p. 45 • HS: LS1.A: Structure and Function, p. 80 **IIC: *NSTA Quick Reference Guide to the NGSS K–12:*** Disciplinary Core Ideas Column • Grade 4: LS1.D: Information Processing, p. 107 • MS: LS1.D: Information Processing, p. 129 • HS: LS1.A: Structure and Function, p. 154
III. Curriculum, Instruction, and Formative Assessment	**IIIA: *Disciplinary Core Ideas: Reshaping Teaching and Learning*** • Ch. 6: 3–5: LS1.D: Information Processing, p. 111 • Ch. 6: MS: LS1.D: Information Processing, p. 114 • Ch. 6: HS: LS1.A: Structure and Function, p. 114; LS1.D: Information Processing, p. 115
IV. Research on Commonly Held Ideas	**IVA: *Benchmarks for Science Literacy:*** Chapter 15 Research • 6C: Nervous System, p. 345 **IVB: *Making Sense of Secondary Science: Research Into Children's Ideas*** • Ch. 4: The Nervous System, pp. 46–47
V. K–12 Articulation and Connections	**VA: *Next Generation Science Standards:*** Appendices: Progression • Appendix E: LS1.D: Information Processing, p. 4 **VB: *NSTA Quick Reference Guide to the NGSS K–12:*** Progression • LS1.D: Information Processing, p. 69 **VC: *The NSTA Atlas of the Three Dimensions:*** Map Page • 4.1: The Structure and Function of Organisms (LS1.A, LS1.B, & LS1.D)
VI. Assessment Expectation	**VIA: *State Standards*** • Examine your state's standards **VIB: *Next Generation Science Standards:*** Performance Expectations • Grade 4: 4-LS1-1, 4-LS1-2, p. 25 • MS: MS-LS1-3, MS-LS1-8, p. 45 • HS: HS-LS1-2, p. 80

Section and Outcome	Selected Sources and Readings for Study and Reflection Read and examine **related** parts of
	VIC: *NSTA Quick Reference Guide to the NGSS K–12:* Performance Expectations Column • Grade 4: 4-LS1-2, p. 107 • MS: MS-LS1-8, p. 129 • HS: HS-LS1-2, p. 154

Review Chapter 2 instructions on how to use this guide.

Visit curriculumtopicstudy2.org for more information about CTS and additional resources.

NOTE: *Atlas* page numbers have not been provided because *The NSTA Atlas of the Three Dimensions* was produced concurrently with this edition. Titles and map codes are accurate.

Additional Readings:

Available for download at www.curriculumtopicstudy2.org

Copyright © 2020 by Corwin Press, Inc. All rights reserved. Reprinted from *Science Curriculum Topic Study: Bridging the Gap Between Three-Dimensional Standards, Research, and Practice* (2nd ed.) by Page Keeley and Joyce Tugel. Thousand Oaks, CA: Corwin, www.corwin.com. Reproduction authorized for educational use by educators, local school sites, and/or noncommercial or nonprofit entities that have purchased the book.

Cell Division and Differentiation
Grades 9–12 Standards- and Research-Based Study of a Curricular Topic

Section and Outcome	Selected Sources and Readings for Study and Reflection Read and examine **related** parts of
I. Content Knowledge	**IA:** *Science for All Americans* • Ch. 5: Cells, pp. 63–64 (last two paragraphs) **IB:** *Science Matters* (2009 edition) • Ch. 16: Sex—A Good Idea, pp. 285–287; Gene Regulation and Differentiation, pp. 290–291 **IC:** *Framework for K–12 Science Education:* Narrative Section • Ch. 6: LS1.B: Growth and Development of Organisms, pp. 145–146 **ID:** *Disciplinary Core Ideas: Reshaping Teaching and Learning* • Ch. 6: LS1.B: Growth and Development of Organisms, pp. 101–103 **IE:** *The NSTA Atlas of the Three Dimensions:* Narrative Page • 4.1: The Structure and Function of Organisms (LS1.B)
II. Concepts, Core Ideas, or Practices	**IIA:** *Framework for K–12 Science Education:* Grade Band Endpoints • Ch. 6: LS1.B: Growth and Development of Organisms, pp. 146–147 **IIB:** *Next Generation Science Standards:* Disciplinary Core Ideas Column • HS: LS1.B: Growth and Development of Organisms, p. 85 **IIC:** *NSTA Quick Reference Guide to the NGSS K–12:* Disciplinary Core Ideas Column • HS: LS1.B: Growth and Development of Organisms, p. 154
III. Curriculum, Instruction, and Formative Assessment	**IIIA:** *Disciplinary Core Ideas: Reshaping Teaching and Learning* • Ch. 6: MS: Expectations for Grades 6–8; LS1.B: Growth and Development of Organisms, pp. 112–113 • Ch. 6: HS: Expectations for Grades 9–12; LS1.B: Growth and Development of Organisms, p. 114 **IIIB:** *Uncovering Student Ideas:* Assessment Probe and Suggestions for Instruction • USI.3: Sam's Puppy, pp. 125, 129–130
IV. Research on Commonly Held Ideas	**IVA:** *Making Sense of Secondary Science: Research Into Children's Ideas* • Ch. 3: The Meaning of "Growth," p. 37 **IVB:** *Uncovering Student Ideas:* Related Research • USI.3: Sam's Puppy, p. 129 **IVC:** *Disciplinary Core Ideas: Reshaping Teaching and Learning* • Ch. 6: Students' Commonly Held Ideas, p. 117 (second column)
V. K–12 Articulation and Connections	**VA:** *Next Generation Science Standards:* Appendices: Progression • Appendix E: LS1.B: Growth and Development of Organisms, p. 4 **VB:** *NSTA Quick Reference Guide to the NGSS K–12:* Progression • LS1.B: Growth and Development of Organisms, p. 68 **VC:** *The NSTA Atlas of the Three Dimensions:* Map Page • 4.1: The Structure and Function of Organisms (LS1.A, LS1.B, & LS1.D)

Section and Outcome	Selected Sources and Readings for Study and Reflection Read and examine **related** parts of
VI. Assessment Expectation	**VIA:** *State Standards* • Examine your state's standards **VIB:** *Next Generation Science Standards:* Performance Expectations • HS: HS-LS1-4, p. 85 **VIC:** *NSTA Quick Reference Guide to the NGSS K–12:* Performance Expectations Column • HS: HS-LS1-4, p. 154

Review Chapter 2 instructions on how to use this guide.

Visit curriculumtopicstudy2.org for more information about CTS and additional resources.

NOTE: *Atlas* page numbers have not been provided because *The NSTA Atlas of the Three Dimensions* was produced concurrently with this edition. Titles and map codes are accurate.

Additional Readings:

Available for download at www.curriculumtopicstudy2.org

Copyright © 2020 by Corwin Press, Inc. All rights reserved. Reprinted from *Science Curriculum Topic Study: Bridging the Gap Between Three-Dimensional Standards, Research, and Practice* (2nd ed.) by Page Keeley and Joyce Tugel. Thousand Oaks, CA: Corwin, www.corwin.com. Reproduction authorized for educational use by educators, local school sites, and/or noncommercial or nonprofit entities that have purchased the book.

Cells and Biomolecules: Structure and Function
Grades 6–12 Standards- and Research-Based Study of a Curricular Topic

Section and Outcome	Selected Sources and Readings for Study and Reflection Read and examine **related** parts of
I. **Content Knowledge**	**IA:** *Science for All Americans* • Ch. 5: Cells, pp. 62–64 **IB:** *Science Matters* (2009 edition) • Ch. 15: Molecules of Life, pp. 252–259; The Chemical Factories of Life, pp. 259–266 **IC:** *Framework for K–12 Science Education:* Narrative Section • Ch. 6: LS1.A: Structure and Function, p. 143 **ID:** *Disciplinary Core Ideas: Reshaping Teaching and Learning* • Ch. 6: LS1.A: Structure and Function, pp. 100–101 **IE:** *The NSTA Atlas of the Three Dimensions:* Narrative Page • 4.1: The Structure and Function of Organisms (LS1.A)
II. **Concepts, Core Ideas, or Practices**	**IIA:** *Framework for K–12 Science Education:* Grade Band Endpoints • Ch. 6: LS1.A: Structure and Function, pp. 144–145 **IIB:** *Next Generation Science Standards:* Disciplinary Core Ideas Column • MS: LS1.A: Structure and Function, p. 45 • HS: LS1.A: Structure and Function, p. 80 **IIC:** *NSTA Quick Reference Guide to the NGSS K–12:* Disciplinary Core Ideas Column • MS: LS1.B: Structure and Function, p. 128 • HS: LS1.A: Structure and Function, p. 154
III. **Curriculum, Instruction, and Formative Assessment**	**IIIA:** *Uncovering Student Ideas:* Assessment Probe and Suggestions for Instruction • USI.1: Is It Made of Cells? pp. 143, 148–149; Human Body Basics, pp. 151, 155–156 • USI.2: Whale and Shrew, pp. 137, 141 • USI.3: Cells and Size, pp. 117, 121–122 • USI.LS: Pond Water, pp. 33, 37; DNA, Genes, and Chromosomes, pp. 129, 132 **IIIB:** *Disciplinary Core Ideas: Reshaping Teaching and Learning* • Ch. 6: How Does Student Understanding of This Disciplinary Core Idea Develop Over Time? pp. 108–109 • Expectations for Grades 6–8, LS1.A: Structure and Function, p. 112 • Expectations for Grades 9–12, LS1.A: Structure and Function, p. 114 • What Approaches Can We Use to Teach About This Disciplinary Core Idea? pp. 118–119
IV. **Research on Commonly Held Ideas**	**IVA:** *Benchmarks for Science Literacy:* Chapter 15 Research • 5C: Cells, p. 342 **IVB:** *Making Sense of Secondary Science: Research Into Children's Ideas* • Ch. 1: Cell Theory, p. 25 **IVC:** *Uncovering Student Ideas:* Related Research • USI.1: Is It Made of Cells? p. 147; Human Body Basics, p. 155 • USI.2: Whale and Shrew, pp. 140–141

Section and Outcome	Selected Sources and Readings for Study and Reflection Read and examine **related** parts of
	• USI.3: Cells and Size, p. 121 • USI.LS: Pond Water, p. 36; DNA, Genes, and Chromosomes, p. 132 **IVD: Disciplinary Core Ideas: Reshaping Teaching and Learning** • Students' Commonly Held Ideas, pp. 115–117
V. **K–12 Articulation and Connections**	**VA: Next Generation Science Standards:** Appendices: Progression • Appendix E: LS1.A: Structure and Function, p. 4 **VB: NSTA Quick Reference Guide to the NGSS K–12:** Progression • LS1.A: Structure and Function, p. 68 **VC: The NSTA Atlas of the Three Dimensions:** Map Page • 4.1: The Structure and Function of Organisms (LS1.A, LS1.B, & LS1.D)
VI. **Assessment Expectation**	**VIA: State Standards** • Examine your state's standards **VIB: Next Generation Science Standards:** Performance Expectations • MS: MS-LS1-1, MS-LS1-2, MS-LS1-3, p. 45 • HS: HS-LS1-1, HS-LS1-2, HS-LS1-3, p. 80 **VIC: NSTA Quick Reference Guide to the NGSS K–12:** Performance Expectations Column • MS: MS-LS1-1, MS-LS1-2, MS-LS1-3, p. 128 • HS: HS-LS1-1, HS-LS1-2, HS-LS1-3, p. 154

Review Chapter 2 instructions on how to use this guide.
Visit curriculumtopicstudy2.org for more information about CTS and additional resources.

NOTE: *Atlas* page numbers have not been provided because *The NSTA Atlas of the Three Dimensions* was produced concurrently with this edition. Titles and map codes are accurate.

Additional Readings:

Available for download at www.curriculumtopicstudy2.org

Copyright © 2020 by Corwin Press, Inc. All rights reserved. Reprinted from *Science Curriculum Topic Study: Bridging the Gap Between Three-Dimensional Standards, Research, and Practice* (2nd ed.) by Page Keeley and Joyce Tugel. Thousand Oaks, CA: Corwin, www.corwin.com. Reproduction authorized for educational use by educators, local school sites, and/or noncommercial or nonprofit entities that have purchased the book.

Characteristics of Living Things
Grades K–5 Standards- and Research-Based Study of a Curricular Topic

Section and Outcome	Selected Sources and Readings for Study and Reflection Read and examine **related** parts of
I. **Content Knowledge**	**IA:** *Framework for K–12 Science Education:* Narrative Section • Core Idea LS1: From Molecules to Organisms: Structures and Processes, p. 143 **IB:** *Disciplinary Core Ideas: Reshaping Teaching and Learning* • Ch. 6: Core Idea LS1, From Molecules to Organisms: Structures and Processes, pp. 99–100 **IC:** *The NSTA Atlas of the Three Dimensions:* Narrative Page • 4.1: The Structure and Function of Organisms (LS1.A, LS1.B, & LS1.D), p. 58 • 4.2: Flow of Matter and Energy in Living Systems (LS1.C & LS2.B)
II. **Concepts, Core Ideas, or Practices**	**IIA:** *Framework for K–12 Science Education:* Grade Band Endpoints • Ch. 6: LS1.A: Structure and Function, p. 144 • Ch. 6: LS1.B: Growth and Development of Organisms, p. 146 • Ch. 6: LS1.C: Organization for Matter and Energy Flow in Organisms, pp. 147–148 • Ch. 6: LS1.D: Information Processing, p. 149 **IIB:** *Next Generation Science Standards:* Disciplinary Core Ideas Column • K: LS1.C: Organization for Matter and Energy Flow in Organisms, p. 6 • Grade 1: LS1.A: Structure and Function; LS1.B, Growth and Development of Organisms; LS1.D, Information Processing, p. 10 • Grade 3: LS1.B: Growth and Development of Organisms, p. 20 • Grade 4: LS1.A: Structure and Function; LS1.D: Information Processing, p. 25 • Grade 5: LS1.C: Organization for Matter and Energy Flow in Organisms, p. 29 **IIC:** *NSTA Quick Reference Guide to the NGSS K–12:* Disciplinary Core Ideas Column • K: LS1.C: Organization for Matter and Energy Flow in Organisms, p. 92 • Grade 1: LS1.A: Structure and Function; LS1.B: Growth and Development of Organisms; LS1.D: Information Processing, p. 94 • Grade 3: LS1.B: Growth and Development of Organisms, p. 104 • Grade 4: LS1.A: Structure and Function; LS1.D: Information Processing, p. 107 • Grade 5: LS1.C: Organization for Matter and Energy Flow in Organisms, p. 111
III. **Curriculum, Instruction, and Formative Assessment**	**IIIA:** *Disciplinary Core Ideas: Reshaping Teaching and Learning* • K–2: Expectations for Grades K–2, p. 110 • 3–5: Expectations for Grades 3–5, pp. 110–112 **IIIB:** *Uncovering Student Ideas:* Assessment Probe and Suggestions for Instruction • USI.1: Is It Living? pp. 133, 139–141; Functions of Living Things, pp. 157, 163–164 • USI.LS: Cucumber Seeds, pp. 9, 13–14 • USI.K-2: Is It Living? pp. 3, 6–7; Do They Need Air? pp. 31, 33–34; Senses, pp. 35, 37–38
IV. **Research on Commonly Held Ideas**	**IVA:** *Benchmarks for Science Literacy:* Chapter 15 Research • 5A: Living and Nonliving, p. 341 **IVB:** *Making Sense of Secondary Science: Research Into Children's Ideas* • Ch. 1: The Concept of Living, pp. 17–21 • Ch. 3: Growth as a Criterion of Life, p. 36 • Ch. 4: Behavior as a Criterion of Life, p. 41 • Ch. 5: Reproduction as a Criterion of Life, p. 48

Section and Outcome	Selected Sources and Readings for Study and Reflection Read and examine **related** parts of
	IVC: *Uncovering Student Ideas:* Related Research • USI.1: Is It Living? pp. 137–139; Functions of Living Things, pp. 162–163 • USI.LS: Cucumber Seeds, pp. 12–13 • USI.K-2: Is It Living? p. 6; Do They Need Air? p. 33; Senses, p. 37 **IVD:** *Disciplinary Core Ideas: Reshaping Teaching and Learning* • Students' Commonly Held Ideas, pp. 115–116
V. K–12 Articulation and Connections	**VA:** *Next Generation Science Standards:* Appendices: Progression • Appendix E: LS1.A: Structure and Function; LS1.B: Growth and Development of Organisms; LS1.C: Organization for Matter and Energy Flow in Organisms; LS1.D: Information Processing, p. 4 **VB:** *NSTA Quick Reference Guide to the NGSS K–12:* Progression • LS1: From Molecules to Organisms: Structures and Processes, pp. 68–69 **VC:** *The NSTA Atlas of the Three Dimensions:* Map Page • 4.1: The Structure and Function of Organisms (LS1.A, LS1.B, & LS1.D), p. 59 • 4.2: Flow of Matter and Energy in Living Systems (LS1.C & LS2.B)
VI. Assessment Expectation	**VIA:** *State Standards* • Examine your state's standards **VIB:** *Next Generation Science Standards:* Performance Expectations • K: K-LS1-1, p. 6 • Grade 1: 1-LS1-1, 1-LS1-2, p. 10 • Grade 3: 3-LS1-1, p. 20 • Grade 4: 4-LS1-1, 4-LS1-2, p. 25 • Grade 5: 5-LS1-1, p. 29 **VIC:** *NSTA Quick Reference Guide to the NGSS K–12:* Performance Expectations Column • K: K-LS1-1, p. 92 • Grade 1: 1-LS1-1, 1-LS1-2, p. 94 • Grade 3: 3-LS1-1, p. 104 • Grade 4: 4-LS1-1, 4-LS1-2, p. 107 • Grade 5: 5-LS1-1, p. 111

Review Chapter 2 instructions on how to use this guide.

Visit curriculumtopicstudy2.org for more information about CTS and additional resources.

NOTE: *Atlas* page numbers have not been provided because *The NSTA Atlas of the Three Dimensions* was produced concurrently with this edition. Titles and map codes are accurate.

Additional Readings:

 Available for download at www.curriculumtopicstudy2.org

Copyright © 2020 by Corwin Press, Inc. All rights reserved. Reprinted from *Science Curriculum Topic Study: Bridging the Gap Between Three-Dimensional Standards, Research, and Practice* (2nd ed.) by Page Keeley and Joyce Tugel. Thousand Oaks, CA: Corwin, www.corwin.com. Reproduction authorized for educational use by educators, local school sites, and/or noncommercial or nonprofit entities that have purchased the book.

Energy Extraction, Food, and Nutrition
Grades K–12 Standards- and Research-Based Study of a Curricular Topic

Section and Outcome	Selected Sources and Readings for Study and Reflection Read and examine **related** parts of
I. Content Knowledge	**IA:** *Science for All Americans* • Ch. 5: Cells, p. 62 • Ch. 5: Flow of Matter and Energy, p. 66 • Ch. 6: Basic Functions, p. 77 **IB:** *Science Matters* (2009 edition) • Ch. 15: The Power Plant, pp. 263–266 **IC:** *Framework for K–12 Science Education:* Narrative Section • Ch. 6: LS1.C: Organization for Matter and Energy Flow in Organisms, p. 147 **ID:** *Disciplinary Core Ideas: Reshaping Teaching and Learning* • Ch. 6: LS1.C: Organization for Matter and Energy Flow in Organisms, pp. 103–105 **IE:** *The NSTA Atlas of the Three Dimensions:* Narrative Page • 4.2: Flow of Matter and Energy in Living Systems (LS1.C & LS2.B)
II. Concepts, Core Ideas, or Practices	**IIA:** *Framework for K–12 Science Education:* Grade Band Endpoints • Ch. 6: LS1.C: Organization for Matter and Energy Flow in Organisms, pp. 147–148 **IIB:** *Next Generation Science Standards:* Disciplinary Core Ideas Column • K: LS1: Organization for Matter and Energy Flow in Organisms, p. 6 • Grade 5: LS1.C: Organization for Matter and Energy Flow in Organisms; PS3.D: Energy in Chemical Processes and Everyday Life, p. 29 • MS: LS1.C: Organization for Matter and Energy Flow in Organisms; PS3.D: Energy in Chemical Processes and Everyday Life, p. 47 • HS: LS1.C: Organization for Matter and Energy Flow in Organisms, p. 81 **IIC:** *NSTA Quick Reference Guide to the NGSS K–12:* Disciplinary Core Ideas Column • K: LS1.C: Organization for Matter and Energy Flow in Organisms, p. 92 • Grade 5: LS1.C: Organization for Matter and Energy Flow in Organisms, p. 111; PS3.D: Energy in Chemical Processes and Everyday Life, p. 113 • MS: LS1.C: Organization for Matter and Energy Flow in Organisms, p. 129; PS3.D: Energy in Chemical Processes and Everyday Life, p. 129 • HS: LS1.C: Organization for Matter and Energy Flow in Organisms, p. 154–155
III. Curriculum, Instruction, and Formative Assessment	**IIIA:** *Disciplinary Core Ideas: Reshaping Teaching and Learning* • Ch. 6: How Does Student Understanding of This Disciplinary Core Idea Develop Over Time? p. 109 • Ch. 6: Expectations for Grades K–2, LS1.C, p. 110 • Ch. 6: Expectations for Grades 3–5, LS1.C, p. 111 • Ch. 6: Expectations for Grades 6–8, LS1.C, p. 113 • Ch. 6: Expectations for Grades 9–12, LS1.C, pp. 114–115 **IIIB:** *Uncovering Student Ideas:* Assessment Probe and Suggestions for Instruction • USI.2: Is It Food for Plants? pp. 113, 118 • USI.3: Respiration, pp. 131, 135–136 • USI.4: Is It Food? pp. 91, 96; Digestive System, pp. 131, 135 • USI.LS: Food for Corn, pp. 69, 73–74

Section and Outcome	Selected Sources and Readings for Study and Reflection Read and examine **related** parts of
IV. **Research on Commonly Held Ideas**	**IVA:** *Benchmarks for Science Literacy:* Chapter 15 Research • 5E: Food, p. 342; Plant and Animal Nutrition, pp. 342–343 • 6C: Digestive System, p. 345 **IVB:** *Making Sense of Secondary Science: Research Into Children's Ideas* • Ch. 2: Nutrition, pp. 27–34 • Ch. 7: Respiration, pp. 66–68 **IVC:** *Uncovering Student Ideas:* Related Research • USI.2: Is It Food for Plants? p. 117 • USI.3: Respiration, p. 135 • USI.4: Is It Food? p. 95; Digestive System, pp. 134–135 • USI.LS: Food for Corn, p. 72–73 **IVD:** *Disciplinary Core Ideas: Reshaping Teaching and Learning* • Ch. 6: Students' Commonly Held Ideas, pp. 117–118 (start with last paragraph)
V. **K–12 Articulation and Connections**	**VA:** *Next Generation Science Standards:* Appendices: Progression • Appendix E: LS1.C: Organization for Matter and Energy Flow in Organisms, p. 4 **VB:** *NSTA Quick Reference Guide to the NGSS K–12:* Progression • LS1.C: Organization for Matter and Energy Flow in Organisms, p. 69 **VC:** *The NSTA Atlas of the Three Dimensions:* Map Page • 4.2: Flow of Matter and Energy in Living Systems (LS1.C & LS2.B)
VI. **Assessment Expectation**	**VIA:** *State Standards* • Examine your state's standards **VIB:** *Next Generation Science Standards:* Performance Expectations • K: K-LS1-1, p. 6 • Grade 5: 5-LS1-1, 5-PS3-1, p. 29 • MS: MS-LS1-6, MS-LS1-7, p. 47 • HS: HS-LS1-5, HS-LS1-6, HS-LS1-7, p. 81 **VIC:** *NSTA Quick Reference Guide to the NGSS K–12:* Performance Expectations • K: K-LS1-1, p. 92 • Grade 5: 5-LS1-1, p. 111 • MS: MS-LS1-6, MS-LS1-7, p. 129 • HS: HS-LS1-5, HS-LS1-6, HS-LS1-7, pp. 154–155

Review Chapter 2 instructions on how to use this guide.
Visit curriculumtopicstudy2.org for more information about CTS and additional resources.

NOTE: This study can be combined with related sections from Energy in Chemical Processes and Everyday Life.

NOTE: *Atlas* page numbers have not been provided because *The NSTA Atlas of the Three Dimensions* was produced concurrently with this edition. Titles and map codes are accurate.

Additional Readings:

 Available for download at www.curriculumtopicstudy2.org

Copyright © 2020 by Corwin Press, Inc. All rights reserved. Reprinted from *Science Curriculum Topic Study: Bridging the Gap Between Three-Dimensional Standards, Research, and Practice* (2nd ed.) by Page Keeley and Joyce Tugel. Thousand Oaks, CA: Corwin, www.corwin.com. Reproduction authorized for educational use by educators, local school sites, and/or noncommercial or nonprofit entities that have purchased the book.

Macroscopic Structure and Function of Organisms
Grades K–5 Standards- and Research-Based Study of a Curricular Topic

Section and Outcome	Selected Sources and Readings for Study and Reflection Read and examine **related** parts of
I. Content Knowledge	**IA:** *Science for All Americans* • Ch. 6: Basic Functions, pp. 76–78 **IB:** *Framework for K–12 Science Education:* Narrative Section • Ch. 6: LS1.A: Structure and Function, p. 143 **IC:** *Disciplinary Core Ideas: Reshaping Teaching and Learning* • Ch. 6: LS1.A: Structure and Function, pp. 100–101 **ID:** *The NSTA Atlas of the Three Dimensions:* Narrative Page • 4.1: The Structure and Function of Organisms (LS1.A, LS1.B, & LS1.D)
II. Concepts, Core Ideas, or Practices	**IIA:** *Framework for K–12 Science Education:* Grade Band Endpoints • Ch. 6: LS1.A: Structure and Function, pp. 144–145 **IIB:** *Next Generation Science Standards:* Disciplinary Core Ideas Column • Grade 1: LS1.A: Structure and Function; p. 10 • Grade 4: LS1.A: Structure and Function, p. 25 **IIC:** *NSTA Quick Reference Guide to the NGSS K–12:* Disciplinary Core Ideas Column • Grade 1: LS1.A: Structure and Function, p. 94 • Grade 4: HS: LS1.A: Structure and Function, p. 107
III. Curriculum, Instruction, and Formative Assessment	**IIIA:** *Disciplinary Core Ideas: Reshaping Teaching and Learning* • Ch. 6: How Does Student Understanding of This Disciplinary Core Idea Develop Over Time? pp. 108–109 • Expectations for Grades K–2, LS1.A, p. 110 • Expectations for Grades 3–5, LS1.A, pp. 110–111 **IIIB:** *Uncovering Student Ideas:* Assessment Probe and Suggestions for Instruction • USI.1: Functions of Living Things, pp. 157, 163–164 • USI.K-2: Is It Made of Parts? pp. 21, 23–24
IV. Research on Commonly Held Ideas	**IVA:** *Benchmarks for Science Literacy:* Chapter 15 Research • 5C: Cells, p. 342 • 6C: Basic Functions, pp. 344–345 **IVB:** *Making Sense of Secondary Science: Research Into Children's Ideas* • Ch. 1: Organization of the Body-Structure and Function, p. 26 • Ch. 2: Human Digestion and Assimilation, pp. 29–30 • Ch. 4: The Nervous System, pp. 46–47; Muscles and Skeleton, p. 47 **IVC:** *Uncovering Student Ideas:* Related Research • USI.1: Functions of Living Things, pp. 162–163 • USI.K-2: Is It Made of Parts? p. 21, 23 **IVD:** *Disciplinary Core Ideas: Reshaping Teaching and Learning* • Ch. 6: Core Idea LS1, Students' Commonly Held Ideas, pp. 115–117

Section and Outcome	Selected Sources and Readings for Study and Reflection Read and examine **related** parts of
V. **K–12 Articulation and Connections**	**VA:** *Next Generation Science Standards:* Appendices: Progression<ul><li>Appendix E: LS1.A: Structure and Function, p. 4</li></ul>**VB:** *NSTA Quick Reference Guide to the NGSS K–12:* Progression<ul><li>LS1.A: Structure and Function, p. 68</li></ul>**VC:** *The NSTA Atlas of the Three Dimensions:* Map Page<ul><li>4.1: The Structure and Function of Organisms (LS1.A)</li></ul>
VI. **Assessment Expectation**	**VIA:** *State Standards*<ul><li>Examine your state's standards</li></ul>**VIB:** *Next Generation Science Standards:* Performance Expectations<ul><li>Grade 1: 1-LS1-1, p. 10</li><li>Grade 4: 4-LS1-1, p. 25</li></ul>**VIC:** *NSTA Quick Reference Guide to the NGSS K–12:* Performance Expectations<ul><li>Grade 1: 1-LS1-1, p. 94</li><li>Grade 4: 4-LS1-1, p. 107</li></ul>

Review Chapter 2 instructions on how to use this guide.

Visit curriculumtopicstudy2.org for more information about CTS and additional resources.

NOTE: *Atlas* page numbers have not been provided because *The NSTA Atlas of the Three Dimensions* was produced concurrently with this edition. Titles and map codes are accurate.

Additional Readings:

online resources Available for download at **www.curriculumtopicstudy2.org**

Copyright © 2020 by Corwin Press, Inc. All rights reserved. Reprinted from *Science Curriculum Topic Study: Bridging the Gap Between Three-Dimensional Standards, Research, and Practice* (2nd ed.) by Page Keeley and Joyce Tugel. Thousand Oaks, CA: Corwin, www.corwin.com. Reproduction authorized for educational use by educators, local school sites, and/or noncommercial or nonprofit entities that have purchased the book.

Organs and Systems of the Human Body
Grades K–12 Standards- and Research-Based Study of a Curricular Topic

Section and Outcome	Selected Sources and Readings for Study and Reflection Read and examine **related** parts of
I. Content Knowledge	**IA:** *Science for All Americans* • Ch. 6: Basic Functions, pp. 76–78 **IB:** *Science Matters* (2009 edition) • Ch. 15: The Organization of Life, p. 266 **IC:** *Framework for K–12 Science Education:* Narrative Section • Ch. 6: LS1.A: Structure and Function, p. 143 **ID:** *Disciplinary Core Ideas: Reshaping Teaching and Learning* • Ch. 6: LS1.A: Structure and Function, pp. 100–101 **IE:** *The NSTA Atlas of the Three Dimensions:* Narrative Page • 4.1: The Structure and Function of Organisms (LS1.A)
II. Concepts, Core Ideas, or Practices	**IIA:** *Framework for K–12 Science Education:* Grade Band Endpoints • Ch. 6: LS1.A: Structure and Function, pp. 144–145 **IIB:** *Next Generation Science Standards:* Disciplinary Core Ideas Column • Grade 1: LS1.A: Structure and Function, p. 10 • Grade 4: LS1.A: Structure and Function, p. 25 • MS: LS1.A: Structure and Function, p. 45 • HS: LS1.A: Structure and Function, p. 80 **IIC:** *NSTA Quick Reference Guide to the NGSS K–12:* Disciplinary Core Ideas Column • Grade 1: LS1.A: Structure and Function, p. 94 • Grade 4: LS1.A: Structure and Function, p. 107 • MS: LS1.A: Structure and Function, p. 128 • HS: LS1.A: Structure and Function, p. 154
III. Curriculum, Instruction, and Formative Assessment	**IIIA:** *Disciplinary Core Ideas: Reshaping Teaching and Learning* • Ch. 6: How Does Student Understanding of This Disciplinary Core Idea Develop Over Time? pp. 108–109 • Ch. 6: Expectations for Grades K–2, LS1.A, p. 110 • Ch. 6: Expectations for Grades 3–5, LS1.A, pp. 110–111 • Ch. 6: Expectations for Grades 6–8, LS1.A, p. 112 • Ch. 6: What Approaches Can We Use to Teach About This Disciplinary Core Idea? pp. 118–119 **IIIB:** *Uncovering Student Ideas:* Assessment Probe and Suggestions for Instruction • USI.1: Human Body Basics, pp. 151, 155–156 • USI.4: Digestive System, pp. 131, 135 • USI.LS: Human Body, pp. 141, 144; Excretory System, pp. 145, 148–149 • USI.K–2: Is It Made of Parts? pp. 21, 23–24
IV. Research on Commonly Held Ideas	**IVA:** *Benchmarks for Science Literacy:* Chapter 15 Research • 5C: Cells, p. 342 • 6C: Basic Functions, pp. 344–345 **IVB:** *Making Sense of Secondary Science: Research Into Children's Ideas* • Ch. 1: Organization of the Body-Structure and Function, p. 26 • Ch. 2: Human Digestion and Assimilation, pp. 29–30 • Ch. 4: The Nervous System, pp. 46–47; Muscles and Skeleton, p. 47

Section and Outcome	Selected Sources and Readings for Study and Reflection Read and examine **related** parts of
	IVC: *Uncovering Student Ideas:* Related Research • USI.1: Human Body Basics, p. 155 • USI.4: Digestive System, pp. 134–135 • USI.LS: Human Body, pp. 143–144; Excretory System, p. 148 • USI.K-2: Is It Made of Parts? p. 23 **IVD: *Disciplinary Core Ideas: Reshaping Teaching and Learning*** • Ch. 6: Core Idea LS1, Students' Commonly Held Ideas, pp. 115–117
V. **K–12 Articulation and Connections**	**VA: *Next Generation Science Standards:*** Appendices: Progression • Appendix E: LS1.A: Structure and Function, p. 4 **VB: *NSTA Quick Reference Guide to the NGSS K–12:*** Progression • LS1.A: Structure and Function, p. 68 **VC: *The NSTA Atlas of the Three Dimensions:*** Map Page • 4.1: The Structure and Function of Organisms (LS1.A, LS1.B, & LS1.D)
VI. **Assessment Expectation**	**VIA: *State Standards*** • Examine your state's standards **VIB: *Next Generation Science Standards:*** Performance Expectations • Grade 1: 1-LS1-1, p. 10 • Grade 4: 4-LS1-1, p. 25 • MS: MS-LS1-3, p. 45 • HS: HS-LS1-2, p. 80 **VIC: *NSTA Quick Reference Guide to the NGSS K–12:*** Performance Expectations Column • Grade 1: 1-LS1-1, p. 94 • Grade 4: 4-LS1-1, p. 107 • MS: MS-LS1-3, p. 128 • HS: HS-LS1-2, p. 154

Review Chapter 2 instructions on how to use this guide.

Visit curriculumtopicstudy2.org for more information about CTS and additional resources.

NOTE: *Atlas* page numbers have not been provided because *The NSTA Atlas of the Three Dimensions* was produced concurrently with this edition. Titles and map codes are accurate.

Additional Readings:

 Available for download at www.curriculumtopicstudy2.org

Copyright © 2020 by Corwin Press, Inc. All rights reserved. Reprinted from *Science Curriculum Topic Study: Bridging the Gap Between Three-Dimensional Standards, Research, and Practice* (2nd ed.) by Page Keeley and Joyce Tugel. Thousand Oaks, CA: Corwin, www.corwin.com. Reproduction authorized for educational use by educators, local school sites, and/or noncommercial or nonprofit entities that have purchased the book.

Photosynthesis and Respiration at the Organism Level
Grades K–12 Standards- and Research-Based Study of a Curricular Topic

Section and Outcome	Selected Sources and Readings for Study and Reflection Read and examine **related** parts of
I. Content Knowledge	**IA:** *Science for All Americans* • Ch. 5: Flow of Matter and Energy, pp. 66–67 **IB:** *Science Matters* (2009 edition) • Ch. 15: The Power Plant, pp. 263–266 **IC:** *Framework for K–12 Science Education:* Narrative Section • Ch. 6: LS1.C: Organization for Matter and Energy Flow in Organisms, p. 147 **ID:** *Disciplinary Core Ideas: Reshaping Teaching and Learning* • Ch. 6: LS1.C: Organization for Matter and Energy Flow in Organisms, pp. 103–105 **IE:** *The NSTA Atlas of the Three Dimensions:* Narrative Page • 4.2: Flow of Matter and Energy in Living Systems (LS1.C)
II. Concepts, Core Ideas, or Practices	**IIA:** *Framework for K–12 Science Education:* Grade Band Endpoints • Ch. 6: LS1.C: Organization for Matter and Energy Flow in Organisms, pp. 147–148 **IIB:** *Next Generation Science Standards:* Disciplinary Core Ideas Column • K: LS1.C: Organization for Matter and Energy Flow in Organisms, p. 6 • Grade 5: LS1.C: Organization for Matter and Energy Flow in Organisms, p. 29 • MS: LS1.C: Organization for Matter and Energy Flow in Organisms, p. 47 • HS: LS1.C: Organization for Matter and Energy Flow in Organisms, p. 81 **IIC:** *NSTA Quick Reference Guide to the NGSS K–12:* Disciplinary Core Ideas Column • K: LS1.C: Organization for Matter and Energy Flow in Organisms, p. 92 • Grade 5: LS1.C: Organization for Matter and Energy Flow in Organisms, p. 111 • MS: LS1.C: Organization for Matter and Energy Flow in Organisms, p. 129; • HS: LS1.C: Organization for Matter and Energy Flow in Organisms, pp. 154–155
III. Curriculum, Instruction, and Formative Assessment	**IIIA:** *Disciplinary Core Ideas: Reshaping Teaching and Learning* • Ch. 6: How Does Student Understanding of This Disciplinary Core Idea Develop Over Time? pp. 108–109 • Ch. 6: Expectations for Grades K–2, LS1.C, p. 110 • Ch. 6: Expectations for Grades 3–5, LS1.C, p. 111 • Ch. 6: Expectations for Grades 6–8, LS1.C, p. 113 • Ch. 6: Expectations for Grades 9–12, LS1.C, pp. 114–115 • Using Simulations and Multiple Representations, pp. 119–120 **IIIB:** *Uncovering Student Ideas:* Assessment Probe and Suggestions for Instruction • USI.1: Functions of Living Things, pp. 157, 163–164 • USI.2: Is It Food for Plants? pp. 113, 118 • USI.3: Respiration, pp. 131, 135–136 • USI.LS: Chlorophyll, pp. 51, 55; Apple Tree, pp. 57, 61; Light and Dark, pp. 63, 66–67; Food for Corn, pp. 69, 73–74
IV. Research on Commonly Held Ideas	**IVA:** *Benchmarks for Science Literacy:* Chapter 15 Research • 5E: Plant and Animal Nutrition, pp. 342–343 **IVB:** *Making Sense of Secondary Science: Research Into Children's Ideas* • Ch. 2: Plant Nutrition, pp. 30–32; Photosynthesis, pp. 32–33; Gas Exchange by Plants, pp. 33–34

Section and Outcome	Selected Sources and Readings for Study and Reflection Read and examine **related** parts of
	• Ch. 7: Gas Exchange and Balance, p. 66; Respiration, pp. 66–68 • Ch. 13: Gases Involved in Life Processes, pp. 110–111 **IVC: *Uncovering Student Ideas:* Related Research** • USI.1: Functions of Living Things, pp. 162–163 • USI.2: Is It Food for Plants? pp. 117–118; Sequoia Tree, p. 126 • USI.3: Respiration, p. 135 • USI.LS: Chlorophyll, pp. 54–55; Apple Tree, pp. 60–61; Light and Dark, p. 66; Food for Corn, pp. 72–73 **IVD: *Disciplinary Core Ideas: Reshaping Teaching and Learning*** • Ch. 6: Students' Commonly Held Ideas, pp. 117–118 (start with last paragraph)
V. **K–12** **Articulation** **and** **Connections**	**VA: *Next Generation Science Standards:* Appendices: Progression** • Appendix E: LS1.C: Organization for Matter and Energy Flow in Organisms, p. 4 **VB: *NSTA Quick Reference Guide to the NGSS K–12:* Progression** • LS1.C: Organization for Matter and Energy Flow in Organisms, p. 69 **VC: *The NSTA Atlas of the Three Dimensions:* Map Page** • 4.2: Flow of Matter and Energy in Living Systems (LS1.C & LS2.B)
VI. **Assessment** **Expectation**	**VIA: *State Standards*** • Examine your state's standards **VIB: *Next Generation Science Standards:* Performance Expectations** • K: K-LS1-1, p. 6 • Grade 5: 5-LS1-1, p. 29 • MS: MS-LS1-6, MS-LS1-7, p. 47 • HS: HS-LS1-5, HS-LS1-6, HS-LS1-7, p. 81 **VIC: *NSTA Quick Reference Guide to the NGSS K–12:* Performance Expectations Column** • K: K-LS1-1, p. 92 • Grade 5: 5-LS1-1, p. 111 • MS: MS-LS1-6, MS-LS1-7, p. 129 • HS: HS-LS1-5, HS-LS1-6, HS-LS1-7, pp. 154–155

Review Chapter 2 instructions on how to use this guide.

Visit curriculumtopicstudy2.org for more information about CTS and additional resources.

NOTE: *Atlas* page numbers have not been provided because *The NSTA Atlas of the Three Dimensions* was produced concurrently with this edition. Titles and map codes are accurate.

Additional Readings:

 Available for download at www.curriculumtopicstudy2.org

Copyright © 2020 by Corwin Press, Inc. All rights reserved. Reprinted from *Science Curriculum Topic Study: Bridging the Gap Between Three-Dimensional Standards, Research, and Practice* (2nd ed.) by Page Keeley and Joyce Tugel. Thousand Oaks, CA: Corwin, www.corwin.com. Reproduction authorized for educational use by educators, local school sites, and/or noncommercial or nonprofit entities that have purchased the book.

Cycling of Matter and Flow of Energy in Ecosystems
Grades 3–12 Standards- and Research-Based Study of a Curricular Topic

Section and Outcome	Selected Sources and Readings for Study and Reflection Read and examine **related** parts of
I. **Content Knowledge**	**IA:** *Science for All Americans* • Ch. 5: Flow of Matter and Energy, pp. 66–67 **IB:** *Science Matters* (2009 edition) • Ch. 19: Nutrients and the Carbon Cycle, 332–333; Energy and the Food Web, 329–332 **IC:** *Framework for K–12 Science Education:* Narrative Section • Ch. 6: LS2.B: Cycles of Matter and Energy Transfer in Ecosystems, pp. 152–153 **ID:** *Disciplinary Core Ideas: Reshaping Teaching and Learning* • Ch. 7: Ecosystem Science: Tracing Matter and Energy Through Ecosystems, pp. 126–128 **IE:** *The NSTA Atlas of the Three Dimensions:* Narrative Page • 4.2: Flow of Matter and Energy in Living Systems (LS2.B)
II. **Concepts, Core Ideas, or Practices**	**IIA:** *Framework for K–12 Science Education:* Grade Band Endpoints • Ch. 6: LS2.B: Cycles of Matter and Energy Transfer in Ecosystems, pp. 153–154 **IIB:** *Next Generation Science Standards:* Disciplinary Core Ideas Column • Grade 5: LS2.B: Cycles of Matter and Energy Transfer in Ecosystems, p. 29 • MS: LS2.B: Cycles of Matter and Energy Transfer in Ecosystems, p. 47 • HS: LS2.B: Cycles of Matter and Energy Transfer in Ecosystems, p. 81 **IIC:** *NSTA Quick Reference Guide to the NGSS K–12:* Disciplinary Core Ideas Column • Grade 5: LS2.B: Cycles of Matter and Energy Transfer in Ecosystems, p. 111 • MS: LS2.B: Cycles of Matter and Energy Transfer in Ecosystems, p. 130 • HS: LS2.B: Cycles of Matter and Energy Transfer in Ecosystems, pp. 156–157
III. **Curriculum, Instruction, and Formative Assessment**	**IIIA:** *Disciplinary Core Ideas: Reshaping Teaching and Learning* • Ch. 7: Elementary School Learning: Connecting Actors and Enablers With Matter and Energy, pp. 131–133 • Ch. 7: Middle and High School Learning: Matter Cycles and Energy Flows, pp. 133–135 • Ch. 7: What Approaches Can We Use to Teach About This Disciplinary Core Idea? pp. 140–142 **IIIB:** *Uncovering Student Ideas:* Assessment Probe and Suggestions for Instruction • USI.2: Giant Sequoia Tree, pp. 121, 126–127 • USI.3: Rotting Apple, pp. 139, 143–144; Earth's Mass, pp. 147, 153 • USI.LS: Food Chain Energy, pp. 91, 94–95; Ecosystem Cycles, pp. 97, 101; No More Plants, pp. 103, 106–107
IV. **Research on Commonly Held Ideas**	**IVA:** *Benchmarks for Science Literacy:* Chapter 15 Research • 5E: Flow of Matter and Energy, pp. 342–343 **IVB:** *Making Sense of Secondary Science: Research Into Children's Ideas* • Ch. 2: Food Chains and Ecological Cycles, pp. 34–35 • Ch. 7: Nutrition and Energy Flow, pp. 59–60; Food Chains and Webs, pp. 60–62; Decay, pp. 63–64; Cycling of Matter Through Ecosystems, p. 65; Gas Exchange and Balance, p. 66

Section and Outcome	Selected Sources and Readings for Study and Reflection Read and examine **related** parts of
	IVC: *Uncovering Student Ideas:* Related Research • USI.2: Giant Sequoia Tree, p. 126 • USI.3: Rotting Apple, p. 143; Earth's Mass, pp. 152–153 • USI.LS: Food Chain Energy, p. 94; Ecosystem Cycles, p. 100; No More Plants, p. 106
V. **K–12 Articulation and Connections**	**VA:** *Next Generation Science Standards:* Appendices: Progression • Appendix E: LS2.B: Cycles of Matter and Energy Transfer in Ecosystems, p. 5 **VB:** *NSTA Quick Reference Guide to the NGSS K–12:* Progression • LS2.B: Cycles of Matter and Energy Transfer in Ecosystems, p. 70 **VC:** *The NSTA Atlas of the Three Dimensions:* Map Page • 4.2: Flow of Matter and Energy in Living Systems (LS1.C & LS2.B)
VI. **Assessment Expectation**	**VIA:** *State Standards* • Examine your state's standards **VIB:** *Next Generation Science Standards:* Performance Expectations • Grade 5: 5-LS2-1, p. 29 • MS: MS-LS2-3, p. 47 • HS: HS-LS2-3, HS-LS2-4, HS-LS2-5, p. 81 **VIC:** *NSTA Quick Reference Guide to the NGSS K–12:* Performance Expectations Column • Grade 5: 5-LS2-1, p. 111 • MS: MS-LS2-3, p. 130 • HS: HS-LS2-3, HS-LS2-4, HS-LS2-5, pp. 156–157

Review Chapter 2 instructions on how to use this guide.

Visit curriculumtopicstudy2.org for more information about CTS and additional resources.

NOTE: This study can be combined with the following crosscutting concept CTS Guide: Energy and Matter: Flows, Cycles, and Conservation.

NOTE: *Atlas* page numbers have not been provided because *The NSTA Atlas of the Three Dimensions* was produced concurrently with this edition. Titles and map codes are accurate.

Additional Readings:

 Available for download at www.curriculumtopicstudy2.org

Copyright © 2020 by Corwin Press, Inc. All rights reserved. Reprinted from *Science Curriculum Topic Study: Bridging the Gap Between Three-Dimensional Standards, Research, and Practice* (2nd ed.) by Page Keeley and Joyce Tugel. Thousand Oaks, CA: Corwin, www.corwin.com. Reproduction authorized for educational use by educators, local school sites, and/or noncommercial or nonprofit entities that have purchased the book.

Cycling of Matter in Ecosystems
Grades 3–12 Standards- and Research-Based Study of a Curricular Topic

Section and Outcome	Selected Sources and Readings for Study and Reflection Read and examine **related** parts of
I. **Content Knowledge**	**IA:** *Science for All Americans* • Ch. 5: Flow of Matter and Energy, pp. 66–67 **IB:** *Science Matters* (2009 edition) • Ch. 19: Nutrients and the Carbon Cycle, 332–333 **IC:** *Framework for K–12 Science Education:* Narrative Section • Ch. 6: LS2.B: Cycles of Matter and Energy Transfer in Ecosystems, pp. 152–153 **ID:** *Disciplinary Core Ideas: Reshaping Teaching and Learning* • Ch. 7: Ecosystem Science: Tracing Matter and Energy Through Ecosystems: Matter Cycles, pp. 126–127 **IE:** *The NSTA Atlas of the Three Dimensions:* Narrative Page • 4.2: Flow of Matter and Energy in Living Systems (LS2.B)
II. **Concepts, Core Ideas, or Practices**	**IIA:** *Framework for K–12 Science Education:* Grade Band Endpoints • Ch. 6: LS2.B: Cycles of Matter and Energy Transfer in Ecosystems, pp. 153–154 **IIB:** *Next Generation Science Standards:* Disciplinary Core Ideas Column • Grade 5: LS2.B: Cycles of Matter and Energy Transfer in Ecosystems, p. 29 • MS: LS2.B: Cycles of Matter and Energy Transfer in Ecosystems, p. 47 • HS: LS2.B: Cycles of Matter and Energy Transfer in Ecosystems, p. 81 **IIC:** *NSTA Quick Reference Guide to the NGSS K–12:* Disciplinary Core Ideas Column • Grade 5: LS2.B: Cycles of Matter and Energy Transfer in Ecosystems, p. 111 • MS: LS2.B: Cycles of Matter and Energy Transfer in Ecosystems, p. 130 • HS: LS2.B: Cycles of Matter and Energy Transfer in Ecosystems, pp. 156–157
III. **Curriculum, Instruction, and Formative Assessment**	**IIIA:** *Uncovering Student Ideas:* Assessment Probe and Suggestions for Instruction • USI.2: Giant Sequoia Tree, pp. 121, 126–127 • USI.3: Rotting Apple, pp. 139, 141, 143–144; Earth's Mass, pp. 147, 149–150, 153 • USI.LS: Is It a Consumer? pp. 85, 86–87, 88; Ecosystem Cycles, pp. 97, 98–99, 101; No More Plants, pp. 103, 104–105, 106–107 **IIIC:** *Disciplinary Core Ideas: Reshaping Teaching and Learning* • Ch. 7: Elementary School Learning: Connecting Actors and Enablers With Matter and Energy, pp. 131–133 • Ch. 7: Middle and High School Learning: Matter Cycles and Energy Flows, pp. 133–135 • Ch. 7: What Approaches Can We Use to Teach About This Disciplinary Core Idea? pp. 140–142
IV. **Research on Commonly Held Ideas**	**IVA:** *Benchmarks for Science Literacy:* Chapter 15 Research • 5E: Flow of Matter and Energy, pp. 342–343 **IVB:** *Making Sense of Secondary Science: Research Into Children's Ideas* • Ch. 2: Food Chains and Ecological Cycles, pp. 34–35 • Ch. 7: Cycling of Matter Through Ecosystems, p. 65

Section and Outcome	Selected Sources and Readings for Study and Reflection Read and examine **related** parts of
	IVC: *Uncovering Student Ideas:* Related Research • USI.2: Giant Sequoia Tree, p. 126 • USI.3: Rotting Apple, pp. 143; Earth's Mass, pp. 152–153 • USI.LS: Is It a Consumer? p. 88; Ecosystem Cycles, p. 100; No More Plants, p. 106
V. **K–12 Articulation and Connections**	**VA:** *Next Generation Science Standards:* Appendices: Progression • Appendix E: LS2.B: Cycles of Matter and Energy Transfer in Ecosystems, p. 5 **VB:** *NSTA Quick Reference Guide to the NGSS K–12:* Progression • LS2.B: Cycles of Matter and Energy Transfer in Ecosystems, p. 70 **VC:** *The NSTA Atlas of the Three Dimensions:* Map Page • 4.2: Flow of Matter and Energy in Living Systems (LS1.C & LS2.B)
VI. **Assessment Expectation**	**VIA:** *State Standards* • Examine your state's standards **VIB:** *Next Generation Science Standards:* Performance Expectations • Grade 5: 3-LS2-1, p. 29 • MS: MS-LS2-3, p. 47 • HS: HS-LS2-3, HS-LS2-4, HS-LS2-5, p. 81 **VIC:** *NSTA Quick Reference Guide to the NGSS K–12:* Performance Expectations Column • Grade 5: 5-LS2-1, p. 111 • MS: MS-LS2-3, p. 130 • HS: HS-LS2-3, HS-LS2-4, HS-LS2-5, pp. 156–157

Review Chapter 2 instructions on how to use this guide.

Visit curriculumtopicstudy2.org for more information about CTS and additional resources.

NOTE: This study can be combined with the following crosscutting concept CTS Guide: Energy and Matter: Flows, Cycles, and Conservation.

NOTE: *Atlas* page numbers have not been provided because *The NSTA Atlas of the Three Dimensions* was produced concurrently with this edition. Titles and map codes are accurate.

Additional Readings:

Available for download at **www.curriculumtopicstudy2.org**

Copyright © 2020 by Corwin Press, Inc. All rights reserved. Reprinted from *Science Curriculum Topic Study: Bridging the Gap Between Three-Dimensional Standards, Research, and Practice* (2nd ed.) by Page Keeley and Joyce Tugel. Thousand Oaks, CA: Corwin, www.corwin.com. Reproduction authorized for educational use by educators, local school sites, and/or noncommercial or nonprofit entities that have purchased the book.

Decomposition and Decay
Grades 3–12 Standards- and Research-Based Study of a Curricular Topic

Section and Outcome	Selected Sources and Readings for Study and Reflection Read and examine **related** parts of
I. **Content Knowledge**	**IA: *Science for All Americans*** • Ch. 5: Flow of Matter and Energy, p. 66 **IB: *Science Matters*** (2009 edition) • Ch. 19: Energy and the Food Web, 329–332 **IC: *Framework for K–12 Science Education:*** Narrative Section • Ch. 6: LS2.B: Cycles of Matter and Energy Transfer in Ecosystems, pp. 152–153 **ID: *The NSTA Atlas of the Three Dimensions:*** Narrative Page • 4.2: Flow of Matter and Energy in Living Systems (LS2.B)
II. **Concepts, Core Ideas, or Practices**	**IIA: *Framework for K–12 Science Education:*** Grade Band Endpoints • Ch. 6: LS2.B: Cycles of Matter and Energy Transfer in Ecosystems, pp. 153–154 **IIB: *Next Generation Science Standards:*** Disciplinary Core Ideas Column • Grade 5: LS2.B: Cycles of Matter and Energy Transfer in Ecosystems, p. 29 • MS: LS2.B: Cycles of Matter and Energy Transfer in Ecosystems, p. 47 • HS: LS2.B: Cycles of Matter and Energy Transfer in Ecosystems, p. 81 **IIC: *NSTA Quick Reference Guide to the NGSS K–12:*** Disciplinary Core Ideas Column • Grade 5: LS2.B: Cycles of Matter and Energy Transfer in Ecosystems, p. 111 • MS: LS2.B: Cycles of Matter and Energy Transfer in Ecosystems, p. 130 • HS: LS2.B: Cycles of Matter and Energy Transfer in Ecosystems, pp. 156–157
III. **Curriculum, Instruction, and Formative Assessment**	**IIIA: *Disciplinary Core Ideas: Reshaping Teaching and Learning*** • K–5: Ch. 7: Elementary School Learning: Connecting Actors and Enablers With Matter and Energy, pp. 131–133 **IIIB: *Uncovering Student Ideas:*** Assessment Probe and Suggestions for Instruction • USI.3: Rotting Apple, pp. 139, 143–144; Earth's Mass, pp. 147, 153
IV. **Research on Commonly Held Ideas**	**IVA: *Benchmarks for Science Literacy:*** Chapter 15 Research • 5E: Decay, p. 343 **IVB: *Making Sense of Secondary Science: Research Into Children's Ideas*** • Ch. 6: Decay and Recycling, p. 57 • Ch. 7: Decay, pp. 63–65; **IVC: *Uncovering Student Ideas:*** Related Research • USI.3: Rotting Apple, p. 143; Earth's Mass, pp. 152–153
V. **K–12 Articulation and Connections**	**VA: *Next Generation Science Standards:*** Appendices: Progression • Appendix E: LS2.B: Cycles of Matter and Energy Transfer in Ecosystems, p. 5 **VB: *NSTA Quick Reference Guide to the NGSS K–12:*** Progression • LS2.B: Cycles of Matter and Energy Transfer in Ecosystems, p. 70 **VC: *The NSTA Atlas of the Three Dimensions:*** Map Page • 4.2: Flow of Matter and Energy in Living Systems (LS1.C & LS2.B)

Section and Outcome	Selected Sources and Readings for Study and Reflection Read and examine **related** parts of
VI. **Assessment** **Expectation**	**VIA:** *State Standards* • Examine your state's standards **VIB:** *Next Generation Science Standards:* Performance Expectations • Grade 5: 5-LS2-1, p. 29 • MS: MS-LS2-3, p. 47 • HS: HS-LS2-3, HS-LS2-4, p. 81 **VIC:** *NSTA Quick Reference Guide to the NGSS K–12:* Performance Expectations Column • Grade 5: 5-LS2-1, p. 111 • MS: MS-LS2-3, p. 130 • HS: HS-LS2-3, HS-LS2-4, p. 156

Review Chapter 2 instructions on how to use this guide.

Visit curriculumtopicstudy2.org for more information about CTS and additional resources.

NOTE: *Atlas* page numbers have not been provided because *The NSTA Atlas of the Three Dimensions* was produced concurrently with this edition. Titles and map codes are accurate.

Additional Readings:

Available for download at www.curriculumtopicstudy2.org

Copyright © 2020 by Corwin Press, Inc. All rights reserved. Reprinted from *Science Curriculum Topic Study: Bridging the Gap Between Three-Dimensional Standards, Research, and Practice* (2nd ed.) by Page Keeley and Joyce Tugel. Thousand Oaks, CA: Corwin, www.corwin.com. Reproduction authorized for educational use by educators, local school sites, and/or noncommercial or nonprofit entities that have purchased the book.

Ecosystem Stability, Disruptions, and Change
Grades 3–12 Standards- and Research-Based Study of a Curricular Topic

Section and Outcome	Selected Sources and Readings for Study and Reflection Read and examine **related** parts of
I. Content Knowledge	**IA:** *Science for All Americans* • Ch. 5: Interdependence of Life, pp. 65–66 (last two paragraphs); Flow of Matter and Energy, p. 67 (last two paragraphs) **IB:** *Science Matters* (2009 edition) • Ch. 19: Ecosystems, pp. 326–329 **IC:** *Framework for K–12 Science Education:* Narrative Section • Ch. 6: LS2.C: Ecosystem Dynamics, Functioning, and Resilience, pp. 154–155 **ID:** *Disciplinary Core Ideas: Reshaping Teaching and Learning* • Community Ecology: Dynamic Processes of Changes in Populations, pp. 128–130 **IE:** *The NSTA Atlas of the Three Dimensions:* Narrative Page • 4.4: Ecosystem Dynamics, Functioning, and Resilience (LS2.C)
II. Concepts, Core Ideas, or Practices	**IIA:** *Framework for K–12 Science Education:* Grade Band Endpoints • Ch. 6: LS2.C: Ecosystem Dynamics, Functioning, and Resilience, pp. 155–156 **IIB:** *Next Generation Science Standards:* Disciplinary Core Ideas Column • Grade 3: LS2.C: Ecosystem Dynamics, Functioning, and Resilience, p. 19 • MS: LS2.C: Ecosystem Dynamics, Functioning, and Resilience, pp. 49 • LS2.A: Interdependent Relationships in Ecosystems, p. 155; • HS: LS2.C: Ecosystem Dynamics, Functioning, and Resilience, p. 83 **IIC:** *NSTA Quick Reference Guide to the NGSS K–12:* Disciplinary Core Ideas Column • Grade 3: LS2.C: Ecosystem Dynamics, Functioning, and Resilience, p. 105 • MS: LS2.C: Ecosystem Dynamics, Functioning, and Resilience, p. 130 • LS2.A: Interdependent Relationships in Ecosystems, p. 155; • HS: LS2.C: Ecosystem Dynamics, Functioning, and Resilience, pp. 156–157
III. Curriculum, Instruction, and Formative Assessment	**IIIA:** *Disciplinary Core Ideas: Reshaping Teaching and Learning* • K–5: Ch. 7: Elementary School Learning: Observing Organisms in Their Environment, pp. 136–137 • 6–12: Ch. 7: Middle and High School Learning: Learning to Reason About Populations and Communities, pp. 137–140 **IIIB:** *Uncovering Student Ideas:* Assessment Probe and Suggestions for Instruction • USI.2: Habitat Change, pp. 143, 147 • USI.LS: No More Plants, pp. 103, 106–107
IV. Research on Commonly Held Ideas	**IVA:** *Benchmarks for Science Literacy:* Chapter 15 Research • 5D: Interdependence of Life, p. 342 **IVB:** *Making Sense of Secondary Science: Research Into Children's Ideas* • Ch. 7: Progression in Reasoning about Ecosystems, p. 59; Communities, Populations, and Competition Between Organisms, pp. 62–63; Environments, p. 63 **IVC:** *Uncovering Student Ideas:* Related Research • USI.2: Habitat Change, p. 147 • USI.LS: No More Plants, p. 106

Section and Outcome	Selected Sources and Readings for Study and Reflection Read and examine **related** parts of
V. K–12 Articulation and Connections	**VA:** *Next Generation Science Standards:* Appendices: Progression • Appendix E: LS2.C: Ecosystem Dynamics, Functioning, and Resilience, p. 5 **VB:** *NSTA Quick Reference Guide to the NGSS K–12:* Progression • LS2.C: Ecosystem Dynamics, Functioning, and Resilience, p. 71 **VC:** *The NSTA Atlas of the Three Dimensions:* Map Page • 4.4: Ecosystem Dynamics, Functioning, and Resilience (LS2.C & LS4.D)
VI. Assessment Expectation	**VIA:** *State Standards* • Examine your state's standards **VIB:** *Next Generation Science Standards:* Performance Expectations • Grade 3: 3-LS4-4, p. 19 • MS: MS-LS2-4, p. 47; MS-LS2-5, p. 49 • HS: HS-LS2-1, HS-LS2-6, HS-LS2-7, p. 83 **VIC:** *NSTA Quick Reference Guide to the NGSS K–12:* Performance Expectations Column • Grade 3: 3-LS4-4, p. 105 • MS: MS-LS2-4, MS-LS2-5, p. 130 • HS: HS-LS2-1, p. 155; HS-LS2-6, HS-LS2-7, pp. 156–157

Review Chapter 2 instructions on how to use this guide.

Visit curriculumtopicstudy2.org for more information about CTS and additional resources.

NOTE: *Atlas* page numbers have not been provided because *The NSTA Atlas of the Three Dimensions* was produced concurrently with this edition. Titles and map codes are accurate.

Additional Readings:

Available for download at www.curriculumtopicstudy2.org

Copyright © 2020 by Corwin Press, Inc. All rights reserved. Reprinted from *Science Curriculum Topic Study: Bridging the Gap Between Three-Dimensional Standards, Research, and Practice* (2nd ed.) by Page Keeley and Joyce Tugel. Thousand Oaks, CA: Corwin, www.corwin.com. Reproduction authorized for educational use by educators, local school sites, and/or noncommercial or nonprofit entities that have purchased the book.

Food Chains and Food Webs (K–8)
Grades K–8 Standards- and Research-Based Study of a Curricular Topic

Section and Outcome	Selected Sources and Readings for Study and Reflection Read and examine **related** parts of
I. Content Knowledge	**IA: *Science for All Americans*** • Ch. 5: Diversity of Life, p. 61 (last paragraph); Flow of Matter and Energy, pp. 66–67 **IB: *Science Matters*** (2009 edition) • Ch. 19: Energy and the Food Web, 329–332 **IC: *Framework for K–12 Science Education:*** Narrative Section • Ch. 6: LS2.A: Interdependent Relationships in Ecosystems, p. 151 • Ch. 6: LS2.B: Cycles of Matter and Energy Transfer in Ecosystems, pp. 152–153 **ID: *Disciplinary Core Ideas: Reshaping Teaching and Learning*** • Ch. 7: Ecosystem Science: Tracing Matter and Energy Through Ecosystems: Matter Cycles, pp. 126–127 **IE: *The NSTA Atlas of the Three Dimensions:*** Narrative Page • 4.2: Flow of Matter and Energy in Living Systems (LS2.B)
II. Concepts, Core Ideas, or Practices	**IIA: *Framework for K–12 Science Education:*** Grade Band Endpoints • Ch. 6: LS2.A: Interdependent Relationships in Ecosystems, pp. 151–152 • Ch. 6: LS2.B: Cycles of Matter and Energy Transfer in Ecosystems, pp. 153–154 **IIB: *Next Generation Science Standards:*** Disciplinary Core Ideas Column • K: LS1.C: Organization for Matter and Energy Flow in Organisms, p. 6 • Grade 5: LS2.A: Interdependent Relationships in Ecosystems; LS2.B: Cycles of Matter and Energy Transfer in Ecosystems, p. 29 • MS: LS2.B: Cycles of Matter and Energy Transfer in Ecosystems, p. 47 **IIC: *NSTA Quick Reference Guide to the NGSS K–12:*** Disciplinary Core Ideas Column • K: LS1.C: Organization for Matter and Energy Flow in Organisms, p. 92 • Grade 5: LS2.A: Interdependent Relationships in Ecosystems, p. 111; LS2.B: Cycles of Matter and Energy Transfer in Ecosystems, p. 111 • MS: LS2.B: Cycles of Matter and Energy Transfer in Ecosystems, p. 130
III. Curriculum, Instruction, and Formative Assessment	**IIIA: *Disciplinary Core Ideas: Reshaping Teaching and Learning*** • Ch. 7: Elementary School Learning: Connecting Actors and Enablers With Matter and Energy, pp. 131–133 • Ch. 7: Middle and High School Learning: Matter Cycles and Energy Flows, pp. 133–135 • Ch. 7: What Approaches Can We Use to Teach About This Disciplinary Core Idea? pp. 140–142 **IIIB: *Uncovering Student Ideas:*** Assessment Probe and Suggestions for Instruction • USI.LS: Is It a Consumer? pp. 85, 88; Food Chain Energy, pp. 91, 94–95; No More Plants, pp. 103, 106–107
IV. Research on Commonly Held Ideas	**IVA: *Benchmarks for Science Literacy:*** Chapter 15 Research • 5E: Flow of Matter and Energy, pp. 342–343 **IVB: *Making Sense of Secondary Science: Research Into Children's Ideas*** • Ch. 2: Food Chains and Ecological Cycles, pp. 34–35 • Ch. 7: Food Chains and Webs, pp. 60–62

Section and Outcome	Selected Sources and Readings for Study and Reflection Read and examine **related** parts of
	IVC: *Uncovering Student Ideas:* Related Research • USI.LS: Is It a Consumer? p. 88; Food Chain Energy, p. 94; No More Plants, p. 106
V. **K–12 Articulation and Connections**	**VA: *Next Generation Science Standards:*** Appendices: Progression • Appendix E: LS2.B: Cycles of Matter and Energy Transfer in Ecosystems, p. 5 **VB: *NSTA Quick Reference Guide to the NGSS K–12:*** Progression • LS2.B: Cycles of Matter and Energy Transfer in Ecosystems, p. 70 **VC: *The NSTA Atlas of the Three Dimensions:*** Map Page • 4.2: Flow of Matter and Energy in Living Systems (LS1.C & LS2.B)
VI. **Assessment Expectation**	**VIA: *State Standards*** • Examine your state's standards **VIB: *Next Generation Science Standards:*** Performance Expectations • K: K-LS1-1, p. 6 • Grade 5: 5-LS2-1, p. 29 • MS: MS-LS2-3, p. 47 **VIC: *NSTA Quick Reference Guide to the NGSS K–12:*** Performance Expectations Column • K: K-LS1-1, p. 92 • Grade 5: 5-LS2-1, p. 111 • MS: MS-LS2-3, p. 130

Review Chapter 2 instructions on how to use this guide.

Visit curriculumtopicstudy2.org for more information about CTS and additional resources.

NOTE: *Atlas* page numbers have not been provided because *The NSTA Atlas of the Three Dimensions* was produced concurrently with this edition. Titles and map codes are accurate.

Additional Readings:

Available for download at **www.curriculumtopicstudy2.org**

Copyright © 2020 by Corwin Press, Inc. All rights reserved. Reprinted from *Science Curriculum Topic Study: Bridging the Gap Between Three-Dimensional Standards, Research, and Practice* (2nd ed.) by Page Keeley and Joyce Tugel. Thousand Oaks, CA: Corwin, www.corwin.com. Reproduction authorized for educational use by educators, local school sites, and/or noncommercial or nonprofit entities that have purchased the book.

Group Behaviors and Social Interactions in Ecosystems
Grades 3–12 Standards- and Research-Based Study of a Curricular Topic

Section and Outcome	Selected Sources and Readings for Study and Reflection Read and examine **related** parts of
I. Content Knowledge	**IA:** *Science for All Americans* • Ch. 5: Interdependence of Life, pp. 64–66 **IB:** *Framework for K–12 Science Education:* Narrative Section • Ch. 6: LS2.D: Social Interactions and Group Behavior, p. 156 **IC:** *The NSTA Atlas of the Three Dimensions:* Narrative Page • 4.3: Interdependent Relationships and Social Interactions in Ecosystems (LS2.D)
II. Concepts, Core Ideas, or Practices	**IIA:** *Framework for K–12 Science Education:* Grade Band Endpoints • Ch. 6: LS2.D: Social Interactions and Group Behavior, pp. 156–157 **IIB:** *Next Generation Science Standards:* Disciplinary Core Ideas Column • Grade 3: LS2.D: Social Interactions and Group Behavior, p. 19 • HS: LS2.D: Social Interactions and Group Behavior, p. 83 **IIC:** *NSTA Quick Reference Guide to the NGSS K–12:* Disciplinary Core Ideas Column • Grade 3: LS2.D: Social Interactions and Group Behavior, p. 104 • HS: LS2.D: Social Interactions and Group Behavior, p. 158
III. Curriculum, Instruction, and Formative Assessment	**IIIA:** *Disciplinary Core Ideas: Reshaping Teaching and Learning* • Ch. 7: 6–12: Middle and High School Learning: Learning to Reason About Populations and Communities, pp. 139–140 (starting with last bullet on p. 139)
IV. Research on Commonly Held Ideas	No available research for this topic.
V. K–12 Articulation and Connections	**VA:** *Next Generation Science Standards:* Appendices: Progression • Appendix E: LS2.D: Social Interactions and Group Behavior, p. 5 **VB:** *NSTA Quick Reference Guide to the NGSS K–12:* Progression • LS2.D: Social Interactions and Group Behavior, p. 71 **VC:** *The NSTA Atlas of the Three Dimensions:* Map Page • 4.3: Interdependent Relationships and Social Interactions in Ecosystems (LS2.A & LS2.D)

Section and Outcome	Selected Sources and Readings for Study and Reflection Read and examine **related** parts of
VI. Assessment Expectation	**VIA:** *State Standards* • Examine your state's standards **VIB:** *Next Generation Science Standards:* Performance Expectations • Grade 3: 3-LS2-1, p. 19 • HS: HS-LS2-8, p. 83 **VIC:** *NSTA Quick Reference Guide to the NGSS K–12:* Performance Expectations Column • Grade 3: 3-LS2-1, p. 104 • HS: HS-LS2-8, p. 158

Review Chapter 2 instructions on how to use this guide.

Visit curriculumtopicstudy2.org for more information about CTS and additional resources.

NOTE: *Atlas* page numbers have not been provided because *The NSTA Atlas of the Three Dimensions* was produced concurrently with this edition. Titles and map codes are accurate.

Additional Readings:

Available for download at **www.curriculumtopicstudy2.org**

Copyright © 2020 by Corwin Press, Inc. All rights reserved. Reprinted from *Science Curriculum Topic Study: Bridging the Gap Between Three-Dimensional Standards, Research, and Practice* (2nd ed.) by Page Keeley and Joyce Tugel. Thousand Oaks, CA: Corwin, www.corwin.com. Reproduction authorized for educational use by educators, local school sites, and/or noncommercial or nonprofit entities that have purchased the book.

Interdependency in Ecosystems
Grades K–12 Standards- and Research-Based Study of a Curricular Topic

Section and Outcome	Selected Sources and Readings for Study and Reflection Read and examine **related** parts of
I. Content Knowledge	**IA:** *Science for All Americans* • Ch. 5: Interdependence of Life, pp. 64–66 **IB:** *Science Matters* (2009 edition) • Ch. 19: Ecosystems, pp. 326–329 **IC:** *Framework for K–12 Science Education:* Narrative Section • Ch. 6: LS2.A: Interdependent Relationships in Ecosystems, p. 151 **ID:** *Disciplinary Core Ideas: Reshaping Teaching and Learning* • Ch. 7: Community Ecology: Dynamic Processes of Changes in Populations, pp. 128–130 **IE:** *The NSTA Atlas of the Three Dimensions:* Narrative Page • 4.3: Interdependent Relationships and Social Interactions in Ecosystems (LS2.A)
II. Concepts, Core Ideas, or Practices	**IIA:** *Framework for K–12 Science Education:* Grade Band Endpoints • Ch. 6: LS2.A: Interdependent Relationships in Ecosystems, pp. 151–152 **IIB:** *Next Generation Science Standards:* Disciplinary Core Ideas Column • Grade 2: LS2.A: Interdependent Relationships in Ecosystems, p. 14 • Grade 5: LS2.A: Interdependent Relationships in Ecosystems, p. 29 • MS: LS2.A: Interdependent Relationships in Ecosystems, pp. 47, 49 • HS: LS2.A: Interdependent Relationships in Ecosystems, p. 83 **IIC:** *NSTA Quick Reference Guide to the NGSS K–12:* Disciplinary Core Ideas Column • Grade 2: LS2.A: Interdependent Relationships in Ecosystems, p. 96 • Grade 5: LS2.A: Interdependent Relationships in Ecosystems, p. 111 • MS: LS2.A: Interdependent Relationships in Ecosystems, p. 130 • HS: LS2.A: Interdependent Relationships in Ecosystems, pp. 155–156
III. Curriculum, Instruction, and Formative Assessment	**IIIA:** *Disciplinary Core Ideas: Reshaping Teaching and Learning* • K–5: Ch. 7: Elementary School Learning: Observing Organisms in Their Environment, pp. 136–137 • 6–12: Ch. 7: Middle and High School Learning: Learning to Reason About Populations and Communities, pp. 137–140 **IIIB:** *Uncovering Student Ideas:* Assessment Probe and Suggestions for Instruction • USI.LS: No More Plants, pp. 103, 106–107
IV. Research on Commonly Held Ideas	**IVA:** *Benchmarks for Science Literacy:* Chapter 15 Research • 5D: Interdependence of Life, p. 342 **IVB:** *Making Sense of Secondary Science: Research Into Children's Ideas* • Ch. 7: Communities, Populations, and Competition Between Organisms, pp. 62–63 **IVC:** *Uncovering Student Ideas:* Related Research • USI.LS: No More Plants, p. 106
V. K–12 Articulation and Connections	**VA:** *Next Generation Science Standards:* Appendices: Progression • Appendix E: LS2.A: Interdependent Relationships in Ecosystems, p. 5 **VB:** *NSTA Quick Reference Guide to the NGSS K–12:* Progression • LS2.A: Interdependent Relationships in Ecosystems, p. 70

Section and Outcome	Selected Sources and Readings for Study and Reflection Read and examine **related** parts of
	VC: *The NSTA Atlas of the Three Dimensions:* Map Page • 4.3: Interdependent Relationships and Social Interactions in Ecosystems (LS2.A & LS2.D)
VI. Assessment Expectation	**VIA:** *State Standards* • Examine your state's standards **VIB:** *Next Generation Science Standards:* Performance Expectations • Grade 2: 2-LS2-1, 2-LS2-2, p. 14 • Grade 5: 5-LS2-1, p. 29 • MS: MS-LS2-1, p. 47; MS-LS2-2, p. 49 • HS: HS-LS2-1, HS-LS2-2, p. 83 **VIC:** *NSTA Quick Reference Guide to the NGSS K–12:* Performance Expectations Column • Grade 2: 2-LS2-1, 2-LS2-2, p. 96 • Grade 5: 5-LS2-1, p. 111 • MS: MS-LS2-1, MS-LS2-2, p. 130 • HS: HS-LS2-1, HS-LS2-2, pp. 155–156

Review Chapter 2 instructions on how to use this guide.

Visit curriculumtopicstudy2.org for more information about CTS and additional resources.

NOTE: *Atlas* page numbers have not been provided because *The NSTA Atlas of the Three Dimensions* was produced concurrently with this edition. Titles and map codes are accurate.

Additional Readings:

 Available for download at www.curriculumtopicstudy2.org

Copyright © 2020 by Corwin Press, Inc. All rights reserved. Reprinted from *Science Curriculum Topic Study: Bridging the Gap Between Three-Dimensional Standards, Research, and Practice* (2nd ed.) by Page Keeley and Joyce Tugel. Thousand Oaks, CA: Corwin, www.corwin.com. Reproduction authorized for educational use by educators, local school sites, and/or noncommercial or nonprofit entities that have purchased the book.

Transfer of Energy in Ecosystems
Grades 6–12 Standards- and Research-Based Study of a Curricular Topic

Section and Outcome	Selected Sources and Readings for Study and Reflection Read and examine **related** parts of
I. Content Knowledge	**IA:** *Science for All Americans* • Ch. 5: Flow of Matter and Energy, pp. 66–67 **IB:** *Science Matters* (2009 edition) • Ch. 19: Energy and the Food Web, 329–332 **IC:** *Framework for K–12 Science Education:* Narrative Section • Ch. 6: LS2.B: Cycles of Matter and Energy Transfer in Ecosystems, pp. 152–153 **ID:** *Disciplinary Core Ideas: Reshaping Teaching and Learning* • Ch. 7: Energy Flows, pp. 127–128 **IE:** *The NSTA Atlas of the Three Dimensions:* Narrative Page • 4.2: Flow of Matter and Energy in Living Systems (LS2.B)
II. Concepts, Core Ideas, or Practices	**IIA:** *Framework for K–12 Science Education:* Grade Band Endpoints • Ch. 6: LS2.B: Cycles of Matter and Energy Transfer in Ecosystems, pp. 153–154 **IIB:** *Next Generation Science Standards:* Disciplinary Core Ideas Column • MS: LS2.B: Cycles of Matter and Energy Transfer in Ecosystems, p. 47 • HS: LS2.B: Cycles of Matter and Energy Transfer in Ecosystems, p. 81 **IIC:** *NSTA Quick Reference Guide to the NGSS K–12:* Disciplinary Core Ideas Column • MS: LS2.B: Cycles of Matter and Energy Transfer in Ecosystems, p. 130 • HS: LS2.B: Cycles of Matter and Energy Transfer in Ecosystems, pp. 156–157
III. Curriculum, Instruction, and Formative Assessment	**IIIA:** *Disciplinary Core Ideas: Reshaping Teaching and Learning* • Ch. 7: Middle and High School Learning: Matter Cycles and Energy Flows, pp. 133–135 • Ch. 7: What Approaches Can We Use to Teach About This Disciplinary Core Idea? pp. 140–142 **IIIB:** *Uncovering Student Ideas:* Assessment Probe and Suggestions for Instruction • USI.LS: Food Chain Energy, pp. 91, 94–95; Ecosystem Cycles, pp. 97, 101; No More Plants, pp. 103, 106–107
IV. Research on Commonly Held Ideas	**IVA:** *Benchmarks for Science Literacy:* Chapter 15 Research • 5E: Flow of Matter and Energy, pp. 342–343 **IVB:** *Making Sense of Secondary Science: Research Into Children's Ideas* • Ch. 2: Food Chains and Ecological Cycles, pp. 34–35 • Ch. 7: Nutrition and Energy Flow, pp. 59–60; Food Chains and Webs, pp. 60–62 **IVC:** *Uncovering Student Ideas:* Related Research • USI.LS: Food Chain Energy, p. 94; Ecosystem Cycles, p. 100; No More Plants, p. 106
V. K–12 Articulation and Connections	**VA:** *Next Generation Science Standards:* Appendices: Progression Chart • Appendix E: LS2.B: Cycles of Matter and Energy Transfer in Ecosystems, p. 5 **VB:** *NSTA Quick Reference Guide to the NGSS K–12:* Progression Chart • LS2.B: Cycles of Matter and Energy Transfer in Ecosystems, p. 70 **VC:** *The NSTA Atlas of the Three Dimensions:* Map Page • 4.2: Flow of Matter and Energy in Living Systems (LS1.C & LS2.B)

Section and Outcome	Selected Sources and Readings for Study and Reflection Read and examine **related** parts of
VI. Assessment Expectation	**VIA:** *State Standards*<ul><li>Examine your state's standards</li></ul>**VIB:** *Next Generation Science Standards:* Performance Expectations<ul><li>MS: MS-LS2-3, p. 47</li><li>HS: HS-LS2-3, HS-LS2-4, HS-LS2-5, p. 81</li></ul>**VIC:** *NSTA Quick Reference Guide to the NGSS K–12:* Performance Expectations Column<ul><li>MS: MS-LS2-3, p. 130</li><li>HS: HS-LS2-3, HS-LS2-4, HS-LS2-5, pp. 156–157</li></ul>

Review Chapter 2 instructions on how to use this guide.

Visit curriculumtopicstudy2.org for more information about CTS and additional resources.

NOTE: This study can be combined with the following crosscutting concept CTS guide: Matter and Energy: Flows, Cycles, and Conservation.

NOTE: *Atlas* page numbers have not been provided because *The NSTA Atlas of the Three Dimensions* was produced concurrently with this edition. Titles and map codes are accurate.

Additional Readings:

Available for download at www.curriculumtopicstudy2.org

Copyright © 2020 by Corwin Press, Inc. All rights reserved. Reprinted from *Science Curriculum Topic Study: Bridging the Gap Between Three-Dimensional Standards, Research, and Practice* (2nd ed.) by Page Keeley and Joyce Tugel. Thousand Oaks, CA: Corwin, www.corwin.com. Reproduction authorized for educational use by educators, local school sites, and/or noncommercial or nonprofit entities that have purchased the book.

Adaptation
Grades 3–12 Standards- and Research-Based Study of a Curricular Topic

Section and Outcome	Selected Sources and Readings for Study and Reflection Read and examine **related** parts of
I. Content Knowledge	**IA:** *Science for All Americans* • Ch. 5: Evolution of Life, pp. 67–69 (paragraphs 4–7) **IB:** *Science Matters* (2009 edition) • Ch. 18: Natural Selection, pp. 310–312 **IC:** *Framework for K–12 Science Education:* Narrative Section • Ch. 6: LS4.C: Adaptation, pp. 164–165 **ID:** *Disciplinary Core Ideas: Reshaping Teaching and Learning* • Ch. 9: Core Idea LS4.C: p. 168 (first column) • Ch. 9: Core Explanatory Mechanisms, pp. 168–169 **IE:** *The NSTA Atlas of the Three Dimensions:* Narrative Page • 4.6 Natural Selection and Adaptation (LS4.C)
II. Concepts, Core Ideas, or Practices	**IIA:** *Framework for K–12 Science Education:* Grade Band Endpoints • Ch. 6: LS4.C: Adaptation, pp. 165–166 **IIB:** *Next Generation Science Standards:* Disciplinary Core Ideas Column • Grade 3: LS4.C: Adaptation, p. 19 • MS: LS4.C: Adaptation, p. 52 • HS: LS4.C: Adaptation, pp. 83, 87 **IIC:** *NSTA Quick Reference Guide to the NGSS K–12:* Disciplinary Core Ideas Column • Grade 3: LS4.C: Adaptation, p. 105 • MS: LS4.C: Adaptation, p. 132 • HS: LS4.C: Adaptation, pp. 159–160
III. Curriculum, Instruction, and Assessment	**IIIA:** *Disciplinary Core Ideas: Reshaping Teaching and Learning* • Ch. 9: How Does Student Understanding Of This Disciplinary Core Idea Develop Over Time? pp. 174–175 • Ch. 9: Building From Student Ideas in Early Grades, pp. 177–178 **IIIB:** *Uncovering Student Ideas:* Assessment Probe and Suggestions for Instruction • USI.2: Habitat Change, pp. 143, 145, 147 • USI.4: Adaptation, pp. 113, 115, 117–118 • USI.LS: Changing Environment, pp. 109, 113–114
IV. Research on Commonly Held Ideas	**IVA:** *Benchmarks for Science Literacy:* Chapter 15 Research • 5D: Habitat, p. 342 • 5F: Adaptation, p. 344 **IVB:** *Making Sense of Secondary Science: Research Into Children's Ideas* • Ch. 5: Adaptation, pp. 52–53 **IVC:** *Uncovering Student Ideas:* Related Research • USI.2: Habitat Change, p. 147 • USI.4: Adaptation, pp. 117 • USI.LS: Changing Environment, p. 113

Section and Outcome	Selected Sources and Readings for Study and Reflection Read and examine **related** parts of
V. **K–12 Articulation and Connections**	**VA:** *Next Generation Science Standards:* Appendices: Progression<ul><li>Appendix E: LS4.C: Adaptation, p. 6</li></ul>**VB:** *NSTA Quick Reference Guide to the NGSS K–12:* Progression<ul><li>LS4.C: Adaptation, p. 73</li></ul>**VC:** *The NSTA Atlas of the Three Dimensions:* Map Page<ul><li>4.6 Natural Selection and Adaptation (LS4.A, LS4.B, & LS4.C)</li></ul>
VI. **Assessment Expectation**	**VIA:** *State Standards*<ul><li>Examine your state's standards</li></ul>**VIB:** *Next Generation Science Standards:* Performance Expectations<ul><li>Grade 3: 3-LS4-4, p. 19</li><li>MS: MS-LS4-6, p. 52</li><li>HS: HS-LS4-2, HS-LS4-3, HS-LS4-4, HS-LS4-5, p. 87</li></ul>**VIC:** *NSTA Quick Reference Guide to the NGSS K–12:* Performance Expectations<ul><li>Grade 3: 3-LS4-4, p. 105</li><li>MS: MS-LS4-6, p. 132</li><li>HS: HS-LS4-2, HS-LS4-3, p. 159; HS-LS4-4, HS-LS4-5, p. 160</li></ul>

Review Chapter 2 instructions on how to use this guide.

Visit curriculumtopicstudy2.org for more information about CTS and additional resources.

NOTE: *Atlas* page numbers have not been provided because *The NSTA Atlas of the Three Dimensions* was produced concurrently with this edition. Titles and map codes are accurate.

Additional Readings:

Available for download at www.curriculumtopicstudy2.org

Copyright © 2020 by Corwin Press, Inc. All rights reserved. Reprinted from *Science Curriculum Topic Study: Bridging the Gap Between Three-Dimensional Standards, Research, and Practice* (2nd ed.) by Page Keeley and Joyce Tugel. Thousand Oaks, CA: Corwin, www.corwin.com. Reproduction authorized for educational use by educators, local school sites, and/or noncommercial or nonprofit entities that have purchased the book.

Biodiversity and Human Impact
Grades 3–12 Standards- and Research-Based Study of a Curricular Topic

Section and Outcome	Selected Sources and Readings for Study and Reflection Read and examine **related** parts of
I. **Content Knowledge**	**IA: *Science for All Americans*** • Chapter 5, Diversity of Life, pp. 60–61 (last paragraph) **IB: *Framework for K–12 Science Education:*** Narrative Section • Chapter 6, LS4.D: Biodiversity and Humans, p. 166 **IC: *Disciplinary Core Ideas: Reshaping Teaching and Learning*** • Ch. 9: The Four Component Ideas of LS4, p. 168 (start at first column, second paragraph) • Ch. 9: Core Explanatory Mechanisms, pp. 170–171 (start with second paragraph) **ID: *The NSTA Atlas of the Three Dimensions:*** Narrative Page • 4.4: Ecosystem Dynamics, Functioning, and Resilience (LS2.C & LS4.D)
II. **Concepts, Core Ideas, or Practices**	**IIA: *Framework for K–12 Science Education:*** Grade Band Endpoints • Chapter 6, LS4.D: Biodiversity and Humans, pp. 166–167 **IIB: *Next Generation Science Standards:*** Disciplinary Core Ideas Column • Grade 3: LS4.D: Biodiversity and Humans, p. 19 • MS: LS4.D: Biodiversity and Humans, p. 49 • HS: LS4.D: Biodiversity and Humans, pp. 83–84 **IIC: *NSTA Quick Reference Guide to the NGSS K–12:*** Disciplinary Core Ideas Column • Grade 3: LS4.D: Biodiversity and Humans, p. 105 • MS: LS4.D: Biodiversity and Humans, p. 130 • HS: LS4.D: Biodiversity and Humans, pp. 157 and 160
III. **Curriculum, Instruction, and Formative Assessment**	**IIIA: *Disciplinary Core Ideas: Reshaping Teaching and Learning*** • Ch. 9: Building From Student Ideas in Early Grades, pp. 177–178 (start with last paragraph) • Ch. 9: Refining Intuitions in Middle and High School, pp. 179–180 (start with last paragraph) **IIIB: *Uncovering Student Ideas:*** Assessment Probe and Suggestions for Instruction • USI.2: Habitat Change, pp. 143, 147 • USI.LS: Changing Environment, pp. 109, 113–114
IV. **Research on Commonly Held Ideas**	**IVA: *Benchmarks for Science Literacy:*** Chapter 15 Research • 5D: Habitat, page 344 **IVB: *Uncovering Student Ideas:*** Related Research • USI2: Habitat Change, p. 147 • USI.LS: Changing Environment, p. 113
V. **K–12 Articulation and Connections**	**VA: *Next Generation Science Standards:*** Appendices: Progression • Appendix E: LS4.D: Biodiversity and Humans, p. 6 **VB: *NSTA Quick Reference Guide to the NGSS K–12:*** Progression • LS4.D: Biodiversity and Humans, p. 74 **VC: *The NSTA Atlas of the Three Dimensions:*** Map Page • 4.4: Ecosystem Dynamics, Functioning, and Resilience (LS2.C & LS4.D)

Section and Outcome	Selected Sources and Readings for Study and Reflection Read and examine **related** parts of
VI. **Assessment** **Expectation**	**VIA: *State Standards*** • Examine your state's standards **VIB: *Next Generation Science Standards:*** Performance Expectations • Grade 3: 3-LS4-4, p. 19 • MS: MS-LS2-5, p. 49 • HS: HS-LS2-7, p. 83 **VIC: *NSTA Quick Reference Guide to the NGSS K–12:*** Performance Expectations • Grade 3: 3-LS4-4, p. 105 • MS: MS-LS2-5, p. 130 • HS: HS-LS2-7, p. 157

Review Chapter 2 instructions on how to use this guide.

Visit curriculumtopicstudy2.org for more information about CTS and additional resources.

NOTE: *Atlas* page numbers have not been provided because *The NSTA Atlas of the Three Dimensions* was produced concurrently with this edition. Titles and map codes are accurate.

Additional Readings:

Available for download at www.curriculumtopicstudy2.org

Copyright © 2020 by Corwin Press, Inc. All rights reserved. Reprinted from *Science Curriculum Topic Study: Bridging the Gap Between Three-Dimensional Standards, Research, and Practice* (2nd ed.) by Page Keeley and Joyce Tugel. Thousand Oaks, CA: Corwin, www.corwin.com. Reproduction authorized for educational use by educators, local school sites, and/or noncommercial or nonprofit entities that have purchased the book.

Biological Evolution
Standards- and Research-Based Study of a Curricular Topic

Section and Outcome	Selected Sources and Readings for Study and Reflection Read and examine **related** parts of
I. Content Knowledge	**IA:** *Science for All Americans* • Ch. 5: Evolution of Life, pp. 67–69 (paras. 4–7) **IB:** *Science Matters* (2009 edition) • Ch. 18: Evolution, pp. 304–325 **IC:** *Framework for K–12 Science Education:* Narrative Section • Ch. 6: LS4: Biological Evolution: Unity and Diversity, p. 161 • Ch. 6: LS4.A: Evidence of Common Ancestry and Diversity, p. 162 • Ch. 6: LS4.B: Natural Selection, p. 163 • Ch. 6: LS4.C: Adaptation, pp. 164–165 • Ch. 6: LS4.D: Biodiversity and Humans, p. 166 **ID:** *Disciplinary Core Ideas: Reshaping Teaching and Learning* • Ch. 9: Core Idea LS4-Biological Evolution: Unity and Diversity, pp. 165–172 **IE:** *The NSTA Atlas of the Three Dimensions:* Narrative Page • 4.6: Natural Selection and Evolution (LS4.A, LS4.B, & LS4.C)
II. Concepts, Core Ideas, or Practices	**IIA:** *Framework for K–12 Science Education:* Grade Band Endpoints • Ch. 6: LS4.A: Evidence of Common Ancestry and Diversity, pp. 162–163 • Ch. 6: LS4.B: Natural Selection, p. 164 • Ch. 6: LS4.C: Adaptation, pp. 165–166 • Ch. 6: LS4.D: Biodiversity and Humans, pp. 166–167 **IIB:** *Next Generation Science Standards:* Disciplinary Core Ideas Column • Grade 3: LS4.A: Evidence of Common Ancestry and Diversity; LS4.C: Adaptation, p. 19; LS4.B: Natural Selection, p. 20 • MS: LS4.A: Evidence of Common Ancestry and Diversity, LS4.C: Adaptation, p. 52; LS4.B: Natural Selection, pp. 50, 52 • HS: LS4.A: Evidence of Common Ancestry and Diversity, LS4.B: Natural Selection, p. 87; LS4.C: Adaptation, pp. 83, 87 **IIC:** *NSTA Quick Reference Guide to the NGSS K–12:* Disciplinary Core Ideas Column • Grade 3: LS4.A: Evidence of Common Ancestry and Diversity, p. 104; LS4.B: Natural Selection, p. 104; LS4.C: Adaptation, p. 105 • MS: LS4.A: Evidence of Common Ancestry and Diversity, pp. 131, 132; LS4.B: Natural Selection, p. 132; LS4.C: Adaptation, p. 132 • HS: LS4.A: Evidence of Common Ancestry and Diversity, p. 159; LS4.B: Natural Selection, p. 159; LS4.C: Adaptation, pp. 159–160; and LS4.D: Biodiversity and Humans, p. 160
III. Curriculum, Instruction, and Formative Assessment	**IIIA:** *Disciplinary Core Ideas: Reshaping Teaching and Learning* • Ch. 9: What Approaches Can We Use to Teach About This Disciplinary Core Idea? p. 176 • Ch. 9: Building From Student Ideas in Early Grades, pp. 177–178 • Ch. 9: Refining Intuitions in Middle and High School, pp. 178–180 **IIIB:** *Uncovering Student Ideas:* Assessment Probe and Suggestions for Instruction • USI2: Habitat Change, pp. 143, 147 • USI4: Biological Evolution, pp. 99, 103–104; Adaptation, pp. 113, 117–118; Is It "Fitter"? pp. 119, 123 • USI.LS: Changing Environment, pp. 109, 113–114

Section and Outcome	Selected Sources and Readings for Study and Reflection Read and examine **related** parts of
IV. **Research on Commonly Held Ideas**	**IVA:** *Benchmarks for Science Literacy:* Chapter 15 Research • 5F: Evolution of Life, pp. 343–344 **IVB:** *Making Sense of Secondary Science: Research Into Children's Ideas* • Ch. 1: The Concept of Species, p. 25; Adaptation, p. 26 • Ch. 5: Adaptation, pp. 52–53 **IVC:** *Uncovering Student Ideas:* Related Research • USI2: Habitat Change, p. 147 • USI4: Biological Evolution, p. 103; Adaptation, p. 117; Is It "Fitter"? p. 123 • USI.LS: Changing Environment, p. 113
V. **K–12 Articulation and Connections**	**VA:** *Next Generation Science Standards:* Appendices: Progression • Appendix E: LS4.A: Evidence of Common Ancestry and Diversity, LS4.B: Natural Selection, LS4.C: Adaptation, p. 6 **VB:** *NSTA Quick Reference Guide to the NGSS K–12:* Progression • LS4, Biological Evolution: Unity and Diversity, pp. 72–74 **VC:** *The NSTA Atlas of the Three Dimensions:* Map Page • 4.6: Natural Selection and Evolution (LS4.A, LS4.B, & LS4.C)
VI. **Assessment Expectation**	**VIA:** *State Standards* • Examine your state's standards **VIB:** *Next Generation Science Standards:* Performance Expectations • Grade 3: 3-LS4-1, 3-LS4-3, p. 19; p. 30; 3-LS4-2, p. 20 • MS-LS4-5, p. 50; MS: MS-LS4-1, MS-LS4-2, MS-LS4-3, MS-LS4-4, MS-LS4-6, p. 52 • HS: HS-LS4-6, p. 83; HS-LS4-1, HS-LS4-2, HS-LS4-3, HS-LS4-4, HS-LS4-5, p. 87 **VIC:** *NSTA Quick Reference Guide to the NGSS K–12:* Performance Expectations • Grade 3: 3-LS4-1, 3-LS4-2, p. 104; 3-LS4-3, 3-LS4-4, p. 105 • MS: MS-LS4-1, MS-LS4-2, MS-LS4-3, MS-LS4-4, MS-LS4-5, MS-LS4-6, pp. 131–132 • HS: HS-LS4-1, HS-LS4-2, HS-LS4-3, p. 159; HS-LS4-4, HS-LS4-5, HS-LS4-6, p. 160

Review Chapter 2 instructions on how to use this guide.

Visit curriculumtopicstudy2.org for more information about CTS and additional resources.

NOTE: *Atlas* page numbers have not been provided because *The NSTA Atlas of the Three Dimensions* was produced concurrently with this edition. Titles and map codes are accurate.

Additional Readings:

 Available for download at www.curriculumtopicstudy2.org

Copyright © 2020 by Corwin Press, Inc. All rights reserved. Reprinted from *Science Curriculum Topic Study: Bridging the Gap Between Three-Dimensional Standards, Research, and Practice* (2nd ed.) by Page Keeley and Joyce Tugel. Thousand Oaks, CA: Corwin, www.corwin.com. Reproduction authorized for educational use by educators, local school sites, and/or noncommercial or nonprofit entities that have purchased the book.

Diversity of Species and Evidence of Common Ancestry
Grades 3–12 Standards- and Research-Based Study of a Curricular Topic

Section and Outcome	Selected Sources and Readings for Study and Reflection Read and examine **related** parts of
I. Content Knowledge	**IA:** *Science for All Americans* • Ch. 5: Evolution of Life, pp. 67–69 (paras. 4–7) • Ch. 10: Explaining the Diversity of Life, pp. 157–159 **IB:** *Science Matters* (2009 edition) • Ch. 18: Evolution, pp. 310–325 **IC:** *Framework for K–12 Science Education:* Narrative Section • Ch. 6: LS4.A: Evidence of Common Ancestry and Diversity, p. 162 **ID:** *Disciplinary Core Ideas: Reshaping Teaching and Learning* • Ch. 9: What Is This Disciplinary Core Idea and Why Is It Important? pp. 165–167 **IE:** *The NSTA Atlas of the Three Dimensions:* Narrative Page • 4.6 Natural Selection and Evolution (LS4.A)
II. Concepts, Core Ideas, or Practices	**IIA:** *Framework for K–12 Science Education:* Grade Band Endpoints • Ch. 6: LS4.A: Evidence of Common Ancestry and Diversity, pp. 162–163 **IIB:** *Next Generation Science Standards:* Disciplinary Core Ideas Column • Grade 3: LS4.A: Evidence of Common Ancestry and Diversity, p. 19 • MS: LS4.A: Evidence of Common Ancestry and Diversity, p. 52 • HS: LS4.A: Evidence of Common Ancestry and Diversity, p. 87 **IIC:** *NSTA Quick Reference Guide to the NGSS K–12:* Disciplinary Core Ideas Column • Grade 3: LS4.A: Evidence of Common Ancestry and Diversity, p. 104 • MS: LS4.A: Evidence of Common Ancestry and Diversity, pp. 131–132 • HS: LS4.A: Evidence of Common Ancestry and Diversity, p. 159
III. Curriculum, Instruction, and Formative Assessment	**IIIA:** *Disciplinary Core Ideas: Reshaping Teaching and Learning* • Ch. 9: What Approaches Can We Use to Teach About This Disciplinary Core Idea? p. 176 • Building From Student Ideas in Early Grades, pp. 177–178 • Refining Intuitions in Middle and High School, pp. 178–180
IV. Research on Commonly Held Ideas	**IVA:** *Benchmarks for Science Literacy:* Chapter 15 Research • 5F: Evolution and Reasoning Ability, p. 344 **IVB:** *Making Sense of Secondary Science: Research Into Children's Ideas* • Ch. 1: The Concept of Species, p. 25
V. K–12 Articulation and Connections	**VA:** *Next Generation Science Standards:* Appendices: Progression • Appendix E: LS4.A: Evidence of Common Ancestry and Diversity, p. 6 **VB:** *NSTA Quick Reference Guide to the NGSS K–12:* Progression • LS4.A: Evidence of Common Ancestry and Diversity, p. 72 **VC:** *The NSTA Atlas of the Three Dimensions:* Map Page • 4.6: Natural Selection and Evolution (LS4.A, LS4.B, & LS4.C)

Section and Outcome	Selected Sources and Readings for Study and Reflection Read and examine **related** parts of
VI. **Assessment** **Expectation**	**VIA:** *State Standards* • Examine your state's standards **VIB:** *Next Generation Science Standards:* Performance Expectations • Grade 3: 3-LS4-1, p. 19 • MS: MS-LS4-1, MS-LS4-2, and MS-LS4-3, p. 52 • HS: HS-LS4-1, p. 87 **VIC:** *NSTA Quick Reference Guide to the NGSS K–12:* Performance Expectations • Grade 3: 3-LS4-1, p. 104 • MS: MS-LS4-1, MS-LS4-2, and MS-LS4-3, pp. 131–132 • HS: HS-LS4-1, p. 159

Review Chapter 2 for instructions on how to use this guide.

Visit curriculumtopicstudy2.org for more information about CTS and additional resources.

NOTE: *Atlas* page numbers have not been provided because *The NSTA Atlas of the Three Dimensions* was produced concurrently with this edition. Titles and map codes are accurate.

Additional Readings:

Available for download at **www.curriculumtopicstudy2.org**

Copyright © 2020 by Corwin Press, Inc. All rights reserved. Reprinted from *Science Curriculum Topic Study: Bridging the Gap Between Three-Dimensional Standards, Research, and Practice* (2nd ed.) by Page Keeley and Joyce Tugel. Thousand Oaks, CA: Corwin, www.corwin.com. Reproduction authorized for educational use by educators, local school sites, and/or noncommercial or nonprofit entities that have purchased the book.

DNA, Genes, and Proteins
Grades 6–12 Standards- and Research-Based Study of a Curricular Topic

Section and Outcome	Selected Sources and Readings for Study and Reflection Read and examine **related** parts of
I. **Content Knowledge**	**IA:** *Science for All Americans* • Ch. 5: Cells, pp. 62–64; Heredity, pp. 61–62 **IB:** *Science Matters* (2009 edition) • Ch. 16: DNA and RNA: Messengers of the Code, pp. 276–285 **IC:** *Framework for K–12 Science Education:* Narrative Section • Ch. 6: LS1.A: Structure and Function, p. 144 • Ch. 6: LS3.A: Inheritance of Traits, p. 158 **1D:** *Disciplinary Core Ideas: Reshaping Teaching and Learning* • Ch. 8: LS3.A: Inheritance of Traits, pp. 146–151 **IE:** *The NSTA Atlas of the Three Dimensions:* Narrative Page • 4.5 Inheritance and Variation of Traits (LS3.A)
II. **Concepts, Core Ideas, or Practices**	**IIA:** *Framework for K–12 Science Education:* Grade Band Endpoints • Ch. 6: LS3.A: Inheritance of Traits, Grade 8–12, pp. 158–159 **IIB:** *Next Generation Science Standards:* Disciplinary Core Ideas Column • MS: LS3.A: Inheritance of Traits, p. 50 • HS: LS1.A: Structure and Function; pp. 80, 85; LS3.A: Inheritance of Traits, p. 85 **IIC:** *NSTA Quick Reference Guide to the NGSS K–12:* Disciplinary Core Ideas Column • MS: LS3.A: Inheritance of Traits, p. 131 • HS: LS1.A: Structure and Function, p. 154; LS3.A: Inheritance of Traits, p. 158
III. **Curriculum, Instruction, and Formative Assessment**	**IIIA:** *Disciplinary Core Ideas: Reshaping Teaching and Learning* • Ch. 8: Middle School Grades, pp. 156–157 • Ch. 8: High School Grades, p. 158 • Ch. 8: What Approaches Can We Use to Teach About This Disciplinary Core Idea? p. 161 **IIIB:** *Uncovering Student Ideas:* Assessment Probe and Suggestions for Instruction • USI.LS: DNA, Genes, and Chromosomes, pp. 129, 132
IV. **Research on Commonly Held Ideas**	**IVA:** *Making Sense of Secondary Science: Research Into Children's Ideas* • Ch. 5: The Mechanism of Inheritance, pp. 51–52 **IVB:** *Uncovering Student Ideas:* Related Research • USI.LS: DNA, Genes, and Chromosomes, p. 132 **IVC:** *Disciplinary Core Ideas: Reshaping Teaching and Learning* • Ch. 8: Challenges to Student Understanding, pp. 158–159
V. **K–12 Articulation and Connections**	**VA:** *Next Generation Science Standards:* Appendices: Progression • Appendix E: LS3.A: Inheritance of Traits, p. 6 **VB:** *NSTA Quick Reference Guide to the NGSS K–12:* Progression • LS1.A: Structure and Function, p. 68; LS3.A: Inheritance of Traits, p. 71 **VC:** *The NSTA Atlas of the Three Dimensions:* Map Page • 4.5: Inheritance and Variation of Traits (LS3.A & LS3.B)

Section and Outcome	Selected Sources and Readings for Study and Reflection Read and examine **related** parts of
VI. **Assessment** **Expectation**	**VIA:** *State Standards* • Examine your state's standards **VIB:** *Next Generation Science Standards:* Performance Expectations • MS: MS-LS3-1, MS-LS3-2, p. 50 • HS: HS-LS1-1, p. 80; HS-LS3-1, p. 85 **VIC:** *NSTA Quick Reference Guide to the NGSS K–12:* Performance Expectations • MS: MS-LS3-1, MS-LS3-2, p. 131 • HS: HS-LS1-1, p. 154; HS-LS3-1, p. 158

Review Chapter 2 instructions on how to use this guide.

Visit curriculumtopicstudy2.org for more information about CTS and additional resources.

NOTE: *Atlas* page numbers have not been provided because *The NSTA Atlas of the Three Dimensions* was produced concurrently with this edition. Titles and map codes are accurate.

Additional Readings:

Available for download at www.curriculumtopicstudy2.org

Copyright © 2020 by Corwin Press, Inc. All rights reserved. Reprinted from *Science Curriculum Topic Study: Bridging the Gap Between Three-Dimensional Standards, Research, and Practice* (2nd ed.) by Page Keeley and Joyce Tugel. Thousand Oaks, CA: Corwin, www.corwin.com. Reproduction authorized for educational use by educators, local school sites, and/or noncommercial or nonprofit entities that have purchased the book.

Inheritance of Traits
Grades K–12 Standards- and Research-Based Study of a Curricular Topic

Section and Outcome	Selected Sources and Readings for Study and Reflection Read and examine **related** parts of
I. **Content Knowledge**	**IA:** *Science for All Americans* • Ch. 5: Heredity, pp. 61–62 **IB:** *Science Matters* (2009 edition) • Ch. 16: The Code of Life, pp. 273–291 **IC:** *Framework for K–12 Science Education:* Narrative Section • Ch. 6: LS3.A: Inheritance of Traits, p. 158 **ID:** *Disciplinary Core Ideas: Reshaping Teaching and Learning* • Ch. 8: LS3.A: Inheritance of Traits, pp. 146–151 **IE:** *The NSTA Atlas of the Three Dimensions:* Narrative Page • 4.3: Inheritance and Variation of Traits (LS3.A & LS3.B)
II. **Concepts, Core Ideas, or Practices**	**IIA:** *Framework for K–12 Science Education:* Grade Band Endpoints • Ch. 6: LS3.A: Inheritance of Traits, pp. 158–159 **IIB:** *Next Generation Science Standards:* Disciplinary Core Ideas Column • Grade 1: LS3.A: Inheritance of Traits, p. 10 • Grade 3: LS3.A: Inheritance of Traits, p. 20 • MS: LS3.A: Inheritance of Traits, p. 50 • HS: LS3.A: Inheritance of Traits, p. 85 **IIC:** *NSTA Quick Reference Guide to the NGSS K–12:* Disciplinary Core Ideas • Grade 1: LS3.A: Inheritance of Traits, p. 94 • Grade 3: LS3.A: Inheritance of Traits, p. 104 • MS: LS3.A: Inheritance of Traits, p. 131 • HS: LS3.A: Inheritance of Traits, p. 158
III. **Curriculum, Instruction, and Formative Assessment**	**IIIA:** *Disciplinary Core Ideas: Reshaping Teaching and Learning* • Ch. 8: How Does Student Understanding of This Disciplinary Core Idea Develop Over Time? p. 155 • Ch. 8: Grades K–2, pp. 155–156 • Ch. 8: Grades 3–5, p. 156 • Ch. 8: Middle School Grades, pp. 156–157 • Ch. 8: High School Grades, p. 158 • What Approaches Can We Use to Teach About This Disciplinary Core Idea? pp. 160–161 **IIIB:** *Uncovering Student Ideas:* Assessment Probe and Suggestions for Instruction • USI.2: Baby Mice, pp. 129, 134–135 • USI.LS: DNA, Genes, and Chromosomes, pp. 129, 132; Eye Color, pp. 135, 138
IV. **Research on Commonly Held Ideas**	**IVA:** *Benchmarks for Science Literacy:* Chapter 15 Research • 5B: Heredity, p. 341 **IVB:** *Making Sense of Secondary Science: Research Into Children's Ideas* • Ch. 5: The Mechanism of Inheritance, pp. 51–52 **IVC:** *Uncovering Student Ideas:* Related Research • USI.2: Baby Mice, pp. 133–134 • USI.LS: DNA, Genes, and Chromosomes, p. 132; Eye Color, p. 138

Section and Outcome	Selected Sources and Readings for Study and Reflection Read and examine **related** parts of
	IVD: *Disciplinary Core Ideas: Reshaping Teaching and Learning* • Ch. 8: Challenges to Student Understanding, pp. 158–159
V. **K–12 Articulation and Connections**	**VA:** *Next Generation Science Standards:* Appendices: Progression • Appendix E: LS3.A: Inheritance of Traits, p. 6 **VB:** *NSTA Quick Reference Guide to the NGSS K–12:* Progression • LS3.A: Inheritance of Traits, p. 71 **VC:** *The NSTA Atlas of the Three Dimensions:* Map Page • 4.3: Inheritance and Variation of Traits (LS3.A)
VI. **Assessment Expectation**	**VIA:** *State Standards* • Examine your state's standards **VIB:** *Next Generation Science Standards:* Performance Expectations • Grade 1: 1-LS3-1, p. 10 • Grade 3: 3-LS3-1, 3-LS3-2, p. 20 • MS: MS-LS3-1, MS-LS3-2, p. 50 • HS: HS-LS3-1, p. 85 **VIC:** *NSTA Quick Reference Guide to the NGSS K–12:* Performance Expectations • Grade 1: 1-LS3-1, p. 94 • Grade 3: 3-LS3-1, 3-LS3-2, p. 104 • MS: MS-LS3-1, MS-LS3-2, p. 131 • HS: HS-LS3-1, p. 158

Review Chapter 2 instructions on how to use this guide.

Visit curriculumtopicstudy2.org for more information about CTS and additional resources.

NOTE: *Atlas* page numbers have not been provided because *The NSTA Atlas of the Three Dimensions* was produced concurrently with this edition. Titles and map codes are accurate.

Additional Readings:

Available for download at www.curriculumtopicstudy2.org

Copyright © 2020 by Corwin Press, Inc. All rights reserved. Reprinted from *Science Curriculum Topic Study: Bridging the Gap Between Three-Dimensional Standards, Research, and Practice* (2nd ed.) by Page Keeley and Joyce Tugel. Thousand Oaks, CA: Corwin, www.corwin.com. Reproduction authorized for educational use by educators, local school sites, and/or noncommercial or nonprofit entities that have purchased the book.

Natural Selection
Grades 3–12 Standards- and Research-Based Study of a Curricular Topic

Section and Outcome	Selected Sources and Readings for Study and Reflection Read and examine **related** parts of
I. **Content Knowledge**	**IA:** *Science for All Americans* • Ch. 5: Evolution of Life, pp. 67–69 (paras. 4–7) **IB:** *Science Matters* (2009 edition) • Ch. 18: Natural Selection, pp. 310–312 **IC:** *Framework for K–12 Science Education:* Narrative Section • Ch. 6: LS4.B: Natural Selection, p. 163 **ID:** *Disciplinary Core Ideas: Reshaping Teaching and Learning* • Ch. 9: The Four Component Ideas of LS4, pp. 168–170 **IE:** *The NSTA Atlas of the Three Dimensions:* Narrative Page • 4.6: Natural Selection and Evolution (LS4.A, LS4.B, & LS4.C)
II. **Concepts, Core Ideas, or Practices**	**IIA:** *Framework for K–12 Science Education:* Grade Band Endpoints • Ch. 6: LS4.B: Natural Selection, p. 164 **IIB:** *Next Generation Science Standards:* Disciplinary Core Ideas Column • Grade 3: LS4.B: Natural Selection, p. 20 • MS: LS4.B: Natural Selection, pp. 50, 52 • HS: LS4.B: Natural Selection, p. 87 **IIC:** *NSTA Quick Reference Guide to the NGSS K–12:* Disciplinary Core Ideas • Grade 3: LS4.B: Natural Selection, p. 104 • MS: LS4.B: Natural Selection, p. 132 • HS: LS4.B: Natural Selection, p. 159
III. **Curriculum, Instruction, and Formative Assessment**	**IIIA:** *Disciplinary Core Ideas: Reshaping Teaching and Learning* • Ch. 9: What Approaches Can We Use to Teach About This Disciplinary Core Idea? p. 176 • Ch. 9: K–5: Building From Student Ideas in Early Grades, pp. 177–178 • Ch. 9: Refining Intuitions in Middle and High School, pp. 178–180 **IIIB:** *Uncovering Student Ideas:* Assessment Probe and Suggestions for Instruction • USI2: Habitat Change, pp. 143, 147 • USI4: Is It "Fitter"? pp. 119, 123
IV. **Research on Commonly Held Ideas**	**IVA:** *Benchmarks for Science Literacy:* Chapter 15 Research • 5F: Adaptation, p. 344 **IVB:** *Making Sense of Secondary Science: Research Into Children's Ideas* • Ch. 1: The Concept of Species, p. 25; Adaptation, p. 26 • Ch. 5: Adaptation, pp. 52–53 **IVC:** *Uncovering Student Ideas:* Related Research • USI2: Habitat Change, p. 147 • USI4: Is It "Fitter"? p. 123 **IVD:** *Disciplinary Core Ideas: Reshaping Teaching and Learning* • Ch. 9: How Does Student Understanding of This Disciplinary Core Idea Develop Over Time? pp. 174–175

Section and Outcome	Selected Sources and Readings for Study and Reflection Read and examine **related** parts of
V. **K-12 Articulation and Connections**	**VA:** *Next Generation Science Standards:* Appendices: Progression • Appendix E: LS4.B: Natural Selection, p. 6 **VB:** *NSTA Quick-Reference Guide to the NGSS K-12:* Progression • LS4.B: Natural Selection, p. 72 **VC:** *The NSTA Atlas of the Three Dimensions:* Map Page • 4.6: Natural Selection and Evolution (LS4.A, LS4.B, & LS4.C)
VI. **Assessment Expectation**	**VIA:** *State Standards* • Examine your state's standards **VIB:** *Next Generation Science Standards:* Performance Expectations • Grade 3: 3-LS4-2, p. 20 • MS: MS-LS4-4, p. 52; MS-LS4-5, p. 50 • HS: HS-LS4-2, HS-LS4-3, p. 87 **VIC:** *NSTA Quick Reference Guide to the NGSS K-12:* Performance Expectations • Grade 3: 3-LS4-2, p. 104 • MS: MS-LS4-4, MS-LS.4-5, p. 132 • HS: HS-LS4-2, HS-LS4-3, p. 159

Review Chapter 2 instructions on how to use this guide.

Visit curriculumtopicstudy2.org for more information about CTS and additional resources.

NOTE: *Atlas* page numbers have not been provided because *The NSTA Atlas of the Three Dimensions* was produced concurrently with this edition. Titles and map codes are accurate.

Additional Readings:

Available for download at www.curriculumtopicstudy2.org

Copyright © 2020 by Corwin Press, Inc. All rights reserved. Reprinted from *Science Curriculum Topic Study: Bridging the Gap Between Three-Dimensional Standards, Research, and Practice* (2nd ed.) by Page Keeley and Joyce Tugel. Thousand Oaks, CA: Corwin, www.corwin.com. Reproduction authorized for educational use by educators, local school sites, and/or noncommercial or nonprofit entities that have purchased the book.

Reproduction, Growth, and Development
Grades K–12 Standards- and Research-Based Study of a Curricular Topic

Section and Outcome	Selected Sources and Readings for Study and Reflection Read and examine **related** parts of
I. Content Knowledge	**IA:** *Science for All Americans* • Ch. 5: Heredity, pp. 61–62 • Ch. 6: Human Development, pp. 73–76 **IB:** *Science Matters* (2009 edition) • Ch. 16: Sex: A Good Idea, pp. 285–289 **IC:** *Framework for K–12 Science Education:* Narrative Section • Ch. 6: LS1.B: Growth and Development of Organisms, pp. 145–146 **ID:** *Disciplinary Core Ideas: Reshaping Teaching and Learning* • Ch. 6: LS1.B: Growth and Development of Organisms, pp. 101–103 **IE: VC:** *The NSTA Atlas of the Three Dimensions:* Narrative Page • 4.1: The Structure and Function of Organisms (LS1.B)
II. Concepts, Core Ideas, or Practices	**IIA:** *Framework for K–12 Science Education:* Grade Band Endpoints • Ch. 6: LS1.B: Growth and Development of Organisms, pp. 146–147 **IIB:** *Next Generation Science Standards:* Disciplinary Core Ideas Column • Grade 1: LS1.B: Growth and Development of Organisms, p. 10 • Grade 3: LS1.B: Growth and Development of Organisms, p. 20 • MS: LS1.B: Growth and Development of Organisms, p. 50 • HS: LS1.B: Growth and Development of Organisms, p. 85 **IIC:** *NSTA Quick Reference Guide to the NGSS K–12:* Disciplinary Core Ideas • Grade 1: LS1.B: Growth and Development of Organisms, p. 94 • Grade 3: LS1.B: Growth and Development of Organisms, p. 104 • MS: LS1.B: Growth and Development of Organisms, pp. 128–129 • HS: LS1.B: Growth and Development of Organisms, p. 154
III. Curriculum, Instruction, and Formative Assessment	**IIIA:** *Disciplinary Core Ideas: Reshaping Teaching and Learning* • Ch. 6: LS1.B: Expectations for Grades K–2, p. 110 • Ch. 6: LS1.B: Expectations for Grades 3–5, p. 111 • Ch. 6: LS1.B: Expectations for Grades 6–8, p. 112–113 • Ch. 6: LS1.B: Expectations for Grades 9–12, p. 114 **IIIB:** *Uncovering Student Ideas:* Assessment Probe and Suggestions for Instruction • USI.3: Sam's Puppy, pp. 125, 129–130; Does It Have a Life Cycle? pp. 111, 115 • USI.LS: Eggs, pp. 117, 120–121; Chrysalis, pp. 123, 126–127
IV. Research on Commonly Held Ideas	**IVA:** *Benchmarks for Science Literacy:* Chapter 15 Research • 6B: Human Development, p. 344 **IVB:** *Making Sense of Secondary Science: Research Into Children's Ideas* • Ch. 3: The Meaning of "Growth," p. 37 • Ch. 5: Reproduction, pp. 48–51 **IVC:** *Uncovering Student Ideas:* Related Research • USI.3: Sam's Puppy, p. 129; Does It Have a Life Cycle? p. 115 • USI.LS: Eggs, p. 120; Chrysalis, p. 126 **IVD:** *Disciplinary Core Ideas: Reshaping Teaching and Learning* • Ch. 6: Students' Commonly Held Ideas, p. 117 (second column)

Section and Outcome	Selected Sources and Readings for Study and Reflection Read and examine **related** parts of
V. **K–12 Articulation and Connections**	**VA:** *Next Generation Science Standards:* Appendices: Progression • Appendix E: LS1.B: Growth and Development of Organisms, p. 4 **VB:** *NSTA Quick Reference Guide to the NGSS K–12:* Progression • LS1.B: Growth and Development of Organisms, p. 68 **VC:** *The NSTA Atlas of the Three Dimensions:* Map Page • 4.1: The Structure and Function of Organisms (LS1.A, LS1.B, LS1.D)
VI. **Assessment Expectation**	**VIA:** *State Standards* • Examine your state's standards **VIB:** *Next Generation Science Standards:* Performance Expectations • Grade 1: 1-LS1-2, p. 10 • Grade 3: 3-LS1-1, p. 20 • MS: MS-LS1-4, MS-LS1-5, p. 50 • HS: HS-LS1-4, p. 85 **VIC:** *NSTA Quick Reference Guide to the NGSS K–12:* Performance Expectations • Grade 1: 1-LS1-2, p. 94 • Grade 3: 3-LS1-1, p. 104 • MS: MS-LS1-4, MS-LS1-5, pp. 128–129 • HS: HS-LS1-4, p. 154

Review Chapter 2 instructions on how to use this guide.

Visit curriculumtopicstudy2.org for more information about CTS and additional resources.

NOTE: *Atlas* page numbers have not been provided because *The NSTA Atlas of the Three Dimensions* was produced concurrently with this edition. Titles and map codes are accurate.

Additional Readings:

Available for download at www.curriculumtopicstudy2.org

Copyright © 2020 by Corwin Press, Inc. All rights reserved. Reprinted from *Science Curriculum Topic Study: Bridging the Gap Between Three-Dimensional Standards, Research, and Practice* (2nd ed.) by Page Keeley and Joyce Tugel. Thousand Oaks, CA: Corwin, www.corwin.com. Reproduction authorized for educational use by educators, local school sites, and/or noncommercial or nonprofit entities that have purchased the book.

Variation of Traits
Grades K–12 Standards- and Research-Based Study of a Curricular Topic

Section and Outcome	Selected Sources and Readings for Study and Reflection Read and examine **related** parts of
I. **Content Knowledge**	**IA:** *Science for All Americans* • Ch. 5: Heredity, pp. 61–62 **IB:** *Science Matters* (2009 edition) • Ch. 18: Natural Selection, pp. 310–313 **IC:** *Framework for K–12 Science Education:* Narrative Section • Ch. 6: LS3.B: Variation of Traits, p. 160 **ID:** *Disciplinary Core Ideas: Reshaping Teaching and Learning* • Ch. 8: LS3.B: Variation of Traits, pp. 151–154 **IE:** *The NSTA Atlas of the Three Dimensions:* Narrative Page • 4.5 Inheritance and Variation of Traits (LS3.B)
II. **Concepts, Core Ideas, or Practices**	**IIA:** *Framework for K–12 Science Education:* Grade Band Endpoints • Ch. 6: LS3.B: Variation of Traits, pp. 160–161 **IIB:** *Next Generation Science Standards:* Disciplinary Core Ideas • Grade 1: LS3.B: Variation of Traits, p. 10 • Grade 3: LS3.B: Variation of Traits, p. 20 • MS: LS3.B: Variation of Traits, p. 50 • HS: LS3.B: Variation of Traits, p. 85 **IIC:** *NSTA Quick Reference Guide to NGSS K–12:* Disciplinary Core Ideas • Grade 1: LS3.B: Variation of Traits, p. 94 • Grade 3: LS3.B: Variation of Traits, p. 104 • MS: LS3.B: Variation of Traits, p. 131 • HS: LS3.B: Variation of Traits, p. 158
III. **Curriculum, Instruction, and Formative Assessment**	**IIIA:** *Disciplinary Core Ideas: Reshaping Teaching and Learning* • Ch. 8: How Does Student Understanding of This Disciplinary Core Idea Develop Over Time? p. 155 • Ch. 8: Grades K–2, pp. 155–156 • Ch. 8: Grades 3–5, p. 156 • Ch. 8: Middle School Grades, pp. 156–157 • Ch. 8: High School Grades, p. 158 **IIIB:** *Uncovering Student Ideas:* Assessment Probe and Suggestions for Instruction • USI.2: Baby Mice, pp. 129, 134–135
IV. **Research on Commonly Held Ideas**	**IVA:** *Benchmarks for Science Literacy:* Chapter 15 Research • 5B: Heredity, p. 341 **IVB:** *Making Sense of Secondary Science: Research Into Children's Ideas* • Ch. 5: Variation and Resemblance, p. 51; Sources of Variation, p. 52 **IVC:** *Uncovering Student Ideas:* Related Research • USI.2: Baby Mice, pp. 133–134

Section and Outcome	Selected Sources and Readings for Study and Reflection Read and examine **related** parts of
V. **K–12 Articulation and Connections**	**VA:** *Next Generation Science Standards:* Appendices: Progression • Appendix E: LS3.B: Variation of Traits, p. 6 **VB:** *NSTA Quick Reference Guide to the NGSS K–12:* Progression • LS3.B: Variation of Traits, p. 72 **VC:** *The NSTA Atlas of the Three Dimensions:* Map Page • 4.5 Inheritance and Variation of Traits (LS3.A & LS3.B)
VI. **Assessment Expectation**	**VIA:** *State Standards* • Examine your state's standards **VIB:** *Next Generation Science Standards:* Performance Expectations • Grade 1: 1-LS3-1, p. 10 • Grade 3: 3-LS3-1, 3-LS3-2, p. 20 • MS: MS-LS3-1, MS-LS3-2, p. 50 • HS: HS-LS3-2, HS-LS3-3, p. 85 **VIC:** *NSTA Quick Reference Guide to the NGSS K–12:* Performance Expectations • Grade 1: 1-LS3-1, p. 94 • Grade 3: 3-LS3-1, 3-LS3-1, p. 104 • MS: MS-LS3-1, MS-LS3-2, p. 131 • HS: HS-LS3-2, HS-LS3-3, p. 158

Review Chapter 2 instructions on how to use this guide.

Visit curriculumtopicstudy2.org for more information about CTS and additional resources.

NOTE: *Atlas* page numbers have not been provided because *The NSTA Atlas of the Three Dimensions* was produced concurrently with this edition. Titles and map codes are accurate.

Additional Readings:

online resources — Available for download at **www.curriculumtopicstudy2.org**

Copyright © 2020 by Corwin Press, Inc. All rights reserved. Reprinted from *Science Curriculum Topic Study: Bridging the Gap Between Three-Dimensional Standards, Research, and Practice* (2nd ed.) by Page Keeley and Joyce Tugel. Thousand Oaks, CA: Corwin, www.corwin.com. Reproduction authorized for educational use by educators, local school sites, and/or noncommercial or nonprofit entities that have purchased the book.

Category B: Physical Science Guides

The twenty-seven CTS guides in this section are divided into two subsections. The guides focus on physical and chemical structures, properties, and processes. They range in scale from subatomic to large systems.

Matter Guides focus on structure, properties, and processes of chemical systems. The alphabetically arranged guides in this section include

- Atoms and Molecules
- Behavior and Characteristics of Gases
- Chemical Bonding
- Chemical Reactions
- Conservation of Matter
- Elements, Compounds, and the Periodic Table
- Mixtures and Solutions
- Nuclear Processes
- Particulate Nature of Matter
- Properties of Matter
- States of Matter

Force, Motion, and Energy Guides focus on interactions between objects and within systems that involve forces or energy. They also include wave concepts and ideas that underlie our modern technologies. The alphabetically arranged guides in this section include

- Concept of Energy
- Conservation of Energy
- Electric Charge and Current
- Energy in Chemical Processes and Everyday Life
- Force and Motion
- Forces Between Objects
- Gravitational Force
- Kinetic and Potential Energy
- Magnetism
- Nuclear Energy
- Relationship Between Energy and Forces

- Sound
- Transfer of Energy
- Visible Light and Electromagnetic Radiation
- Waves and Information Technologies
- Waves and Wave Properties

NOTES FOR USING CATEGORY B GUIDES

Overall

- *Atlas* page numbers have not been provided because *The NSTA Atlas of the Three Dimensions* was produced concurrently with this edition. Titles and map codes are accurate.
- The same resources are not always included in each guide. For example, some guides include readings from *Science Matters* for section I, others do not.
- Eliminate redundancy. Some readings include the exact same information. However, even when the information is the same, there may be an advantage in how the information is presented. Select the reading based on the resources you have available to use for CTS and/or the advantage of using one over the other.

Section I

- Readings from the *Framework* and *Atlas* narrative are exactly the same. Choose one of these resources for this section.
- When reading the *Framework*, stop at Grade Band Endpoints.
- Some *Atlas* maps combine topics. When using the *Atlas* narrative for this section, if there is more than one core idea included in the narrative, focus on the one that is listed on your CTS guide.
- When reading *Disciplinary Core Ideas* for this section, focus on the content that helps you understand this topic. There are also suggestions for instruction embedded in this reading that can be added to CTS section III.

Section II

- Readings from the *Framework*, *NGSS*, and *NSTA Quick Reference Guide* are practically the same. In a few cases, a *Framework* idea was not included in the *NGSS* or was moved to a different grade span. In the *Framework*, goals are described in grade bands K–2, 3–5, 6–8, and 9–12. In the *NGSS* and *NSTA Quick Reference Guide*, they are phrased as disciplinary core ideas and designate a specific grade.
- The readings from the *NGSS* and the *NSTA Quick Reference Guide* are exactly the same. Choose one of these resources. An advantage to using the *NSTA Quick*

Reference Guide is that the disciplinary core idea is matched to the performance expectation (section VI) on the chart.

Section III

- Readings listed for *Disciplinary Core Ideas* are the longest readings. In some guides these have been broken down into subsections. You can read all the designated pages or focus on the subsections that are of interest to you. If doing CTS with a group, you may consider assigning subsections so one person does not have to read all the designated pages.

- Sometimes readings about instructional implications from *Disciplinary Core Ideas* may mention and cite a research study on students' commonly held ideas or difficulties. This can be combined with section IV.

- Several readings may be listed from the *Uncovering Student Ideas in Science* series. This does not mean you need to read them all. Choose ones from the books you have available. The page numbers list the probe that can be used to elicit students' ideas, and the teacher notes contain suggestions for instruction that are designed to address the ideas elicited by the probe. Some of these books also have a section on curricular considerations. This can be added to your reading.

- *Uncovering Student Ideas in Physical Science: Force and Motion* and *Uncovering Student Ideas in Physical Science: Electricity and Magnetism* have an additional front matter discussion that can be used in section III for topics related to forces, motion, electricity, and magnetism.

- There are always new books released in the *Uncovering Student Ideas in Science* series that may include probes that are not listed on the topic guides. The fourth book in the *Uncovering Student Ideas in Physical Science* series focuses on light, sound, and waves. Check the CTS website for a list of new probes published after 2019.

Section IV

- If you are using the research summaries from the *Uncovering Student Ideas* series, they usually include references to the commonly held ideas identified in *Benchmarks* Chapter 15 Research and *Making Sense of Secondary Science* so there is no need to use all three of the resources listed. In addition, the *Uncovering Student Ideas* series includes research published after *Benchmarks* and *Making Sense of Secondary science*.

Section V

- Readings from the *NGSS* Appendix E and the *NSTA Quick Reference Guide* both describe a progression. The difference is the progression is summarized for each grade span in the *NGSS* Appendix E. In the *NSTA Quick Reference Guide*, they are listed by the disciplinary core ideas along with the code for the performance expectation. You can choose one or both of these resources, depending on the level of specificity you desire.

- The *Atlas* includes the same information as the *NSTA Quick Reference Guide* but the visual mapping of the ideas allows you to see precursor ideas and connections, as well as connections to other maps. Be sure to read the front matter of the *Atlas* before using the maps. The information will help you use the maps effectively.

- The *Atlas* map often combines two or more topics. Follow the connections that match the topic you are studying.

Section VI

- The *NGSS* and the *NSTA Quick Reference Guide* provide the exact same information. Clarifications and assessment boundaries are also included in both. Choose one of these resources based on the advantages listed below or the resources you have available.

- An advantage to using the *NSTA Quick Reference Guide* is that the performance expectation is listed next to the disciplinary core idea included in that performance expectation.

- An advantage to using the *NGSS* that is not evident in the *NSTA Quick Reference Guide* is that the scientific and engineering practice and crosscutting concept chart, included below the performance expectations, shows the other two dimensions that are part of the performance expectation.

Atoms and Molecules
Grades 6–12 Standards- and Research-Based Study of a Curricular Topic

Section and Outcome	Selected Sources and Readings for Study and Reflection Read and examine **related** parts of
I. **Content Knowledge**	**IA:** *Science for All Americans* • Ch. 4: Structure of Matter, pp. 46–49 **IB:** *Science Matters* (2009 edition) • Ch. 4: The Atom, p. 67–74 **IC:** *Framework for K–12 Science Education:* Narrative Section • Ch. 5: PS1.A: Structure and Properties of Matter narrative section, pp. 106–107 **ID:** *The NSTA Atlas of the Three Dimensions:* Narrative Page • 3.1: Structure and Properties of Matter (PS1.A)
II. **Concepts, Core Ideas, or Practices**	**IIA:** *Framework for K–12 Science Education:* Grade Band Endpoints • Ch. 5: PS1.A: Structure and Properties of Matter, Grades 8–12, pp. 108–109 **IIB:** *Next Generation Science Standards:* Disciplinary Core Ideas Column • MS: PS1.A: Structure and Properties of Matter, pp. 35, 37 • HS: PS1.A: Structure and Properties of Matter, pp. 68, 70 **IIC:** *NSTA Quick Reference Guide to the NGSS K–12:* Disciplinary Core Ideas Column • MS: PS1.A: Structure and Properties of Matter, pp. 122–123 • HS: PS1.A: Structure and Properties of Matter, p. 147
III. **Curriculum, Instruction, and Formative Assessment**	**IIIA:** *Disciplinary Core Ideas: Reshaping Teaching and Learning* • Ch. 2: 6–8: An Atomic-Molecular Model, pp. 21–24 • Ch. 2: What Approaches Can We Use to Teach About This Disciplinary Idea, Middle School, pp. 28–29 • Ch. 2: 9–12: A Subatomic Model, pp. 25–27 • Ch. 2: What Approaches Can We Use to Teach About This Disciplinary Idea, High School, p. 29 **IIIB:** *Uncovering Student Ideas:* Assessment Probe and Suggestions for Instruction • USI.1 (2018 edition): Is It Made of Molecules? pp. 87, 91–92 • USI.3: Pennies, pp. 17, 21–22 • USI.4: Iron Bar, pp. 17, 21, *Salt Crystals,* pp. 39, 41, 42–43 • USI.PS3: What Do You Know About Atoms and Molecules? pp. 29; 33–34; Neutral Atoms, pp. 119, 122
IV. **Research on Commonly Held Ideas**	**IVA:** *Benchmarks for Science Literacy:* Chapter 15 Research • 4D: The Structure of Matter: Particles, p. 337 **IVB:** *Making Sense of Secondary Science: Research Into Children's Ideas* • Ch. 11: Conceptions of Atomic Size and Mass and Internal Structure of Molecules, pp. 95–96 **IVC:** *Uncovering Student Ideas:* Related Research • USI.1 (2018 edition): Is It Made of Molecules? p. 91 • USI.3: Pennies, pp. 20–21 • USI.4: Iron Bar, pp. 20–21; Salt Crystals, p. 42 • USI.PS3: What Do You Know About Atoms and Molecules? p. 33; Neutral Atoms, p. 122

Section and Outcome	Selected Sources and Readings for Study and Reflection Read and examine **related** parts of
V. **K–12 Articulation and Connections**	**VA:** *Next Generation Science Standards:* Appendices: Progression<ul><li>Appendix E: PS1.A: Structure and Properties of Matter, p. 7</li></ul>**VB:** *NSTA Quick Reference Guide to the NGSS K–12:* Progression<ul><li>PS1.A: Structure and Properties of Matter, p. 61</li></ul>**VC:** *The NSTA Atlas of the Three Dimensions:* Map Page<ul><li>3.1: Structure and Properties of Matter (PS1.A)</li></ul>
VI. **Assessment Expectation**	**VIA:** *State Standards*<ul><li>Examine your state's standards</li></ul>**VIB:** *Next Generation Science Standards:* Performance Expectations<ul><li>MS: MS-PS1-1, MS-PS1-4; p. 35 MS-PS1-5, p. 37</li><li>HS: HS-PS1-1, p. 68; HS-PS1-2, HS-PS1-7, p. 70</li></ul>**VIC:** *NSTA Quick Reference Guide to the NGSS K–12:* Performance Expectations Column<ul><li>MS: MS-PS1-1, p. 122; MS-PS1-4, MS-PS1-5, p. 123</li><li>HS: HS-PS1-1, HS-PS1-2, p. 147; HS-PS1-7, p. 148</li></ul>

Review Chapter 2 instructions on how to use this guide.

Visit curriculumtopicstudy2.org for more information about CTS and additional resources.

NOTE: *Atlas* page numbers have not been provided because *The NSTA Atlas of the Three Dimensions* was produced concurrently with this edition. Titles and map codes are accurate.

Additional Readings:

Available for download at www.curriculumtopicstudy2.org

Copyright © 2020 by Corwin Press, Inc. All rights reserved. Reprinted from *Science Curriculum Topic Study: Bridging the Gap Between Three-Dimensional Standards, Research, and Practice* (2nd ed.) by Page Keeley and Joyce Tugel. Thousand Oaks, CA: Corwin, www.corwin.com. Reproduction authorized for educational use by educators, local school sites, and/or noncommercial or nonprofit entities that have purchased the book.

Behavior and Characteristics of Gases
Grades 3–12 Standards- and Research-Based Study of a Curricular Topic

Section and Outcome	Selected Sources and Readings for Study and Reflection Read and examine **related** parts of
I. Content Knowledge	**IA:** *Science for All Americans* • Ch. 4: Structure of Matter, pp. 47–48 (fifth paragraph) **IB:** *Science Matters* (2009 edition) • Ch. 7: Gases, pp. 117–118 **IC:** *Framework for K–12 Science Education:* Narrative Section • Ch. 5: PS1.A: Structure and Properties of Matter, p. 107 (second paragraph) **ID:** *The NSTA Atlas of the Three Dimensions:* Narrative Page • 3.1: Structure and Properties of Matter (PS1.A)
II. Concepts, Core Ideas, or Practices	**IIA:** *Framework for K–12 Science Education:* Grade Band Endpoints • Ch. 5: PS1.A: Structure and Properties of Matter, pp. 108–109 **IIB:** *Next Generation Science Standards:* Disciplinary Core Ideas Column • Grade 5: PS1.A: Structure and Properties of Matter, p. 28 • MS: PS1.A: Structure and Properties of Matter, p. 35 **IIC:** *NSTA Quick Reference Guide to the NGSS K–12:* Disciplinary Core Ideas Column • Grade 5: PS1.A: Structure and Properties of Matter, p. 112 • MS: PS1.A: Structure and Properties of Matter, p. 123
III. Curriculum, Instruction, and Formative Assessment	**IIIA:** *Disciplinary Core Ideas: Reshaping Teaching and Learning* • Ch. 2: 3–5: The Study of Materials and a Particle Model, pp. 17–21 • Ch. 2: 6–8: An Atomic-Molecular Model, pp. 21–24 **IIIB:** *Uncovering Student Ideas:* Assessment Probe and Suggestions for Instruction • USI.2: What's in the Bubbles? pp. 65, 69–70 • USI.3: Floating Balloon, pp. 39, 43–44; Hot and Cold Balloons, pp. 45, 49 • USIPS.3: Model of Air Inside a Jar, pp. 43, 47; Do They Have Weight and Take Up Space? pp. 59, 62–63
IV. Research on Commonly Held Ideas	**IVA:** *Benchmarks for Science Literacy:* Chapter 15 Research • 4D: Structure of Matter, pp. 336–337 **IVB:** *Making Sense of Secondary Science: Research Into Children's Ideas* • Ch. 9: The Gaseous State, p. 80 • Ch. 11: Particle Ideas about Gases, pp. 93–94 • Ch. 13: Existence of Air, pp. 104–105 **IVC:** *Uncovering Student Ideas:* Related Research • USI.2: What's in the Bubbles? p. 69 • USI.3: Floating Balloon, pp. 42–43; Hot and Cold Balloons, p. 49 • USIPS.3: Model of Air Inside a Jar, pp. 46–47; Do They Have Weight and Take Up Space? p. 6

Section and Outcome	Selected Sources and Readings for Study and Reflection Read and examine **related** parts of
V. **K–12 Articulation and Connections**	**VA:** *Next Generation Science Standards:* Appendices: Progression • Appendix E: PS1.A: Structure and Properties of Matter, p. 7 **VB:** *NSTA Quick Reference Guide to the NGSS K–12:* Progression • PS1.A: Structure and Properties of Matter, p. 61 **VC:** *The NSTA Atlas of the Three Dimensions:* Map Page • 3.1: Structure and Properties of Matter (PS1.A)
VI. **Assessment Expectation**	**VIA:** *State Standards* • Examine your state's standards **VIB:** *Next Generation Science Standards:* Performance Expectations • Grade 5: 5-PS1.2, p. 28 • MS: MS-PS1.4, p. 35 **VIC:** *NSTA Quick Reference Guide to the NGSS K–12:* Performance Expectations Column • Grade 5: 5-PS1.1, p. 112 • MS: MS-PS1.4, p. 123

Review Chapter 2 instructions on how to use this guide.

Visit curriculumtopicstudy2.org/ for more information about CTS and additional resources.

NOTE: *Atlas* page numbers have not been provided because *The NSTA Atlas of the Three Dimensions* was produced concurrently with this edition. Titles and map codes are accurate.

Additional Readings:

Available for download at www.curriculumtopicstudy2.org

Copyright © 2020 by Corwin Press, Inc. All rights reserved. Reprinted from *Science Curriculum Topic Study: Bridging the Gap Between Three-Dimensional Standards, Research, and Practice* (2nd ed.) by Page Keeley and Joyce Tugel. Thousand Oaks, CA: Corwin, www.corwin.com. Reproduction authorized for educational use by educators, local school sites, and/or noncommercial or nonprofit entities that have purchased the book.

Chemical Bonding
Grades 6–12 Standards- and Research-Based Study of a Curricular Topic

Section and Outcome	Selected Sources and Readings for Study and Reflection Read and examine **related** parts of
I. **Content Knowledge**	**IA:** *Science for All Americans* • Ch. 4: Structure of Matter, pp. 46–48 **IB:** *Science Matters* (2009 edition) • Ch. 6: Chemical Bonding, pp. 94–114 **IC:** *Framework for K–12 Science Education:* Narrative Section • Ch. 5: PS1.A: Structure and Properties of Matter, pp. 106–107 • Ch. 5: PS1.B: Chemical Reactions, pp. 109–110 **ID:** *The NSTA Atlas of the Three Dimensions:* Narrative Page • 3.1: Structure and Properties of Matter (PS1.A) • 3.2: Chemical Reactions and Nuclear Processes (PS1.B)
II. **Concepts, Core Ideas, or Practices**	**IIA:** *Framework for K–12 Science Education:* Grade Band Endpoints • Ch. 5: PS1.A: Structure and Properties of Matter, pp. 108–109 • Ch. 5: PS1.B: Chemical Reactions, p. 111 **IIB:** *Next Generation Science Standards:* Disciplinary Core Ideas Column • MS: PS1.A: Structure and Properties of Matter, p. 35; PS1.B: Chemical Reactions, p. 37 • HS: PS1.A: Structure and Properties of Matter, p. 68; PS1.B: Chemical Reactions, p. 70 **IIC:** *NSTA Quick Reference Guide to the NGSS K–12:* Disciplinary Core Ideas Column • MS: PS1.A: Structure and Properties of Matter, p. 122; PS1.B: Chemical Reactions, p. 123 • HS: PS1.A: Structure and Properties of Matter; PS1.B: Chemical Reactions, p. 147
III. **Curriculum, Instruction, and Formative Assessment**	**IIIA:** *Disciplinary Core Ideas: Reshaping Teaching and Learning* • Ch. 2: 6–8: An Atomic-Molecular Model, pp. 21–23 • Ch. 2: 9–12: A Subatomic Model, p. 25 **IIIB:** *Uncovering Student Ideas:* Assessment Probe and Suggestions for Instruction • USI.2: Chemical Bonds, pp. 71, 74–75 • USI.4: Salt Crystals, pp. 39, 42–43 • USI.PS3: Energy and Chemical Bonds, pp. 189, 192–193
IV. **Research on Commonly Held Ideas**	**IVA:** *Benchmarks for Science Literacy:* Chapter 15 Research • 4D: Chemical Changes, p. 337 **IVB:** *Making Sense of Secondary Science: Research Into Children's Ideas* • Ch. 10: Chemical Change, pp. 85–87 • Ch. 11: Conceptions of the Internal Structure of Molecules, p. 96; Particle Models of Giant Ionic Lattices, p. 97 **IVC:** *Uncovering Student Ideas:* Related Research • USI.2: Chemical Bonds, p. 74 • USI.4: Salt Crystals, p. 42 • USI.PS3: Energy and Chemical Bonds, p. 192

Section and Outcome	Selected Sources and Readings for Study and Reflection Read and examine **related** parts of
V. **K–12** **Articulation** **and** **Connections**	**VA:** *Next Generation Science Standards:* Appendices: Progression<ul><li>Appendix E: PS1.A: Structure and Properties of Matter, p. 7</li><li>Appendix E: PS1.B: Chemical Reactions, p. 7</li></ul>**VB:** *NSTA Quick Reference Guide to the NGSS K–12:* Progression<ul><li>PS1.A: Structure and Properties of Matter, p. 61</li><li>PS1.B: Chemical Reactions, p. 62</li></ul>**VC:** *The NSTA Atlas of the Three Dimensions:* Map Page<ul><li>3.1: Structure and Properties of Matter (PS1.A)</li><li>3.2: Chemical Reactions and Nuclear Processes (PS1.B & PS1.C)</li></ul>
VI. **Assessment** **Expectation**	**VIA:** *State Standards*<ul><li>Examine your state's standards</li></ul>**VIB:** *Next Generation Science Standards:* Performance Expectations<ul><li>MS: MS-PS1-1, p, 35; MS-PS1-2, p. 37</li><li>HS: HS-PS1-1, HS-PS1-3, p. 68; HS-PS1-2, HS-PS1-4, p. 70</li></ul>**VIC:** *NSTA Quick Reference Guide to the NGSS K–12:* Performance Expectations Column<ul><li>MS: MS-PS1-1, MS-PS1-2, p. 122</li><li>HS: HS-PS1-1, HS-PS1-2, HS-PS1-3, HS-PS1-4, p. 147</li></ul>

Review Chapter 2 instructions on how to use this guide.

Visit curriculumtopicstudy2.org for more information about CTS and additional resources.

NOTE: *Atlas* page numbers have not been provided because *The NSTA Atlas of the Three Dimensions* was produced concurrently with this edition. Titles and map codes are accurate.

Additional Readings:

Available for download at **www.curriculumtopicstudy2.org**

Copyright © 2020 by Corwin Press, Inc. All rights reserved. Reprinted from *Science Curriculum Topic Study: Bridging the Gap Between Three-Dimensional Standards, Research, and Practice* (2nd ed.) by Page Keeley and Joyce Tugel. Thousand Oaks, CA: Corwin, www.corwin.com. Reproduction authorized for educational use by educators, local school sites, and/or noncommercial or nonprofit entities that have purchased the book.

Chemical Reactions
Grades K–12 Standards- and Research-Based Study of a Curricular Topic

Section and Outcome	Selected Sources and Readings for Study and Reflection Read and examine **related** parts of
I. Content Knowledge	**IA:** *Science for All Americans*<ul><li>Ch. 4: Structure of Matter, pp. 46–48</li><li>Ch. 10: Understanding Fire, pp. 153–155</li></ul>**IB:** *Science Matters* (2009 edition)<ul><li>Ch. 7: Chemical Reactions, pp. 121–124</li></ul>**IC:** *Framework for K–12 Science Education:* Narrative Section<ul><li>Ch. 5: PS1.B: Chemical Reactions, pp. 109–110</li></ul>**ID:** *The NSTA Atlas of the Three Dimensions:* Narrative Page<ul><li>3.2: Chemical Reactions and Nuclear Processes (PS1.B)</li></ul>
II. Concepts, Core Ideas, or Practices	**IIA:** *Framework for K–12 Science Education:* Grade Band Endpoints<ul><li>Ch. 5: PS1.A: PS1.B: Chemical Reactions, pp. 110–111</li></ul>**IIB:** *Next Generation Science Standards:* Disciplinary Core Ideas Column<ul><li>Grade 2: PS1.B: Chemical Reactions, p. 13</li><li>Grade 5: PS1.B: Chemical Reactions, p. 28</li><li>MS: PS1.B: Chemical Reactions, pp. 35, 37</li><li>HS: PS1.B: Chemical Reactions, p. 70</li></ul>**IIC:** *NSTA Quick Reference Guide to the NGSS K–12:* Disciplinary Core Ideas Column<ul><li>Grade 2: PS1.B: Chemical Reactions, p. 97</li><li>Grade 5: PS1.B: Chemical Reactions, pp. 112–113</li><li>MS: PS1.B: Chemical Reactions, pp. 122–124</li><li>HS: PS1.B: Chemical Reactions, pp. 147–148</li></ul>
III. Curriculum, Instruction, and Formative Assessment	**IIIA:** *Disciplinary Core Ideas: Reshaping Teaching and Learning*<ul><li>K–2: Core Idea PS1, p. 17</li><li>3–5: Core Idea PS1, pp. 20–21</li><li>MS: Core Idea PS1, pp. 23–24, 28</li><li>HS: Core Idea PS1, pp. 26–27</li></ul>**IIIB:** *Uncovering Student Ideas:* Assessment Probe and Suggestions for Instruction<ul><li>USI.1 (2018 edition): Rusty Nails, pp. 93, 98</li><li>USI.4: Burning Paper, pp. 23, 28; Nails in a Jar, pp. 31, 36</li><li>USI.K-2: Back and Forth, pp. 63, 66</li><li>USI.PS3: Will It Form a New Substance? pp. 131, 135–136; What Is the Result of a Chemical Change? pp. 137, 141; What Happens to Atoms During a Chemical Reaction? pp. 143, 146–147</li></ul>
IV. Research on Commonly Held Ideas	**IVA:** *Benchmarks for Science Literacy:* Chapter 15 Research<ul><li>4D: Chemical Changes, p. 337</li></ul>**IVB:** *Making Sense of Secondary Science: Research Into Children's Ideas*<ul><li>Ch. 10: Chemical Change, pp. 85–91</li><li>Ch. 13: Composition of Air and Chemical Interactions of Air, p. 110</li></ul>

Section and Outcome	Selected Sources and Readings for Study and Reflection Read and examine **related** parts of
	IVC: *Uncovering Student Ideas:* Related Research<ul><li>USI.1 (2018 edition): Rusty Nails, pp. 96–97</li><li>USI.4: Burning Paper, pp. 27–28; Nails in a Jar, pp. 35–36</li><li>USI.K–2: Back and Forth, p. 65</li><li>USI.PS3: Will It Form a New Substance? pp. 134–135; What Is the Result of a Chemical Change? pp. 140–141; What Happens to Atoms During a Chemical Reaction? p. 14</li></ul>
V. **K–12 Articulation and Connections**	**VA:** *Next Generation Science Standards:* Appendices: Progression<ul><li>Appendix E: PS1.B: Chemical Reactions, p. 7</li></ul>**VB:** *NSTA Quick Reference Guide to the NGSS K–12:* Progression<ul><li>PS1.B: Chemical Reactions, p. 62</li></ul>**VC:** *The NSTA Atlas of the Three Dimensions:* Map Page<ul><li>3.2: Chemical Reactions and Nuclear Processes (PS1.B & PS1.C)</li></ul>
VI. **Assessment Expectation**	**VIA:** *State Standards*<ul><li>Examine your state's standards</li></ul>**VIB:** *Next Generation Science Standards:* Performance Expectations<ul><li>Grade 2: 2-PS1-4, p. 13</li><li>Grade 5: 5-PS1-2, 5-PS1-4, p. 28</li><li>MS: MS-PS1-2, MS-PS1-5, MS-PS1-6, p. 37; MS-PS1-3, p. 35</li><li>HS: HS-PS1-2, HS-PS1-4, HS-PS1-5, HS-PS1-6, HS-PS1-7, p. 70</li></ul>**VIC:** *NSTA Quick Reference Guide to the NGSS K–12:* Performance Expectations Column<ul><li>Grade 2: 2-PS1-4, p. 97</li><li>Grade 5: 5-PS1-2, 5-PS1-4, pp. 112–113</li><li>MS: MS-PS1-2, MS-PS1-3, MS-PS1-5, MS-PS1-6, pp. 122–124</li><li>HS: HS-PS1-2, HS-PS1-4, HS-PS1-5, HS-PS1-6, HS-PS1-7, pp. 147–148</li></ul>

Review Chapter 2 instructions on how to use this guide.

Visit curriculumtopicstudy2.org for more information about CTS and additional resources.

NOTE: *Atlas* page numbers have not been provided because *The NSTA Atlas of the Three Dimensions* was produced concurrently with this edition. Titles and map codes are accurate.

Additional Readings:

 Available for download at www.curriculumtopicstudy2.org

Copyright © 2020 by Corwin Press, Inc. All rights reserved. Reprinted from *Science Curriculum Topic Study: Bridging the Gap Between Three-Dimensional Standards, Research, and Practice* (2nd ed.) by Page Keeley and Joyce Tugel. Thousand Oaks, CA: Corwin, www.corwin.com. Reproduction authorized for educational use by educators, local school sites, and/or noncommercial or nonprofit entities that have purchased the book.

Conservation of Matter
Grades 3–12 Standards- and Research-Based Study of a Curricular Topic

Section and Outcome	Selected Sources and Readings for Study and Reflection Read and examine **related** parts of
I. Content Knowledge	**IA:** *Science for All Americans* • Ch. 10: Understanding Fire, pp. 153–154 (paras. 1–4) **IB:** *Framework for K–12 Science Education:* Narrative Section • Ch. 4: Energy and Matter: Flows, Cycles, and Conservation, pp. 94–95 (first and second paragraphs) • Ch. 5: PS1.B: Chemical Reactions, p. 109 (first paragraph) **IC:** *The NSTA Atlas of the Three Dimensions:* Narrative Page • 3.1: Structure and Properties of Matter (PS1.A) • 3.2: Chemical Reactions and Nuclear Processes (PS1.B & PS1.C)
II. Concepts, Core Ideas, or Practices	**IIA:** *Framework for K–12 Science Education:* Grade Band Endpoints • Ch. 5: PS1.A: Structure and Properties of Matter, p. 108 • Ch. 5: PS1.B: Chemical Reactions, pp. 110–11 **IIB:** *Next Generation Science Standards:* Disciplinary Core Ideas Column • Grade 5: PS1.A: Structure and Properties of Matter; PS1.B: Chemical Reactions, p. 28 • MS: PS1.B: Chemical Reactions, p. 37 • HS: PS1.B: Chemical Reactions, p. 70 **IIIC1:** *NSTA Quick Reference Guide to the NGSS K–12:* Disciplinary Core Ideas Column • Grade 5: PS1.A: Structure and Properties of Matter; PS1.B: Chemical Reactions, p. 112 • MS: PS1.B: Chemical Reactions, p. 123 • HS: PS1.B: Chemical Reactions, p. 148 **IIIC2:** *NSTA Quick Reference Guide to the NGSS K–12:* Crosscutting Concepts Chart • Energy and Matter: Flows, Cycles, and Conservation, p. 60
III. Curriculum, Instruction, and Formative Assessment	**IIIA:** *Disciplinary Core Ideas: Reshaping Teaching and Learning* • Ch. 2: 3–5: PS1.B: Chemical Changes, pp. 20–21 **IIIB:** *Uncovering Student Ideas:* Assessment Probe and Suggestions for Instruction • USI.1 (2018 edition): Cookie Crumbles, pp. 61, 65; Ice Cubes in a Bag, pp. 45, 51–52; Lemonade, pp. 53, 58–59 • USI.3: Hot and Cold Balloons, pp. 45, 49 • USI.4: Burning Paper, pp. 23, 25–26, 28; Nails in a Jar, pp. 31, 32–33, 36 • USI.K-2: Snap Blocks, pp. 59, 60, 61–62 • USI.PS3: What Does Conservation of Matter Mean? pp. 65, 68–69; Salt in Water, pp. 71, 75–76; What Happens to Atoms During a Chemical Reaction? pp. 143, 146–147
IV. Research on Commonly Held Ideas	**IVA:** *Benchmarks for Science Literacy:* Chapter 15 Research • 4D: Structure of Matter: Conservation of Matter, pp. 336–337 **IVB:** *Making Sense of Secondary Science: Research into Children's Ideas* • Ch. 8: Conservation of Matter, p. 77 • Ch. 9: Solids, Liquids, Gases, Melting, Evaporation, and Dissolving, pp. 79–84 • Ch. 10: Conservation of Matter Through Change, pp. 88–89.

Section and Outcome	Selected Sources and Readings for Study and Reflection Read and examine **related** parts of
	IVC: *Uncovering Student Ideas:* Related Research • USI.1 (2018 edition): Cookie Crumbles, p. 65; Ice Cubes in a Bag, pp. 49–50; Lemonade, pp. 57–58 • USI.3: USI.3: Hot and Cold Balloons, p. 49 • USI.4: Burning Paper, pp. 27–28; Nails in a Jar, pp. 35–36 • USI.K-2, Snap Blocks, p. 61 • USI.PS3: What Does Conservation of Matter Mean? p. 68; Salt in Water, pp. 74–75; What Happens to Atoms During a Chemical Reaction? p. 146
V. K–12 Articulation and Connections	**VA: *Next Generation Science Standards:*** Appendices: Progression • Appendix E: PS1.A: Structure of Matter, p. 7 • Appendix E: PS1.B: Chemical Reactions, p. 7 • Appendix G: Energy and Matter: Flows, Cycles, and Conservation, p. 9 **VB: *NSTA Quick Reference Guide to the NGSS K–12:*** Progression • PS1.A: Structure and Properties of Matter, p. 61 • PS1.B: Chemical Reactions, p. 62 • Energy and Matter: Flows, Cycles, and Conservation, p. 60 **VC: *The NSTA Atlas of the Three Dimensions:*** Map Page • 3.1: Structure and Properties of Matter (PS1.A) • 3.2: Chemical Reactions and Nuclear Processes (PS1.B & PS1.C)
VI. Assessment Expectation	**VIA: *State Standards*** • Examine your state's standards **VIB: *Next Generation Science Standards:*** Performance Expectations • Grade 5: 5-PS1.2, p. 28 • MS: MS-PS1.5, p. 37 • HS: HS-PS1.7, p. 70 **VIC: *NSTA Quick Reference Guide to the NGSS K–12:*** Performance Expectations Column • Grade 5: 5-PS1.2, p. 112 • MS: MS-PS1.5, p. 123 • HS: HS-PS1.7, p. 148

Review Chapter 2 instructions on how to use this guide.

Visit curriculumtopicstudy2.org for more information about CTS and additional resources.

NOTE: This study can be combined with the following crosscutting concept guide: Energy and Matter: Flows, Cycles, and Conservation.

NOTE: *Atlas* page numbers have not been provided because *The NSTA Atlas of the Three Dimensions* was produced concurrently with this edition. Titles and map codes are accurate.

Additional Readings:

 Available for download at www.curriculumtopicstudy2.org

Copyright © 2020 by Corwin Press, Inc. All rights reserved. Reprinted from *Science Curriculum Topic Study: Bridging the Gap Between Three-Dimensional Standards, Research, and Practice* (2nd ed.) by Page Keeley and Joyce Tugel. Thousand Oaks, CA: Corwin, www.corwin.com. Reproduction authorized for educational use by educators, local school sites, and/or noncommercial or nonprofit entities that have purchased the book.

Elements, Compounds, and the Periodic Table
Grades 3–12 Standards- and Research-Based Study of a Curricular Topic

Section and Outcome	Selected Sources and Readings for Study and Reflection Read and examine **related** parts of
I. Content Knowledge	**IA:** *Science for All Americans* • Ch. 4: Structure of Matter, pp. 46–47 **IIB.** *Science Matter* (2009 edition) • Ch. 4: The Periodic Table of Chemical Elements, pp. 74–77 **IC:** *Framework for K–12 Science Education:* Narrative Section • Ch. 5: PS1.A: Structure and Properties of Matter, pp. 106–107; PS1.B: Chemical Reactions, pp. 109–110 **ID:** *The NSTA Atlas of the Three Dimensions:* Narrative Page • 3.1: Structure and Properties of Matter (PS1.A) • 3.2: Chemical Reactions and Nuclear Processes (PS1.B)
II. Concepts, Core Ideas, or Practices	**IIA:** *Framework for K–12 Science Education:* Grade Band Endpoints • Ch. 5: PS1.A: Structure and Properties of Matter, pp. 108–109, PS1.B: Chemical Reactions, pp. 110–111 **IIB:** *Next Generation Science Standards:* Disciplinary Core Ideas Column • Grade 5: PS1.B: Chemical Reactions, p. 28 • MS: PS1.A: Structure and Properties of Matter, p. 35; PS1.B: Chemical Reactions, p. 37 • HS: PS1.A: Structure and Properties of Matter, pp. 68, 70; PS1.B: Chemical Reactions, p. 70 **IIC:** *NSTA Quick Reference Guide to the NGSS K–12:* Disciplinary Core Ideas Column • Grade 5: PS1.B: Chemical Reactions, p. 113 • MS: PS1.A: Structure and Properties of Matter and Chemical Reactions, p. 122 • HS: PS1.A: Structure and Properties of Matter and Chemical Reactions, p. 147
III. Curriculum, Instruction, and Formative Assessment	**IIIA:** *Disciplinary Core Ideas: Reshaping Teaching and Learning* • Ch. 2: 6–8: An Atomic-Molecular Model, pp. 21–24 • Ch. 2: 9–12: A Subatomic Model, pp. 25–27 **IIIB:** *Uncovering Student Ideas:* Assessment Probe and Suggestions for Instruction • USI.PS3: Classifying Water, pp. 109, 112; Graphite and Diamonds, pp. 113, 116–117; What Is a Substance? pp. 125, 128–129
IV. Research on Commonly Held Ideas	**IVA:** *Benchmarks for Science Literacy:* Chapter 15 Research • 4D: Nature of Matter, p. 336 **IVB:** *Making Sense of Secondary Science: Research Into Children's Ideas* • Ch. 8: Elements, pp. 76–77 **IVC:** *Uncovering Student Ideas:* Related Research • USI.PS3: Classifying Water, pp. 111–112; Graphite and Diamonds, p. 116; What Is a Substance? p. 128
V. K–12 Articulation and Connections	**VA:** *Next Generation Science Standards:* Appendices: Progression • Appendix E: PS1.A: Structure of Matter and PS1.B: Chemical Reactions, p. 7 **VB:** *NSTA Quick Reference Guide to the NGSS K–12:* Progression • PS1.A: Structure and Properties of Matter and PS1.B: Chemical Reactions, pp. 61–62

Section and Outcome	Selected Sources and Readings for Study and Reflection Read and examine **related** parts of
	VC: *The NSTA Atlas of the Three Dimensions:* Map Page • 3.1: Structure and Properties of Matter (PS1.A) • 3.2: Chemical Reactions and Nuclear Processes (PS1.B & PS1.C)
VI. Assessment Expectation	**VIA: *State Standards*** • Examine your state's standards **VIB: *Next Generation Science Standards:*** Performance Expectations • Grade 5: 5-PS1-4, p. 28 • MS: MS-PS1-1, p. 35; MS-PS1-2, p. 37 • HS: HS-PS1-1, p. 68; HS-PS1-2, p. 70 **VIC: *NSTA Quick Reference Guide to the NGSS K–12:*** Performance Expectations • Grade 5: 5-PS1-4, p. 113 • MS: MS-PS1-1, MS-PS1-2, p. 122 • HS: HS-PS1-1, HS-PS1-2, p. 147

Review Chapter 2 for abbreviations key and instructions on how to use this guide.

Visit curriculumtopicstudy2.org/ for more information about CTS and additional resources.

NOTE: *Atlas* page numbers have not been provided because *The NSTA Atlas of the Three Dimensions* was produced concurrently with this edition. Titles and map codes are accurate.

Additional Readings:

 Available for download at **www.curriculumtopicstudy2.org**

Copyright © 2020 by Corwin Press, Inc. All rights reserved. Reprinted from *Science Curriculum Topic Study: Bridging the Gap Between Three-Dimensional Standards, Research, and Practice* (2nd ed.) by Page Keeley and Joyce Tugel. Thousand Oaks, CA: Corwin, www.corwin.com. Reproduction authorized for educational use by educators, local school sites, and/or noncommercial or nonprofit entities that have purchased the book.

Mixtures and Solutions
Grades K–12 Standards- and Research-Based Study of a Curricular Topic

Section and Outcome	Selected Sources and Readings for Study and Reflection Read and examine **related** parts of
I. **Content Knowledge**	**IA: *Science for All Americans*** • Ch. 4: Structure of Matter, pp. 46–48 **IB: *Science Matters*** (2009 edition) • Ch. 7: Physical Properties, pp. 124–136 **IC: *Framework for K–12 Science Education:*** Narrative Section • Ch. 5: PS1.A: Structure and Properties of Matter, pp. 106–107 **ID: *The NSTA Atlas of the Three Dimensions:*** Narrative Page • 3.1: Structure and Properties of Matter (PS1.A) • 3.2: Chemical Reactions and Nuclear Processes (PS1.B)
II. **Concepts, Core Ideas, or Practices**	**IIA: *Framework for K–12 Science Education:*** Grade Band Endpoints • Ch. 5: PS1.A: Structure and Properties of Matter, pp. 108–109 **IIB: *Next Generation Science Standards:*** Disciplinary Core Ideas Column • Grade 2: PS1.A: Structure and Properties of Matter, p. 13 • Grade 5: PS1.A: Structure and Properties of Matter; PS1.B: Chemical Reactions, p. 28 • MS: PS1.A: Structure and Properties of Matter, p. 37 • HS: PS1.A: Structure and Properties of Matter, p. 68 **IIC: *NSTA Quick Reference Guide to the NGSS K–12:*** Disciplinary Core Ideas Column • Grade 2: PS1.A: Structure and Properties of Matter, p. 96 • Grade 5: PS1.A: Structure and Properties of Matter, pp. 112–113; PS1.B: Chemical Reactions, pp. 112–113 • MS: PS1.A: Structure and Properties of Matter, pp. 122–123 • HS: PS1.A: Structure and Properties of Matter, p. 147
III. **Curriculum, Instruction, and Formative Assessment**	**IIIA: *Disciplinary Core Ideas: Reshaping Teaching and Learning*** • Ch. 9: K–2: PS1.B, p. 17 • Ch. 9: 3–5: PS1.A, p. 18 • Ch. 9: 6–8: PS1.B, p. 24 **IIIB: *Uncovering Student Ideas:*** Assessment Probe and Suggestions for Instruction • USI.1 (2018 edition): Lemonade, pp. 53, 58–59 • USI.4: Sugar Water, pp. 11, 15–16 • USI.PS3: Salt in Water, pp. 71, 75–76; What Is a Substance? pp. 125, 128–129; Does It Have New Properties? pp. 155, 158
IV. **Research on Commonly Held Ideas**	**IVA: *Benchmarks for Science Literacy:*** Chapter 15 Research • 4D: Structure of Matter, pp. 336–337 **IVB: *Making Sense of Secondary Science: Research Into Children's Ideas*** • Ch. 8: Mixtures and Substances, pp. 74–75 • Ch. 9: Dissolving, pp. 83–84 • Ch. 10: Mixtures of Substances, p. 85 • Ch. 11: Particle Ideas About Solutions, p. 95 • Ch. 12: Dissolving Substances in Water, pp. 100–101 • Ch. 13: Composition of Air, p. 110

Section and Outcome	Selected Sources and Readings for Study and Reflection Read and examine **related** parts of
	IVC: *Uncovering Student Ideas:* Related Research • USI.1 (2018 edition): Lemonade, p. 58 • USI.4: Sugar Water, p. 15 • USI.PS3: Salt in Water, pp. 74–75; What Is a Substance? p. 128; Does It Have New Properties? p. 158
V. **K–12** **Articulation** **and** **Connections**	**VA:** *Next Generation Science Standards:* Appendices: Progression • Appendix E: PS1.A: Structure and Properties of Matter, p. 7 • Appendix E: PS1.B: Chemical Reactions, p. 7 **VB:** *NSTA Quick Reference Guide to the NGSS K–12:* Progression • PS1.A: Structure and Properties of Matter, p. 61 • PS1.B: Chemical Reactions, p. 62 **VC:** *The NSTA Atlas of the Three Dimensions:* Map Page • 3.1: Structure and Properties of Matter (PS1.A) • 3.2: Chemical Reactions and Nuclear Processes (PS1.B)
VI. **Assessment** **Expectation**	**VIA:** *State Standards* • Examine your state's standards **VIB:** *Next Generation Science Standards:* Performance Expectations • Grade 2: 2-PS1-1, p. 13 • Grade 5: 5-PS1-1, 5-PS1-2, 5-PS1-3, 5-PS1-4, p. 28 • MS: MS-PS1-2, p. 37 • HS: HS-PS1-3, p. 68 **VIC:** *NSTA Quick Reference Guide to the NGSS K–12:* Performance Expectations • Grade 2: 2-PS1-1, p. 96 • Grade 5: 5-PS1-1, 5-PS1-2, 5-PS1-3, 5-PS1-4, pp. 112–113 • MS: MS-PS1-2, p. 122 • HS: HS-PS1-3, p. 147

Review Chapter 2 instructions on how to use this guide.

Visit curriculumtopicstudy2.org for more information about CTS and additional resources.

NOTE: *Atlas* page numbers have not been provided because *The NSTA Atlas of the Three Dimensions* was produced concurrently with this edition. Titles and map codes are accurate.

Additional Readings:

 Available for download at www.curriculumtopicstudy2.org

Copyright © 2020 by Corwin Press, Inc. All rights reserved. Reprinted from *Science Curriculum Topic Study: Bridging the Gap Between Three-Dimensional Standards, Research, and Practice* (2nd ed.) by Page Keeley and Joyce Tugel. Thousand Oaks, CA: Corwin, www.corwin.com. Reproduction authorized for educational use by educators, local school sites, and/or noncommercial or nonprofit entities that have purchased the book.

Nuclear Processes
Grades 9–12 Standards- and Research-Based Study of a Curricular Topic

Section and Outcome	Selected Sources and Readings for Study and Reflection Read and examine related parts of
I. Content Knowledge	**IA:** *Science for All Americans* • Ch. 4: Structure of Matter, p. 49 (last two paragraphs) • Ch. 10: Splitting the Atom, pp. 155–157 **IB:** *Science Matters* (2009 edition) • Ch. 8: The Nucleus, pp. 137–152 **IC:** *Framework for K–12 Science Education:* Narrative Section • Ch. 2: PS1.C. Nuclear Processes, pp. 111–112 **ID:** *The NSTA Atlas of the Three Dimensions:* Narrative Page • 3.2: Chemical Reactions and Nuclear Processes (PS1.C)
II. Concepts, Core Ideas, or Practices	**IIA:** *Framework for K–12 Science Education:* Grade Band Endpoints • Ch. 2: PS1.C. Nuclear Processes, p. 113 **IIB:** *Next Generation Science Standards:* Disciplinary Core Ideas Column • HS: PS1.C. Nuclear Processes, p. 68 **IIC:** *NSTA Quick Reference Guide to the NGSS K–12:* Disciplinary Core Ideas Column • HS: PS1.C. Nuclear Processes, p. 148
III. Curriculum, Instruction, and Formative Assessment	**IIIA:** *Disciplinary Core Ideas: Reshaping Teaching and Learning* • HS: Ch. 2: Core Idea PS1.C: Nuclear Processes, p. 27 **IIIB:** *Uncovering Student Ideas:* Assessment Probe and Suggestions for Instruction • USI.PS3: Are They Safe to Eat? pp. 165, 169; Radish Seeds, pp. 171, 174–175
IV. Research on Commonly Held Ideas	**IVA:** *Uncovering Student Ideas:* Related Research • USI.PS3: Are They Safe to Eat? pp. 168–169; Radish Seeds, p. 174
V. K–12 Articulation and Connections	**VA:** *Next Generation Science Standards:* Appendices: Progression • Appendix E: PS1.C. Nuclear Processes: No progression **VB:** *NSTA Quick Reference Guide to the NGSS K–12:* Progression • PS1.C: Nuclear Processes, p. 62 **VC:** *The NSTA Atlas of the Three Dimensions:* Map Page • 3.2: Chemical Reactions and Nuclear Processes (PS1.B & PS1.C)
VI. Assessment Expectation	**VIA:** *State Standards* • Examine your state's standards **VIB:** *Next Generation Science Standards:* Performance Expectations • HS: HS-PS1-8, p. 68 **VIC:** *NSTA Quick Reference Guide to the NGSS K–12:* Performance Expectations • HS: HS-PS1-8, p. 148

Section and Outcome	Selected Sources and Readings for Study and Reflection Read and examine **related** parts of

Review Chapter 2 instructions on how to use this guide.

Visit curriculumtopicstudy2.org for more information about CTS and additional resources.

NOTE: *Atlas* page numbers have not been provided because *The NSTA Atlas of the Three Dimensions* was produced concurrently with this edition. Titles and map codes are accurate.

Additional Readings:

Available for download at **www.curriculumtopicstudy2.org**

Copyright © 2020 by Corwin Press, Inc. All rights reserved. Reprinted from *Science Curriculum Topic Study: Bridging the Gap Between Three-Dimensional Standards, Research, and Practice* (2nd ed.) by Page Keeley and Joyce Tugel. Thousand Oaks, CA: Corwin, www.corwin.com. Reproduction authorized for educational use by educators, local school sites, and/or noncommercial or nonprofit entities that have purchased the book.

Particulate Nature of Matter
Grades K–12 Standards- and Research-Based Study of a Curricular Topic

Section and Outcome	Selected Sources and Readings for Study and Reflection Read and examine **related** parts of
I. **Content Knowledge**	**IA:** *Science for All Americans* • Ch. 4: Structure of Matter, pp. 46–49 **IB:** *Science Matters* (2009 edition) • Ch. 4: The Atom, pp. 67–74; Ch. 7: Atomic Architecture, pp. 115–116 **IC:** *Framework for K–12 Science Education:* Narrative Section • Ch. 5: PS1.A: Structure and Properties of Matter, pp. 106–107 **ID:** *The NSTA Atlas of the Three Dimensions:* Narrative Page • 3.1: Structure and Properties of Matter (PS1.A)
II. **Concepts, Core Ideas, or Practices**	**IIA:** *Framework for K–12 Science Education:* Grade Band Endpoints • Ch. 5: PS1.A: Structure and Properties of Matter, pp. 108–109 **IIB:** *Next Generation Science Standards:* Disciplinary Core Ideas Column • Grade 2: PS1.A: Structure and Properties of Matter, p. 13 • Grade 5: PS1.A: Structure and Properties of Matter, p. 28 • MS: PS1.A: Structure and Properties of Matter, p. 35 • HS: PS1.A: Structure and Properties of Matter, pp. 68, 70 **IIC:** *NSTA Quick Reference Guide to the NGSS K–12:* Disciplinary Core Ideas Column • Grade 2: Structure and Properties of Matter, pp. 96, 97 • Grade 5: Structure and Properties of Matter, p. 112 • MS: Structure and Properties of Matter, pp. 122, 123 • HS: Structure and Properties of Matter, p. 147
III. **Curriculum, Instruction, and Formative Assessment**	**IIA:** *Disciplinary Core Ideas: Reshaping Teaching and Learning* • Ch. 2: 3–5: The Study of Materials and a Particle Model, pp. 17–21 • Ch. 2: 6–8: Atomic-Molecular Model, pp. 21–25 • Ch. 2: 9–12: A Subatomic Model, pp. 25–27 **IIIB:** *Uncovering Student Ideas:* Assessment Probe and Suggestions for Instruction • USI.1 (2018 edition): Is It Made of Molecules? pp. 87, 91–92 • USI.4: Iron Bar, pp. 17, 21; Salt Crystals, pp. 39, 42–43 • USI.K-2: Is It Matter? pp. 53, 55–56 • USI.PS3: What Do You Know About Atoms and Molecules? pp. 29, 33–34; Model of Air Inside a Jar, pp. 43, 47; What If You Could Remove All the Atoms? pp. 49, 52–53
IV. **Research on Commonly Held Ideas**	**IVA:** *Benchmarks for Science Literacy:* Chapter 15 Research • 4D: The Structure of Matter: Particles, p. 337 **IVB:** *Making Sense of Secondary Science: Research Into Children's Ideas* • Ch. 11: Particles, pp. 92–97 **IVC:** *Uncovering Student Ideas:* Related Research • USI.1 (2018 edition): Is It Made of Molecules? p. 91 • USI.4: Iron Bar, pp. 20–21; Salt Crystals, p. 42; • USI.K-2: Is It Matter? p. 55 • USI.PS3: What Do You Know About Atoms and Molecules? p. 33; Model of Air Inside a Jar, pp. 46–47; What If You Could Remove All the Atoms? p. 52

Section and Outcome	Selected Sources and Readings for Study and Reflection Read and examine **related** parts of
V. **K–12 Articulation and Connections**	**VA:** *Next Generation Science Standards:* Appendices: Progression • Appendix E: PS1.A: Structure and Properties of Matter, p. 7 **VB:** *NSTA Quick Reference Guide to the NGSS K–12:* Progression • PS1.A: Structure and Properties of Matter, p. 61 **VC:** *The NSTA Atlas of the Three Dimensions:* Map Page • 3.1: Structure and Properties of Matter (PS1.A)
VI. **Assessment Expectation**	**VIA:** *State Standards* • Examine your state's standards **VIB:** *Next Generation Science Standards:* Performance Expectations • Grade 2: 2-PS1-3, p. 13 • Grade 5: 5-PS1-1, p. 28 • MS: MS-PS1-1, p. 35 • HS: HS-PS1-1, HS-PS1-3, p. 68; HS-PS1-2, p. 70 **VIC:** *NSTA Quick Reference Guide to the NGSS K–12:* Performance Expectations • Grade 2: 2-PS1-3, p. 97 • Grade 5: 5-PS1-1, p. 112 • MS: MS-PS1-1, p. 122 • HS: HS-PS1-1, HS-PS1-2, HS-PS1-3, p. 147

Review Chapter 2 instructions on how to use this guide.

Visit curriculumtopicstudy2.org for more information about CTS and additional resources.

NOTE: *Atlas* page numbers have not been provided because *The NSTA Atlas of the Three Dimensions* was produced concurrently with this edition. Titles and map codes are accurate.

Additional Readings:

Available for download at www.curriculumtopicstudy2.org

Copyright © 2020 by Corwin Press, Inc. All rights reserved. Reprinted from *Science Curriculum Topic Study: Bridging the Gap Between Three-Dimensional Standards, Research, and Practice* (2nd ed.) by Page Keeley and Joyce Tugel. Thousand Oaks, CA: Corwin, www.corwin.com. Reproduction authorized for educational use by educators, local school sites, and/or noncommercial or nonprofit entities that have purchased the book.

Properties of Matter
Grades K–12 Standards- and Research-Based Study of a Curricular Topic

Section and Outcome	Selected Sources and Readings for Study and Reflection Read and examine **related** parts of
I. Content Knowledge	**IA:** *Science for All Americans* • Ch. 4: Structure of Matter, pp. 46–48 **IB:** *Science Matters* (2009 edition) • Ch. 7: Physical Properties, pp. 124–136 **IC:** *Framework for K–12 Science Education:* Narrative Section • Ch. 5: PS1.A: Structure and Properties of Matter, pp. 106–107 **ID:** *The NSTA Atlas of the Three Dimensions:* Narrative Page • 3.1: Structure and Properties of Matter (PS1.A)
II. Concepts, Core Ideas, or Practices	**IIA:** *Framework for K–12 Science Education:* Grade Band Endpoints • Ch. 5: PS1.A: Structure and Properties of Matter, pp. 108–109 **IIB:** *Next Generation Science Standards:* Disciplinary Core Ideas Column • Grade 2: PS1.A: Structure and Properties of Matter, p. 13 • Grade 5: PS1.A: Structure and Properties of Matter, p. 28 • MS: PS1.A: Structure and Properties of Matter, pp. 37 • HS: PS1.A: Structure and Properties of Matter, pp. 68, 70 **IIC:** *NSTA Quick Reference Guide to the NGSS K–12:* Disciplinary Core Ideas Column • Grade 2: PS1.A: Structure and Properties of Matter, pp. 96–97 • Grade 5: PS1.A: Structure and Properties of Matter, pp. 112–113 • MS: PS1.A: Structure and Properties of Matter, pp. 122–123 • HS: PS1.A: Structure and Properties of Matter, p. 147
III. Curriculum, Instruction, and Formative Assessment	**IIIA:** *Disciplinary Core Ideas: Reshaping Teaching and Learning* • Ch. 2: K–2: PS1.A: Structure and Properties of Matter, pp. 16–17 • Ch. 2: 3–5: PS1.A: Structure and Properties of Matter, pp. 18–19 • Ch. 2: MS: PS1.A: Structure and Properties of Matter, pp. 22–23 • Ch. 2: HS: PS1.A: Structure and Properties of Matter, pp. 25–26 **IIIB:** *Uncovering Student Ideas:* Assessment Probe and Suggestions for Instruction • USI.2: Comparing Cubes, pp. 19, 23–24; Solids and Holes, pp. 41, 45–46; Turning the Dial, pp. 47, 51–52; Boiling Time and Temperature, pp. 53, 57–58; Freezing Ice, pp. 59, 63–64 • USI.3: Pennies, pp. 17, 21–22 • USI.PS3: Do They Have Weight and Take Up Space? pp. 59, 62–63; Mass, Volume, and Density, pp. 83, 87; Measuring Mass, pp. 89, 92; Do They Have the Same Properties? pp. 93, 96–97; Are They the Same Substance? pp. 99, 102
IV. Research on Commonly Held Ideas	**IVA:** *Benchmarks for Science Literacy:* Chapter 15 Research • 4D: The Structure of Matter, pp. 336–337 **IVB:** *Making Sense of Secondary Science: Research Into Children's Ideas* • Ch. 8: Materials, pp. 73–78 • Ch. 9: Solids, Liquids, and Gases, pp. 79–84 • Ch. 12: Water, pp. 98–103 • Ch. 13: Summary of Ideas about the Physical Properties of Air, pp. 107–110

Section and Outcome	Selected Sources and Readings for Study and Reflection Read and examine **related** parts of
	IVC: Uncovering Student Ideas: Related Research • USI.2: Comparing Cubes, p. 19, 23–24; Solids and Holes, pp. 44–45; Turning the Dial, p. 51; Boiling Time and Temperature, p. 57; Freezing Ice, pp. 62–63 • USI.3: Pennies, pp. 20–21 • Do They Have Weight and Take Up Space? p. 62; Mass, Volume, and Density, p. 86; Measuring Mass, pp. 93; Do They Have the Same Properties? p. 96; Are They the Same Substance? p. 102
V. **K–12 Articulation and Connections**	**VA: Next Generation Science Standards:** Appendices: Progression • Appendix E: PS1.A: Structure and Properties of Matter, p. 7 **VB: NSTA Quick Reference Guide to the NGSS K–12:** Progression • PS1.A: Structure and Properties of Matter, p. 61 **VC: The NSTA Atlas of the Three Dimensions:** Map Page • 3.1: Structure and Properties of Matter (PS1.A)
VI. **Assessment Expectation**	**VIA: State Standards** • Examine your state's standards **VIB: Next Generation Science Standards:** Performance Expectations • Grade 2: 2-PS1-1, 2-PS1-2, p. 13 • Grade 5: 5-PS1-3, 5-PS1-4, p. 28 • MS: MS-PS1-2, p. 37 • HS: HS-PS1-1, p. 68; HS-PS1-2, p. 70 **VC: NSTA Quick Reference Guide to the NGSS K–12:** Performance Expectations • Grade 2: 2-PS1-1, 2-PS1-2, pp. 96–97 • Grade 5: 5-PS1-3, 5-PS1-4, pp. 112–113 • MS: MS-PS1-2, p. 122 • HS: HS-PS1-1, HS-PS1-2, p. 147

Review Chapter 2 for instructions on how to use this guide.

Visit curriculumtopicstudy2.org/ for more information about CTS and additional resources.

NOTE: *Atlas* page numbers have not been provided because *The NSTA Atlas of the Three Dimensions* was produced concurrently with this edition. Titles and map codes are accurate.

Additional Readings:

 Available for download at www.curriculumtopicstudy2.org

Copyright © 2020 by Corwin Press, Inc. All rights reserved. Reprinted from *Science Curriculum Topic Study: Bridging the Gap Between Three-Dimensional Standards, Research, and Practice* (2nd ed.) by Page Keeley and Joyce Tugel. Thousand Oaks, CA: Corwin, www.corwin.com. Reproduction authorized for educational use by educators, local school sites, and/or noncommercial or nonprofit entities that have purchased the book.

States of Matter
Grades K–8 Standards- and Research-Based Study of a Curricular Topic

Section and Outcome	Selected Sources and Readings for Study and Reflection Read and examine **related** parts of
I. **Content Knowledge**	**IA:** *Science for All Americans* • Ch. 4: Structure of Matter, pp. 47–48 (fifth paragraph) **IB:** *Science Matters* (2009 edition) • Ch. 7: The States of Matter, pp. 116–121 **IC:** *Framework for K–12 Science Education:* Narrative Section • Ch. 5: PS1.A: Structure and Properties of Matter, p. 107 (second paragraph) **ID:** *The NSTA Atlas of the Three Dimensions:* Narrative Page • 3.1: Structure and Properties of Matter (PS1.A)
II. **Concepts, Core Ideas, or Practices**	**IIA:** *Framework for K–12 Science Education:* Grade Band Endpoints • Ch. 5: PS1.A: Structure and Properties of Matter, pp. 108–109 **IIB:** *Next Generation Science Standards:* Disciplinary Core Ideas Column • Grade 2: PS1.A: Structure and Properties of Matter, p. 13 • Grade 5: PS1.A: Structure and Properties of Matter, p. 28 • MS: PS1.A: Structure and Properties of Matter, p. 35 **IIC:** *NSTA Quick Reference Guide to the NGSS K–12:* Disciplinary Core Ideas Column • Grade 2: Structure and Properties of Matter, p. 96 • Grade 5: Structure and Properties of Matter, p. 112 • MS: Structure and Properties of Matter, p. 123
III. **Curriculum, Instruction, and Formative Assessment**	**IIIA:** *Disciplinary Core Ideas: Reshaping Teaching and Learning* • Ch. 2: K–2: PS1.A: Structure and Properties of Matter, p. 17 • Ch. 2: 3–5: PS1.A: Structure and Properties of Matter, pp. 18–20 • MS: Core Idea PS1.B: Chemical Reactions. pp. 23–24 **IIIB:** *Uncovering Student Ideas:* Assessment Probe and Suggestions for Instruction • USI.1 (2018 edition): Is It Melting? pp. 73, 77–78 • USI.2: What's in the Bubbles? pp. 65, 69–70 • USI.3: Is It a Solid? pp. 25, 29–30; Floating Balloon, pp. 39, 43–44 • USI.4: Ice Water, pp. 45, 49–50 • USI.PS3: Solids, Liquids, and Gases, pp. 23, 26–27; Model of Air Inside a Jar, pp. 43, 47
IV. **Research on Commonly Held Ideas**	**IVA:** *Benchmarks for Science Literacy:* Chapter 15 Research • 4D: Structure of Matter, pp. 336–337 **IVB:** *Making Sense of Secondary Science: Research Into Children's Ideas* • Ch. 9: Solids, Liquids, and Gases, pp. 79–84 • Ch. 11: Particle Ideas About Change of State, pp. 94–95 • Ch. 13: Existence of Air, pp. 104–105 **IVC:** *Uncovering Student Ideas:* Related Research • USI.1 (2018 edition): Is It Melting? p. 77 • USI.2: What's in the Bubbles? p. 69 • USI.3: Is It a Solid? pp. 28–29; Floating Balloon, pp. 42–43 • USI.4: Ice Water, p. 48 • USI.PS3: Solids, Liquids, and Gases, p. 26; Model of Air Inside a Jar, pp. 46–47

Section and Outcome	Selected Sources and Readings for Study and Reflection Read and examine **related** parts of
V. **K–12** **Articulation** **and** **Connections**	**VA:** *Next Generation Science Standards:* Appendices: Progression • Appendix E: PS1.A: Structure of Matter, p. 7 **VB:** *NSTA Quick Reference Guide to the NGSS K–12:* Progression • PS1.A: Structure and Properties of Matter, p. 61 **VC:** *The NSTA Atlas of the Three Dimensions:* Map Page • 3.1: Structure and Properties of Matter (PS1.A)
VI. **Assessment** **Expectation**	**VIA:** *State Standards* • Examine your state's standards **VIB:** *Next Generation Science Standards:* Performance Expectations • Grade 2: 2-PS1-1, p. 13 • Grade 5: 5-PS1.2, p. 28 • MS: MS-PS1.4, p. 35 **VIC:** *NSTA Quick Reference Guide to the NGSS K–12:* Performance • Grade 2: 2-PS1.1, p. 96 • Grade 5: 5-PS1.1, p. 112 • MS: MS-PS1.4, p. 123

Review Chapter 2 instructions on how to use this guide.

Visit curriculumtopicstudy2.org for more information about CTS and additional resources.

NOTE: *Atlas* page numbers have not been provided because *The NSTA Atlas of the Three Dimensions* was produced concurrently with this edition. Titles and map codes are accurate.

Additional Readings:

Available for download at www.curriculumtopicstudy2.org

Copyright © 2020 by Corwin Press, Inc. All rights reserved. Reprinted from *Science Curriculum Topic Study: Bridging the Gap Between Three-Dimensional Standards, Research, and Practice* (2nd ed.) by Page Keeley and Joyce Tugel. Thousand Oaks, CA: Corwin, www.corwin.com. Reproduction authorized for educational use by educators, local school sites, and/or noncommercial or nonprofit entities that have purchased the book.

Concept of Energy
Grades 3–12 Standards- and Research-Based Study of a Curricular Topic

Section and Outcome	Selected Sources and Readings for Study and Reflection Read and examine **related** parts of
I. Content Knowledge	**IA:** *Science for All Americans* • Ch. 4: Energy Transformation, pp. 49–52 **IB:** *Science Matters* (2009 edition) • Ch. 2: Energy, pp. 26–31 **IC:** *Framework for K–12 Science Education:* Narrative Section • Ch. 5: PS3.A: Definitions of Energy, pp. 120–122 (stop at Grade Band Endpoints) **ID:** *Disciplinary Core Ideas: Reshaping Teaching and Learning* • Ch. 4: PS3: Defining Energy as a Scientific Idea, pp. 59–60 **IE:** *The NSTA Atlas of the Three Dimensions:* Narrative Page • 3.5: Definitions of Energy (PS3.A)
II. Concepts, Core Ideas, or Practices	**IIA:** *Framework for K–12 Science Education:* Grade Band Endpoints • Ch. 5: PS3.A: Definitions of Energy, pp. 122–124 **IIB:** *Next Generation Science Standards:* Disciplinary Core Ideas Column • Grade 4: PS3.A: Definitions of Energy, p. 23 • MS: PS3.A: Definitions of Energy, p. 40 • HS: PS3.A: Definitions of Energy, p. 74 **IIC:** *NSTA Quick Reference Guide to the NGSS K–12:* Disciplinary Core Ideas Column • Grade 4: PS3.A: Definitions of Energy, pp. 108–109 • MS: PS3.A: Definitions of Energy, pp. 125–126 • HS: PS3.A: Definitions of Energy, pp. 150–151
III. Curriculum, Instruction, and Formative Assessment	**IIIA:** *Uncovering Student Ideas:* Assessment Probe and Suggestions for Instruction • USI.4: Warming Water, p. 53, 58 • USI.PS3: Describing Energy, p. 177, 181–182; Which Has More Energy? pp. 213, 217 **IIIB:** *Disciplinary Core Ideas: Reshaping Teaching and Learning* • Ch. 4: By the End of Grade 5, p. 60 • Ch. 4: By the End of Grade 8, p. 60 • Ch. 4: By the End of Grade 12, p. 61
IV. Research on Commonly Held Ideas	**IVA:** *Benchmarks for Science Literacy:* Chapter 15 Research • 4E: Energy Forms and Energy Transformation, p. 338 **IVB:** *Making Sense of Secondary Science: Research Into Children's Ideas* • Ch. 19: Heat and Temperature, pp. 138–141 • Ch. 20: Energy, pp. 143–147 **IVC:** *Uncovering Student Ideas:* Related Research • USI.4: Warming Water, pp. 57–58 • USI.PS3: Describing Energy, pp. 180–181; Which Has More Energy? pp. 216–217 **IVD:** *Disciplinary Core Ideas: Reshaping Teaching and Learning* • Ch. 4: PS3.A: Definitions of Energy-Common Prior Conceptions, p. 59

Section and Outcome	Selected Sources and Readings for Study and Reflection Read and examine **related** parts of
V. **K–12 Articulation and Connections**	**VA:** *Next Generation Science Standards:* Appendices: Progression • Appendix E: PS3.A: Definitions of Energy, p. 7 **VB:** *NSTA Quick Reference Guide to the NGSS K–12:* Progression • PS3.A: Definitions of Energy, p. 64 **VC:** *The NSTA Atlas of the Three Dimensions:* Map Page • 3.5: Definitions of Energy (PS3.A)
VI. **Assessment Expectation**	**VIA:** *State Standards* • Examine your state's standards **VIB:** *Next Generation Science Standards:* Performance Expectations • Grade 4: 4-PS3-1, 4-PS3-2, 4-PS3-3, p. 23 • MS: MS-PS1-1, MS-PS1-2, MS-PS1-3, MS-PS1-4, p. 40 • HS: HS-PS1-1, HS-PS1-2, HS-PS1-3, p. 74 **VIC:** *NSTA Quick Reference Guide to the NGSS K–12:* Performance Expectations • Grade 4: 4-PS3-1, 4-PS3-2, 4-PS3-3, pp. 108–109 • MS: MS-PS3-1, MS-PS3-2, MS-PS3-3, MS-PS3-4, pp. 125–126 • HS: HS-PS3-1, HS-PS3-2, HS-PS3-3, pp. 150–151

Review Chapter 2 instructions on how to use this guide.

Visit curriculumtopicstudy2.org for more information about CTS and additional resources.

NOTE: This study can be combined with the following crosscutting concept guide: Energy and Matter.

NOTE: *Atlas* page numbers have not been provided because *The NSTA Atlas of the Three Dimensions* was produced concurrently with this edition. Titles and map codes are accurate.

Additional Readings:

Available for download at www.curriculumtopicstudy2.org

Copyright © 2020 by Corwin Press, Inc. All rights reserved. Reprinted from *Science Curriculum Topic Study: Bridging the Gap Between Three-Dimensional Standards, Research, and Practice* (2nd ed.) by Page Keeley and Joyce Tugel. Thousand Oaks, CA: Corwin, www.corwin.com. Reproduction authorized for educational use by educators, local school sites, and/or noncommercial or nonprofit entities that have purchased the book.

Conservation of Energy
Grades 6–12 Standards-and Research-Based Study of a Curricular Topic

Section and Outcome	Selected Sources and Readings for Study and Reflection Read and examine **related** parts of
I. **Content Knowledge**	**IA:** *Science for All Americans* • Ch. 4: Energy Transformation, pp. 49–52 • Ch. 5: Flow of Matter and Energy, pp. 66–67 (first 3 paragraphs) **IB:** *Science Matters* (2009 edition) • Ch. 2: Energy Is Conserved, p. 27; Good News: The First Law, pp. 31–33 **IC:** *Framework for K–12 Science Education:* – Narrative Section • Ch. 5: PS3.B: Conservation of Energy and Energy Transfer, pp. 124–125 **ID:** *Disciplinary Core Ideas: Reshaping Teaching and Learning* • Ch. 4: PS3: Energy, pp. 55–57 **IE:** *The NSTA Atlas of the Three Dimensions:* Narrative Page • 3.6: Conservation of Energy and Energy Transfer (PS3.B and PS3.C)
II. **Concepts, Core Ideas, or Practices**	**IIA:** *Framework for K–12 Science Education:* Grade Band Endpoints • Ch. 5: PS3.B: Conservation of Energy and Energy Transfer, pp. 125–126 **IIB:** *Next Generation Science Standards:* Disciplinary Core Ideas Column • MS: PS3.B: Conservation of Energy and Energy Transfer, p. 40 • HS: PS3.B: Conservation of Energy and Energy Transfer, p. 74 **IIC:** *NSTA Quick Reference Guide to the NGSS K–12:* Disciplinary Core Ideas Column • MS: PS3.B: Conservation of Energy and Energy Transfer, p. 126 • HS: PS3.B: Conservation of Energy and Energy Transfer, pp. 150–151
III. **Curriculum, Instruction, and Formative Assessment**	**IIIA:** *Disciplinary Core Ideas: Reshaping Teaching and Learning* • By the End of Grade 8: p. 64 • By the End of Grade 12: p. 64 **IIIB:** *Uncovering Student Ideas:* Assessment Probe and Suggestions for Instruction • USI.PS3: Matter and Energy, pp. 183, 187–188; Hot Soup, pp. 195, 199–200
IV. **Research on Commonly Held Ideas**	**IVA:** *Benchmarks for Science Literacy:* Chapter 15 Research • 4E: Energy Conservation, p. 338 **IVB:** *Making Sense of Secondary Science: Research into Children's Ideas* • Ch. 20: Conservation of Energy, pp. 146–147 **IVC:** *Uncovering Student Ideas:* Related Research • USI.PS3: Matter and Energy, p. 187; Hot Soup, p. 199
V. **K–12 Articulation and Connections**	**VA:** *Next Generation Science Standards:* Appendices: Progression • Appendix E: PS3.B: Conservation of Energy and Energy Transfer, p. 7 **VB:** *NSTA Quick Reference Guide to the NGSS K–12:* Progression • Appendix E: PS3.B: Conservation of Energy and Energy Transfer, p. 65 **VC:** *The NSTA Atlas of the Three Dimensions:* Map Page • 3.6: Conservation of Energy and Energy Transfer (PS3.B and PS3.C)

Section and Outcome	Selected Sources and Readings for Study and Reflection Read and examine **related** parts of
VI. Assessment Expectation	**VIA:** *State Standards* • Examine your state's standards **VIB:** *Next Generation Science Standards*: Performance Expectations • MS: MS-PS3-5, p. 40 • HS: HS-PS3-1, HS-PS3-4, p. 74 **VIC:** *NSTA Quick Reference Guide to the NGSS K–12*: Performance Expectations • MS: MS-PS3-5, p. 126 • HS: HS-PS3-1, HS-PS3-4, pp. 150–151

Review Chapter 2 instructions on how to use this guide.
Visit curriculumtopicstudy2.org for more information about CTS and additional resources.

SEE ALSO: Crosscutting Concept Guide: Energy and Matter: Flows, Cycles, and Conservation.

NOTE: Atlas page numbers have not been provided because *The NSTA Atlas of the Three Dimensions* was produced concurrently with this edition. Titles and map codes are accurate.

Additional Readings:

Available for download at www.curriculumtopicstudy2.org

Copyright © 2020 by Corwin Press, Inc. All rights reserved. Reprinted from *Science Curriculum Topic Study: Bridging the Gap Between Three-Dimensional Standards, Research, and Practice* (2nd ed.) by Page Keeley and Joyce Tugel. Thousand Oaks, CA: Corwin, www.corwin.com. Reproduction authorized for educational use by educators, local school sites, and/or noncommercial or nonprofit entities that have purchased the book.

Electric Charge and Current

Grades 3–12 Standards- and Research-Based Study of a Curricular Topic

Section and Outcome	Selected Sources and Readings for Study and Reflection Read and examine **related** parts of
I. Content Knowledge	**IA:** *Science for All Americans* • Ch. 4: Forces of Nature, pp. 55–57 **IB:** *Science Matters* (2009 edition) • Ch. 3: Electrical Charge and Coulomb's Law, pp. 46–48 **IC:** *Framework for K–12 Science Education:* Narrative Section • Ch. 5: PS2.B: Types of Interactions, pp. 116–117 **ID:** *Disciplinary Core Ideas: Reshaping Teaching and Learning* • Ch. 3: PS2.B: Types of Interactions, p. 41 **IE:** *The NSTA Atlas of the Three Dimensions:* Narrative Page • 3.4: Types of Interactions (PS2.B)
II. Concepts, Core Ideas, or Practices	**IIA:** *Framework for K–12 Science Education:* Grade Band Endpoints • Ch. 5: PS2.B: Types of Interactions, pp. 117–118 **IIB:** *Next Generation Science Standards:* Disciplinary Core Ideas Column • Grade 3: PS2.B: Types of Interactions, p. 18 • Grade 4: PS3.A: Definitions of Energy; PS3.B: Conservation of Energy and Energy Transfer, p. 23 • MS: PS3.A: Definitions of Energy, p. 40; PS2.B: Types of Interactions, p. 38; PS3.C: Relationship Between Energy and Forces, p. 40 • HS: PS2.B: Types of Interactions, p. 95; PS3.A: Definitions of Energy, p. 95, 97–98; PS3.C: Relationship Between Energy and Forces, p. 99 **IIC:** *NSTA Quick Reference Guide to the NGSS K–12:* Disciplinary Core Ideas Column • Grade 3: PS2.B: Types of Interactions, p. 106 • Grade 4: PS3.A: Definitions of Energy, p. 109; PS3.B: Conservation of Energy and Energy Transfer, p. 109 • MS: PS3.A: Definitions of Energy, p. 125; PS2.B: Types of Interactions, pp. 124–125; PS3.C: Relationship Between Energy and Forces, p. 125 • HS: PS2.B: Types of Interactions, pp 68, 72; PS3.A: Definitions of Energy, pp. 72, 74; PS3.C: Relationship Between Energy and Forces, p. 94
III. Curriculum, Instruction, and Formative Assessment	**IIIA:** *Disciplinary Core Ideas: Reshaping Teaching and Learning* • Ch. 3: By the End of Grade 5, PS2.B, p. 47; PS3.A, p. 60; PS3.B, p. 64 • Ch. 3: By the End of Grade 8, PS2.B, p. 49 • Ch. 3: By the End of Grade 12, PS2.B, p. 52 **IIIB:** *Uncovering Student Ideas:* Assessment Probe and Suggestions for Instruction • USI.PS2: Too many to list. Sections 1 and 2 of this book contain twenty-one probes on electric charge and current electricity with suggestions for instruction
IV. Research on Commonly Held Ideas	**IVA:** *Making Sense of Secondary Science: Research Into Children's Ideas* • Ch. 15: Electricity, pp. 117–125 **IVB:** *Uncovering Student Ideas:* Related Research • USI.PS2: Too many to list. Sections 1 and 2 of this book contain twenty-one probes with research summaries on electric charge and current electricity

Section and Outcome	Selected Sources and Readings for Study and Reflection Read and examine **related** parts of
V. **K–12 Articulation and Connections**	**VA:** *Next Generation Science Standards:* Appendices: Progression • Appendix E: PS2.B: Types of Interactions, p. 7; PS3.A: Definitions of Energy, p. 7; PS3.B: Conservation of Energy and Energy Transfer, p. 7; PS3.C: Relationship Between Energy and Forces, p. 8 **VB:** *NSTA Quick Reference Guide to the NGSS K–12:* Progression • PS2.B: Types of Interactions, p. 63; PS3.A: Definitions of Energy, p. 64; PS3.B: Conservation of Energy and Energy Transfer, p. 65; PS3.C: Relationship Between Energy and Forces, p. 65 **VC:** *The NSTA Atlas of the Three Dimensions:* Map Page • 3.4: Types of Interactions (PS2.B)
VI. **Assessment Expectation**	**VIA:** *State Standards* • Examine your state's standards **VIB:** *Next Generation Science Standards:* Performance Expectations • Grade 3: 3-PS2-3, p. 18 • Grade 4: 4-PS3-2, 4-PS3-4, p. 23 • MS: MS-PS2-3, MS-PS2-5, p. 38; MS-PS3-2, p. 40 • HS: HS-PS2-4, HS-PS2-5, p. 72; HS-PS2-6, p. 68; HS-PS3-1, HS-PS3-2, p. HS-PS3-5, p. 74 **VIC:** *NSTA Quick Reference Guide to the NGSS K–12:* Performance Expectations Column • Grade 3: 3-PS2-3, p. 106 • Grade 4: 4-PS3-2, 4-PS3-4, p. 109 • MS: MS-PS2-3, p. 124; MS-PS2.5, MS-PS3-2, p. 125 • HS: HS-PS2-4, p. 149; HS-PS2-5, HS-PS2-6, HS-PS3-1, p. 150; HS-PS3-2, p. 151; HS-PS3-5, p. 152

Review Chapter 2 instructions on how to use this guide.

Visit curriculumtopicstudy2.org for more information about CTS and additional resources.

NOTE: *Atlas* page numbers have not been provided because *The NSTA Atlas of the Three Dimensions* was produced concurrently with this edition. Titles and map codes are accurate.

Additional Readings:

Available for download at www.curriculumtopicstudy2.org

Copyright © 2020 by Corwin Press, Inc. All rights reserved. Reprinted from *Science Curriculum Topic Study: Bridging the Gap Between Three-Dimensional Standards, Research, and Practice* (2nd ed.) by Page Keeley and Joyce Tugel. Thousand Oaks, CA: Corwin, www.corwin.com. Reproduction authorized for educational use by educators, local school sites, and/or noncommercial or nonprofit entities that have purchased the book.

Energy in Chemical Processes and Everyday Life
Grades 3–12 Standards- and Research-Based Study of a Curricular Topic

Section and Outcome	Selected Sources and Readings for Study and Reflection Read and examine **related** parts of
I. **Content Knowledge**	**IA:** *Science for All Americans*<ul><li>Ch. 4: Energy Transformation, pp. 49–52</li><li>Ch. 5: Flow of Matter and Energy, pp. 66–67</li><li>Ch. 8: Energy Sources and Use, pp, 114–115</li></ul>**IB:** *Science Matters* (2009 edition)<ul><li>Ch. 2: Potential Energy, p. 29; Frontiers: New Energy Sources, pp. 41–43</li><li>Ch. 15: Power Plant, pp. 263–266</li></ul>**IC:** *Framework for K–12 Science Education:* Narrative Section<ul><li>Ch. 5: PS3.D: Energy in Chemical Processes and Everyday Life, pp. 128–129</li></ul>**ID:** *Disciplinary Core Ideas: Reshaping Teaching and Learning*<ul><li>Ch. 4: PS3.D: Energy in Chemical Processes and Everyday Life, pp. 67–69</li></ul>**IE:** *The NSTA Atlas of the Three Dimensions:* Narrative Page<ul><li>3.7: Energy in Chemical Processes and Everyday Life (PS3.D)</li></ul>
II. **Concepts, Core Ideas, or Practices**	**IIA:** *Framework for K–12 Science Education:* Grade Band Endpoints<ul><li>Ch. 5: PS3.D: Energy in Chemical Processes and Everyday Life, pp. 129–130</li></ul>**IIB:** *Next Generation Science Standards:* Disciplinary Core Ideas Column<ul><li>Grade 4: PS3.D: Energy in Chemical Processes and Everyday Life, p. 23</li><li>Grade 5: PS3.D: Energy in Chemical Processes and Everyday Life, p. 29</li><li>MS: PS3.D: Energy in Chemical Processes and Everyday Life, p. 47</li><li>HS: PS3.D: Energy in Chemical Processes and Everyday Life, pp. 74, 81</li></ul>**IIC:** *NSTA Quick Reference Guide to the NGSS K–12:* Disciplinary Core Ideas Column<ul><li>Grade 4: PS3.D: Energy in Chemical Processes and Everyday Life, p. 109</li><li>Grade 5: PS3.D: Energy in Chemical Processes and Everyday Life, p. 113</li><li>MS: PS3.D: Energy in Chemical Processes and Everyday Life, p. 129</li><li>HS: PS3.C: PS3.D: Energy in Chemical Processes and Everyday Life, pp. 151, 157</li></ul>
III. **Curriculum, Instruction, and Formative Assessment**	**IIIA:** *Disciplinary Core Ideas: Reshaping Teaching and Learning*<ul><li>Ch. 4: PS3.D: By the End of Grade 5, p. 69</li><li>Ch. 4: PS3.D: By the End of Grade 8, p. 69</li><li>Ch. 4: PS3.D: By the End of Grade 12, pp. 69–70</li></ul>**IIIB:** *Uncovering Student Ideas:* Assessment Probe and Suggestions for Instruction<ul><li>USI.4: Digestive System, pp. 131, 135; Where Does Oil Come From? pp. 151, 155–156</li><li>USI.LS1: Food for Corn, pp. 69, 73–74</li></ul>
IV. **Research on Commonly Held Ideas**	**IVA:** *Benchmarks for Science Literacy:* Chapter 15 Research<ul><li>4E: Heat Transfer, p. 337; Energy Forms and Energy Transformation, p. 338</li></ul>**IVB:** *Making Sense of Secondary Science: Research Into Children's Ideas*<ul><li>Ch. 7: Nutrition and Energy Flow, pp. 59–60</li><li>Ch. 20: Energy and Living Things, pp. 143–144; Energy Stores, p. 144; Energy as a Fuel, pp. 145–146</li></ul>

Section and Outcome	Selected Sources and Readings for Study and Reflection Read and examine **related** parts of
	IVC: *Uncovering Student Ideas:* Related Research • USI.4: Digestive System, pp. 134–135; Where Does Oil Come From? p. 155 • USI.LS1: Food for Corn, pp. 72–73 **IVD:** *Disciplinary Core Ideas: Reshaping Teaching and Learning* • Ch. 4: PS3.D: Energy in Chemical Processes and Everyday Life: Common Prior Conceptions, p. 67
V. **K–12 Articulation and Connections**	**VA:** *Next Generation Science Standards:* Appendices: Progression • Appendix E: PS3.D: Energy in Chemical Processes and Everyday Life, p. 8 **VB:** *NSTA Quick Reference Guide to the NGSS K–12:* Progression • PS3.D: Energy in Chemical Processes and Everyday Life, p. 66 **VC:** *The NSTA Atlas of the Three Dimensions:* Map Page • 3.7: Energy in Chemical Processes and Everyday Life (PS3.D)
VI. **Assessment Expectation**	**VIA:** *State Standards* • Examine your state's standards **VIB:** *Next Generation Science Standards:* Performance Expectations • Grade 4: 4-PS3-4, p. 23 • Grade 5: 5-PS3-1, p. 29 • MS: MS-LS1-6, MS-LS1-7, p. 47 • HS: HS-PS3-3, HS-PS3-4, p. 74; HS-LS2-5, p. 81 **VIC:** *NSTA Quick Reference Guide to the NGSS K–12:* Performance Expectations • Grade 4: 4-PS3-4, p. 109 • Grade 5: 5-PS3-1, p. 113 • MS: MS-LS1-6, MS-LS1-7, p. 129 • HS: HS-PS3-3, HS-PS3-4, p. 151; HS-LS2-5, p. 157

Review Chapter 2 instructions on how to use this guide.

Visit curriculumtopicstudy2.org for more information about CTS and additional resources.

SEE ALSO: Crosscutting Concept Guide: Energy and Matter: Flows, Cycles, and Conservation.

NOTE: *Atlas* page numbers have not been provided because *The NSTA Atlas of the Three Dimensions* was produced concurrently with this edition. Titles and map codes are accurate.

Additional Readings:

 Available for download at **www.curriculumtopicstudy2.org**

Copyright © 2020 by Corwin Press, Inc. All rights reserved. Reprinted from *Science Curriculum Topic Study: Bridging the Gap Between Three-Dimensional Standards, Research, and Practice* (2nd ed.) by Page Keeley and Joyce Tugel. Thousand Oaks, CA: Corwin, www.corwin.com. Reproduction authorized for educational use by educators, local school sites, and/or noncommercial or nonprofit entities that have purchased the book.

Force and Motion
Grades K–12 Standards- and Research-Based Study of a Curricular Topic

Section and Outcome	Selected Sources and Readings for Study and Reflection Read and examine **related** parts of
I. **Content Knowledge**	**IA: *Science for All Americans*** • Ch. 4: Motion, pp. 52–55 • Ch. 4: Forces of Nature, pp. 55–57 **IB: *Science Matters*** (2009 edition) • Ch. 1: The Clockwork Universe, pp. 7–18 **IC: *Framework for K–12 Science Education:*** Narrative Section • Ch. 5: PS2.A: Forces and Motion, pp. 114–115 **ID: *Disciplinary Core Ideas: Reshaping Teaching and Learning*** • Ch. 3: PS2.A: Forces and Motion, pp. 34–40 **IE: *The NSTA Atlas of the Three Dimensions:*** Narrative Page • 3.3: Forces and Motion (PS2.A)
II. **Concepts, Core Ideas, or Practices**	**IIA: *Framework for K–12 Science Education:*** Grade Band Endpoints • Ch. 5: PS2.A: Forces and Motion, pp. 115–116 **IIB: *Next Generation Science Standards:*** Disciplinary Core Ideas Column • K: PS2.A: Forces and Motion, p. 5 • Grade 3: PS2.A: Forces and Motion, p. 18 • MS: PS2.A: Forces and Motion, p. 38 • HS: PS2.A: Forces and Motion, p. 72 **IIC: *NSTA Quick Reference Guide to the NGSS K–12:*** Disciplinary Core Ideas Column • K: PS2.A: Forces and Motion, p. 93 • Grade 3: PS2.A: Forces and Motion, pp. 105, 106 • MS: PS2.A: Forces and Motion, p. 124 • HS: PS2.A: Forces and Motion, p. 149
III. **Curriculum, Instruction, and Formative Assessment**	**IIIA: *Disciplinary Core Ideas: Reshaping Teaching and Learning*** • Ch. 3: By the End of Grade 2, PS2.A: Forces and Motion, p. 45 • Ch. 3: By the End of Grade 5, PS2.A: Forces and Motion, pp. 46–47 • Ch. 3: By the End of Grade 8, PS2.A: Forces and Motion, pp. 48–49 • Ch. 3: By the End of Grade 12, PS2.A: Forces and Motion, p. 51 **IIIB: *Uncovering Student Ideas:*** Assessment Probe and Suggestions for Instruction • USI.PS1: Talking About Forces, pp. 71, 73–74; Force and Motion Ideas, pp. 79, 82; Equal and Opposite, pp. 131, 133–134
IV. **Research on Commonly Held Ideas**	**IVA: *Benchmarks for Science Literacy:*** Chapter 15 Research • 4F: The Concept of Force, p. 339; Newton's Laws of Motion, pp. 339–340 • 4G: Forces of Nature, p. 340 **IVB: *Making Sense of Secondary Science: Research Into Children's Ideas*** • Ch. 21: Forces, pp. 148–153 • Ch. 22: Horizontal Motion, pp. 154–162 **IVC: *Uncovering Student Ideas:*** Related Research • USI.PS1: Talking About Forces, p. 73; Force and Motion Ideas, p. 81; Equal and Opposite, p. 133

Section and Outcome	Selected Sources and Readings for Study and Reflection Read and examine **related** parts of
V. K–12 Articulation and Connections	**VA:** *Next Generation Science Standards:* Appendices: Progression • Appendix E, PS2.A: Forces and Motion, p. 7 **VB:** *NSTA Quick Reference Guide to the NGSS K–12:* Progression • PS2.A: Forces and Motion, p. 63 **VC:** *The NSTA Atlas of the Three Dimensions:* Map Page • 3.3: Forces and Motion (PS2.A)
VI. Assessment Expectation	**VIA:** *State Standards* • Examine your state's standards **VIB:** *Next Generation Science Standards:* Performance Expectations • K: K-PS2-1, K-PS2-2, p. 5 • Grade 3: 3-PS2-1, 3-PS2-2, p. 18 • MS: MS-PS2-1, MS-PS2.2, p. 38 • HS: HS-PS2-1, HS-PS2-2, HS-PS2-3, p. 72 **VIC:** *NSTA Quick Reference Guide to the NGSS K–12:* Performance Expectations • K: K-PS2-1, K-PS2-2, p. 93 • Grade 3: 3-PS2-1, 3-PS2-2, pp. 105–106 • MS: MS-PS2-1, MS-PS2.2, p. 124 • HS: HS-PS2-1, HS-PS2-2, HS-PS2-3, p. 149

Review Chapter 2 instructions on how to use this guide.

Visit curriculumtopicstudy2.org for more information about CTS and additional resources.

NOTE: *Atlas* page numbers have not been provided because *The NSTA Atlas of the Three Dimensions* was produced concurrently with this edition. Titles and map codes are accurate.

Additional Readings:

Available for download at www.curriculumtopicstudy2.org

Copyright © 2020 by Corwin Press, Inc. All rights reserved. Reprinted from *Science Curriculum Topic Study: Bridging the Gap Between Three-Dimensional Standards, Research, and Practice* (2nd ed.) by Page Keeley and Joyce Tugel. Thousand Oaks, CA: Corwin, www.corwin.com. Reproduction authorized for educational use by educators, local school sites, and/or noncommercial or nonprofit entities that have purchased the book.

Forces Between Objects
Grades K–12 Standards- and Research-Based Study of a Curricular Topic

Section and Outcome	Selected Sources and Readings for Study and Reflection Read and examine **related** parts of
I. Content Knowledge	**IA:** *Science for All Americans* • Ch. 4: Motion, pp. 52–55 • Ch. 4: Forces of Nature, pp. 55–57 **IB:** *Science Matters* (2009 edition) • Ch. 1: Gravity, pp. 13–15 • Ch. 3: Electricity and Magnetism, Coulomb's Law, pp. 44–51 **IC:** *Framework for K–12 Science Education:* Narrative Section • Ch. 5: PS2.B: Types of Interactions, pp. 116–117 **ID:** *Disciplinary Core Ideas: Reshaping Teaching and Learning* • Ch. 3: PS2.B: Types of Interactions, pp. 40–43 **IE:** *The NSTA Atlas of the Three Dimensions:* Narrative Page • 3.4: Types of Interactions (PS2.B)
II. Concepts, Core Ideas, or Practices	**IIA:** *Framework for K–12 Science Education:* Grade Band Endpoints • Ch. 5: PS2.B: Types of Interactions, pp. 117–118 **IIB:** *Next Generation Science Standards:* Disciplinary Core Ideas Column • K: PS2.B: Types of Interactions, p. 5 • Grade 3: PS2.B: Types of Interactions, p. 18 • Grade 5: PS2.B: Types of Interactions, p. 31 • MS: PS2.B: Types of Interactions, p. 38 • HS: PS2.B: Types of Interactions, pp. 68, 72 **IIC:** *NSTA Quick Reference Guide to the NGSS K–12:* Disciplinary Core Ideas Column • K: PS2.B: Types of Interactions, p. 93 • Grade 3: PS2.B: Types of Interactions, p. 106 • Grade 5: PS2.B: Types of Interactions, p. 113 • MS: PS2.B: Types of Interactions, pp. 124–125 • HS: PS2.B: Types of Interactions, pp. 147, 149–150
III. Curriculum, Instruction, and Formative Assessment	**IIIA:** *Disciplinary Core Ideas: Reshaping Teaching and Learning* • Ch. 3: By the End of Grade 2, PS2.B, pp. 45–46 • Ch. 3: By the End of Grade 5, PS2.B, p. 47 • Ch. 3: By the End of Grade 2, PS2.B, pp. 49–50 • Ch. 3: By the End of Grade 2, PS2.B, pp. 51–53 **IIIB:** *Uncovering Student Ideas:* Assessment Probe and Suggestions for Instruction • USI.PS1: Does It Have to Touch? pp. 75, 77–78; Experiencing Gravity, p. 157, 160–161; The Tower Drop, p. 177, 179–180 • USI.PS2: Do the Objects Need to Touch? pp. 15, 17; What Happens When You Bring a Balloon Near a Wall? pp. 27, 29; Can Magnets Push or Pull Without Touching? pp. 111, 113; How Can You Represent a Magnetic Field? pp. 159, 161–162
IV. Research on Commonly Held Ideas	**IVA:** *Benchmarks for Science Literacy:* Chapter 15 Research • 4G: Forces of Nature, p. 340 **IVB:** *Making Sense of Secondary Science: Research into Children's Ideas* • Ch. 16: Magnetism and Gravity, p. 126 • Ch. 23: Gravity, pp. 163–167

Section and Outcome	Selected Sources and Readings for Study and Reflection Read and examine **related** parts of
	IVC: *Uncovering Student Ideas:* Related Research • USI.PS1: Does It Have to Touch? p. 77; Experiencing Gravity, p. 160; The Tower Drop, p. 179 • USI.PS2: Do the Objects Need to Touch? p. 15; What Happens When You Bring a Balloon Near a Wall? p, 29; Can Magnets Push or Pull Without Touching? pp. 112–113; How Can You Represent a Magnetic Field? p. 161
V. K–12 Articulation and Connections	**VA: *Next Generation Science Standards:*** Appendices: Progression • Appendix E: PS2.B: Types of Interactions, p. 7 **VB: *NSTA Quick Reference Guide to the NGSS K–12:*** Progression • PS2.B: Types of Interactions, p. 63 **VC: *The NSTA Atlas of the Three Dimensions:*** Map Page • 3.4: Types of Interactions (PS2.B)
VI. Assessment Expectation	**VIA: *State Standards*** • Examine your state's standards **VIB: *Next Generation Science Standards:*** Performance Expectations • K: K-PS2-1, p. 5 • Grade 3: 3-PS2-1, 3-PS2-3, 3-PS2-4, p. 18 • Grade 5: 5-PS2-1, p. 31 • MS: MS-PS2-3, MS-PS2.4, MS-PS2.5, p. 38 • HS: HS-PS1-3, p. 68; HS-PS2-2, HS-PS2-4, p. 72; HS-PS2-6, p. 68 **VIC: *NSTA Quick Reference Guide to the NGSS K–12:*** Performance Expectations • K: K-PS2-1, p. 93 • Grade 3: 3-PS2-1, 3-PS2-3, 3-PS2-4, pp. 105–106 • Grade 5: 5-PS2-1, p. 113 • MS: MS-PS2-3, MS-PS2.4, MS-PS2.5, pp. 124–125 • HS: HS-PS1-3, p. 147; HS-PS2-2, HS-PS2-4, p. 149, HS-PS2-6, p. 150

Review Chapter 2 instructions on how to use this guide.

Visit curriculumtopicstudy2.org for more information about CTS and additional resources.

NOTE: *Atlas* page numbers have not been provided because *The NSTA Atlas of the Three Dimensions* was produced concurrently with this edition. Titles and map codes are accurate.

Additional Readings:

 Available for download at www.curriculumtopicstudy2.org

Copyright © 2020 by Corwin Press, Inc. All rights reserved. Reprinted from *Science Curriculum Topic Study: Bridging the Gap Between Three-Dimensional Standards, Research, and Practice* (2nd ed.) by Page Keeley and Joyce Tugel. Thousand Oaks, CA: Corwin, www.corwin.com. Reproduction authorized for educational use by educators, local school sites, and/or noncommercial or nonprofit entities that have purchased the book.

Gravitational Force
Grades 3–12 Standards- and Research-Based Study of a Curricular Topic

Section and Outcome	Selected Sources and Readings for Study and Reflection Read and examine **related** parts of
I. Content Knowledge	**IA: *Science for All Americans*** • Ch. 4: Forces of Nature, pp. 55–57 **IB: *Science Matters*** (2009 edition) • Ch. 1: Gravity, pp. 13–15 **IC: *Framework for K–12 Science Education:*** Narrative Section • Ch. 5: PS2.B: Types of Interactions, pp. 116–117 • Ch. 5: Relationship Between Energy and Forces, pp. 126–127 **ID: *Disciplinary Core Ideas: Reshaping Teaching and Learning*** • Ch. 3: PS2.B: Types of Interactions, p. 42 **IE: *The NSTA Atlas of the Three Dimensions:*** Narrative Page • 3.4: Types of Interactions (PS2.B)
II. Concepts, Core Ideas, or Practices	**IIA: *Framework for K–12 Science Education:*** Grade Band Endpoints • Ch. 5: PS2.B: Types of Interactions, pp. 117–118 • Ch. 5: Relationship Between Energy and Forces, p. 127 **IIB: *Next Generation Science Standards:*** Disciplinary Core Ideas Column • Grade 3: PS2.A: Forces and Motion, p. 18 • Grade 5: PS2.B: Types of Interactions, p. 31 • MS: PS2.B: Types of Interactions, p. 38 • HS: PS2.B: Types of Interactions, p. 72 **IIC: *NSTA Quick Reference Guide to the NGSS K–12:*** Disciplinary Core Ideas Column • Grade 3: PS2.A: Forces and Motion, p. 106 • Grade 5: PS2.B: Types of Interactions, p. 113 • MS: PS2.B: Types of Interactions, p. 125 • HS: PS2.B: Types of Interactions, pp. 149–150
III. Curriculum, Instruction, and Formative Assessment	**IIIA: *Disciplinary Core Ideas: Reshaping Teaching and Learning*** • Ch. 3: By the End of Grade 5: PS2.B, p. 47 • Ch. 3: By the End of Grade 8: PS2.B, pp. 49–50 • Ch. 3: By the End of Grade 12: PS2.B, pp. 51–52 **IIIB: *Uncovering Student Ideas:*** Assessment Probe and Suggestions for Instruction • USI.1: Talking About Gravity, pp. 99, 104 • USI.PS1: Does It Have to Touch? pp. 75, 77–78; Experiencing Gravity, pp. 157, 160–161; Apple on the Ground, pp. 163, 166; The Tower Drop, pp. 177, 179–180
IV. Research on Commonly Held Ideas	**IVA: *Benchmarks for Science Literacy:*** Chapter 15 Research • 4G: Forces of Nature, p. 340 **IVB: *Making Sense of Secondary Science: Research Into Children's Ideas*** • Ch. 16: Magnetism and Gravity, p. 126 • Ch. 23: Gravity, pp. 163–167 **IVC: *Uncovering Student Ideas:*** Related Research • USI.1: Talking About Gravity, p. 103 • USI.PS1: Does It Have to Touch? p. 77; Experiencing Gravity, p. 160; Apple on the Ground, pp. 165–166; The Tower Drop, p. 179

Section and Outcome	Selected Sources and Readings for Study and Reflection Read and examine **related** parts of
V. **K–12 Articulation and Connections**	**VA:** *Next Generation Science Standards:* Appendices: Progression • Appendix E: PS2.B: Types of Interactions, p. 7 **VB:** *NSTA Quick Reference Guide to the NGSS K–12:* Progression • PS2.B: Types of Interactions, p. 63 **VC:** *The NSTA Atlas of the Three Dimensions:* Map Page • 3.4: Types of Interactions (PS2.B)
VI. **Assessment Expectation**	**VIA:** *State Standards* • Examine your state's standards **VIB:** *Next Generation Science Standards:* Performance Expectations • Grade 3: 3-PS2-3, 3-PS2-4, p. 18 • Grade 5: 5-PS2-1, p. 31 • MS: MS-PS2.4, MS-PS2.5, p. 38 • HS: HS-PS2-4, p. 72 **VIC:** *NSTA Quick Reference Guide to the NGSS K–12:* Performance Expectations • Grade 3: 3-PS2-3, 3-PS2-4, p. 106 • Grade 5: 5-PS2-1, p. 113 • MS: MS-PS2.4, MS-PS2.5, pp. 124–125 • HS: HS-PS2-4, p. 149

Review Chapter 2 instructions on how to use this guide.

Visit curriculumtopicstudy2.org for more information about CTS and additional resources.

NOTE: *Atlas* page numbers have not been provided because *The NSTA Atlas of the Three Dimensions* was produced concurrently with this edition. Titles and map codes are accurate.

Additional Readings:

Available for download at www.curriculumtopicstudy2.org

Copyright © 2020 by Corwin Press, Inc. All rights reserved. Reprinted from *Science Curriculum Topic Study: Bridging the Gap Between Three-Dimensional Standards, Research, and Practice* (2nd ed.) by Page Keeley and Joyce Tugel. Thousand Oaks, CA: Corwin, www.corwin.com. Reproduction authorized for educational use by educators, local school sites, and/or noncommercial or nonprofit entities that have purchased the book.

Kinetic and Potential Energy
Grades 6–12 Standards- and Research-Based Study of a Curricular Topic

Section and Outcome	Selected Sources and Readings for Study and Reflection Read and examine **related** parts of
I. Content Knowledge	**IA: *Science for All Americans*** • Ch. 4: Energy Transformation, pp. 49–52 **IB: *Science Matters*** (2009 edition) • Ch. 2: Potential Energy, p. 29; Kinetic Energy, pp. 29–31 **IC: *Framework for K–12 Science Education:*** Narrative Section • Ch. 5: PS3.A: Definitions of Energy, pp. 120–122 (stop at Grade Band Endpoints) **ID: *Disciplinary Core Ideas: Reshaping Teaching and Learning*** • Ch. 4: PS3: Defining Energy as a Scientific Idea, p. 60 (first column), p. 62 (pendulum example) **VC: *The NSTA Atlas of the Three Dimensions:*** Narrative Page • 3.5: Definitions of Energy (PS3.A)
II. Concepts, Core Ideas, or Practices	**IIA: *Framework for K–12 Science Education:*** Grade Band Endpoints • Ch. 5: PS3.A: Definitions of Energy, pp. 122–124 **IIB: *Next Generation Science Standards:*** Disciplinary Core Ideas Column • MS: PS3.A: Definitions of Energy, p. 40 • HS: PS3.A: Definitions of Energy, p. 74 **IIC: *NSTA Quick Reference Guide to the NGSS K–12:*** Disciplinary Core Ideas Column • MS: PS3.A: Definitions of Energy, pp. 125–126 • HS: PS3.A: Definitions of Energy, pp. 150–151
III. Curriculum, Instruction, and Formative Assessment	**IIIA: *Disciplinary Core Ideas: Reshaping Teaching and Learning*** • Ch. 4: By the End of Grade 5, PS3.A, p. 60 • Ch. 4: By the End of Grade 8, PS3.A, p. 60 • Ch. 4: By the End of Grade 12, PS3.A, p. 61 **IIIB: *Uncovering Student Ideas:*** Assessment Probe and Suggestions for Instruction • USI.PS1: Skate Park, pp. 19, 21; Gravity Rocks, pp. 171, 174
IV. Research on Commonly Held Ideas	**IVA: *Benchmarks for Science Literacy:*** Chapter 15 Research • 4E: Energy Forms and Energy Transformation, p. 338 **IVB: *Making Sense of Secondary Science: Research Into Children's Ideas*** • Ch. 19: Energy, Movement, and Force, pp. 144–145 **IVC: *Uncovering Student Ideas:*** Related Research • USI.PS1: Skate Park, p. 21; Gravity Rocks, pp. 173–174
V. K–12 Articulation and Connections	**VA: *Next Generation Science Standards:*** Appendices: Progression • Appendix E: PS3.A: Definitions of Energy, p. 7 **VB: *NSTA Quick Reference Guide to the NGSS K–12:*** Progression • PS3.A: Definitions of Energy, p. 64 **VC: *The NSTA Atlas of the Three Dimensions:*** Map Page • 3.5: Definitions of Energy (PS3.A)

Section and Outcome	Selected Sources and Readings for Study and Reflection Read and examine **related** parts of
VI. **Assessment** **Expectation**	**VIA:** *State Standards* • Examine your state's standards **VIB:** *Next Generation Science Standards:* Performance Expectations • MS: MS-PS3-1, MS-PS3-2, MS-PS3-3, MS-PS3-4, p. 40 • HS: HS-PS3-1, HS-PS3-2, HS-PS3-3, p. 74 **VIC:** *NSTA Quick Reference Guide to the NGSS K–12:* Performance Expectations • MS: MS-PS3-1, MS-PS3-2, MS-PS3-3, MS-PS3-4, pp. 125–126 • HS: HS-PS3-1, HS-PS3-2, HS-PS3-3, pp. 150–151

Review Chapter 2 instructions on how to use this guide.

Visit curriculumtopicstudy2.org for more information about CTS and additional resources.

NOTE: *Atlas* page numbers have not been provided because *The NSTA Atlas of the Three Dimensions* was produced concurrently with this edition. Titles and map codes are accurate.

Additional Readings:

Available for download at **www.curriculumtopicstudy2.org**

Copyright © 2020 by Corwin Press, Inc. All rights reserved. Reprinted from *Science Curriculum Topic Study: Bridging the Gap Between Three-Dimensional Standards, Research, and Practice* (2nd ed.) by Page Keeley and Joyce Tugel. Thousand Oaks, CA: Corwin, www.corwin.com. Reproduction authorized for educational use by educators, local school sites, and/or noncommercial or nonprofit entities that have purchased the book.

Magnetism
Grades 3–12 Standards- and Research-Based Study of a Curricular Topic

Section and Outcome	Selected Sources and Readings for Study and Reflection Read and examine **related** parts of
I. **Content Knowledge**	**IA: *Science for All Americans*** • Ch. 4: Forces of Nature, pp. 55–57 **IB: *Science Matters*** (2009 edition) • Ch. 3: Magnetism, pp. 48–54 **IC: *Framework for K–12 Science Education:*** Narrative Section • Ch. 5: PS2.B: Types of Interactions, pp. 116–117 **ID: *Disciplinary Core Ideas: Reshaping Teaching and Learning*** • Ch. 3: PS2.B: Types of Interactions, p. 42 **IE: *The NSTA Atlas of the Three Dimensions:*** Narrative Page • 3.4: Types of Interactions (PS2.B)
II. **Concepts, Core Ideas, or Practices**	**IIA: *Framework for K–12 Science Education:*** Grade Band Endpoints • Ch. 5: PS2.B: Types of Interactions, pp. 117–118 • Ch. 5: PS3.C: Relationship Between Energy and Forces, p. 127 **IIB: *Next Generation Science Standards:*** Disciplinary Core Ideas Column • Grade 3: PS2.B: Types of Interactions, p. 18 • MS: PS2.B: Types of Interactions, p. 38; PS3.C: Relationship Between Energy and Forces, p. 40 • HS: PS2.B: Types of Interactions, p. 72; PS3.C: Relationship Between Energy and Forces, p. 74 **IIC: *NSTA Quick Reference Guide to the NGSS K–12:*** Disciplinary Core Ideas Column • Grade 3: PS2.B: Types of Interactions, p. 106 • MS: PS2.B: Types of Interactions, p. 125; PS3.C: Relationship Between Energy and Forces, p. 125 • HS: PS2.B: Types of Interactions, pp. 149–150; PS3.C: Relationship Between Energy and Forces, p. 152
III. **Curriculum, Instruction, and Formative Assessment**	**IIIA: *Disciplinary Core Ideas: Reshaping Teaching and Learning*** • Ch. 3: By the End of Grade 5, PS2.B, p. 47 • Ch. 3: By the End of Grade 8, PS2.B, p. 49 • Ch. 3: By the End of Grade 12, PS2.B, p. 52; PS3.C, p. 67 **IIIB: *Uncovering Student Ideas:*** Assessment Probe and Suggestions for Instruction • USI.PS2: Section 3 in this book includes eighteen probes and suggestions for instruction on magnetism, pp. 111–181
IV. **Research on Commonly Held Ideas**	**IVA: *Benchmarks for Science Literacy:*** Chapter 15 Research • 4G: Forces of Nature, p. 340 **IVB: *Making Sense of Secondary Science: Research Into Children's Ideas*** • Ch. 16: Magnetism, pp. 126–127 **IVC: *Uncovering Student Ideas:*** Related Research • USI.PS2: Section 3 in this book includes eighteen probes and research summaries on magnetism, pp. 111–181

Section and Outcome	Selected Sources and Readings for Study and Reflection Read and examine **related** parts of
V. **K–12 Articulation and Connections**	**VA:** *Next Generation Science Standards:* Appendices: Progression • Appendix E: PS2.B: Types of Interactions, p. 7; PS3.C: Relationship Between Energy and Forces, p. 8 **VB:** *NSTA Quick Reference Guide to the NGSS K–12:* Progression • PS2.B: Types of Interactions, p. 63; PS3.C: Relationship Between Energy and Forces, p. 65 **VC:** *The NSTA Atlas of the Three Dimensions:* Map Page • 3.4: Types of Interactions (PS2.B)
VI. **Assessment Expectation**	**VIA:** *State Standards* • Examine your state's standards **VIB:** *Next Generation Science Standards:* Performance Expectations • Grade 3: 3-PS2-3, 3-PS2-4, p. 18 • MS: MS-PS2-3, MS-PS2.5, p. 38; MS-PS3-2, p. 40 • HS: HS-PS2-5, p. 72; HS-PS3-5, p. 74 **VIC:** *NSTA Quick Reference Guide to the NGSS K–12:* Performance Expectations • Grade 3: 3-PS2-3, 3-PS2-4, p. 106 • MS: MS-PS2-3, p. 124; MS-PS2.5, MS-PS3-2, p. 125 • HS: HS-PS2-5, p. 150 ; HS-PS3-5, p. 152

Review Chapter 2 instructions on how to use this guide.

Visit curriculumtopicstudy2.org for more information about CTS and additional resources.

NOTE: *Atlas* page numbers have not been provided because *The NSTA Atlas of the Three Dimensions* was produced concurrently with this edition. Titles and map codes are accurate.

Additional Readings:

Available for download at www.curriculumtopicstudy2.org

Copyright © 2020 by Corwin Press, Inc. All rights reserved. Reprinted from *Science Curriculum Topic Study: Bridging the Gap Between Three-Dimensional Standards, Research, and Practice* (2nd ed.) by Page Keeley and Joyce Tugel. Thousand Oaks, CA: Corwin, www.corwin.com. Reproduction authorized for educational use by educators, local school sites, and/or noncommercial or nonprofit entities that have purchased the book.

Nuclear Energy
Grades 9–12 Standards- and Research-Based Study of a Curricular Topic

Section and Outcome	Selected Sources and Readings for Study and Reflection Read and examine **related** parts of
I. Content Knowledge	**IA: *Science for All Americans*** • Ch. 4: Energy Transformation, p. 52 • Ch. 8: Energy Sources and Use, pp. 115–116 • Ch. 10: Splitting the Atom, pp. 155–157 **IB: *Science Matters*** (2009 edition) • Ch. 8: Nuclear Energy, pp. 139–145 **IC: *Framework for K–12 Science Education:*** Narrative Section • Ch. 2: PS1.C: Nuclear Processes, pp. 111–112 **VC: *The NSTA Atlas of the Three Dimensions:*** Narrative Page • 3.2: Chemical Reactions and Nuclear Processes (PS1.C) • 3.7: Energy in Chemical Processes and Everyday Life (PS3.D)
II. Concepts, Core Ideas, or Practices	**IIA: *Framework for K–12 Science Education:*** Grade Band Endpoints • Ch. 2: PS1.C: Nuclear Processes, p. 113 • Ch. 5: PS3.D: Energy in Chemical Processes and Everyday Life, Grade 12, p. 130 **IIB: *Next Generation Science Standards:*** Disciplinary Core Ideas Column • HS: PS1.C: Nuclear Processes, p. 68 **IIC: *NSTA Quick Reference Guide to the NGSS K–12:*** Disciplinary Core Ideas Column • HS: PS1.C: Nuclear Processes, p. 148
III. Curriculum, Instruction, and Formative Assessment	**IIIA: *Disciplinary Core Ideas: Reshaping Teaching and Learning*** • HS: Ch. 2: PS1.C: Nuclear Processes, p. 27 • HS: Ch. 4: PS3.D: Energy in Chemical Processes and Everyday Life, By the End of Grade 12, pp. 69–70
IV. Research on Commonly Held Ideas	Not available in the CTS resources
V. K–12 Articulation and Connections	**VA: *Next Generation Science Standards:*** Appendices: Progression • Appendix E: PS1.C: Nuclear Processes: No progression **VB: *NSTA Quick Reference Guide to the NGSS K–12:*** Progression • PS1.C: Nuclear Processes, p. 62 **VC: *The NSTA Atlas of the Three Dimensions:*** Map Page • 3.2: Chemical Reactions and Nuclear Processes (PS1.C) • 3.7: Energy in Chemical Processes and Everyday Life (PS3.D)

Section and Outcome	Selected Sources and Readings for Study and Reflection Read and examine **related** parts of
VI. **Assessment** **Expectation**	**VIA: *State Standards*** • Examine your state's standards **VIB: *Next Generation Science Standards:* Performance Expectations** • HS: Matter and Its Interactions, HS-PS1-8, p. 68 **VIC: *NSTA Quick Reference Guide to the NGSS K–12:* Performance Expectations** • HS: HS-PS1-8, p. 148

Review Chapter 2 instructions on how to use this guide.

Visit curriculumtopicstudy2.org for more information about CTS and additional resources.

NOTE: *Atlas* page numbers have not been provided because *The NSTA Atlas of the Three Dimensions* was produced concurrently with this edition. Titles and map codes are accurate.

Additional Readings:

Available for download at **www.curriculumtopicstudy2.org**

Copyright © 2020 by Corwin Press, Inc. All rights reserved. Reprinted from *Science Curriculum Topic Study: Bridging the Gap Between Three-Dimensional Standards, Research, and Practice* (2nd ed.) by Page Keeley and Joyce Tugel. Thousand Oaks, CA: Corwin, www.corwin.com. Reproduction authorized for educational use by educators, local school sites, and/or noncommercial or nonprofit entities that have purchased the book.

Relationship Between Energy and Forces
Grades K–12 Standards- and Research-Based Study of a Curricular Topic

Section and Outcome	Selected Sources and Readings for Study and Reflection Read and examine **related** parts of
I. Content Knowledge	**IA:** *Science for All Americans* • Ch. 4: Energy Transformation, pp. 49–52 • Ch. 4: Forces of Nature, pp. 55–57 **IB:** *Framework for K–12 Science Education:* Narrative Section • Ch. 5: PS3.C: Relationship Between Energy and Forces, pp. 126–127 **IC:** *Disciplinary Core Ideas: Reshaping Teaching and Learning* • Ch. 4: PS3.C: Forces and Energy Transfer Processes, p. 65 • Ch. 4: PS3.C: Potential Energy and Forces, pp. 65–66 **ID:** *The NSTA Atlas of the Three Dimensions:* Narrative Page • 3.6: Conservation of Energy and Energy Transfer (PS3.C)
II. Concepts, Core Ideas or Practices	**IIA:** *Framework for K–12 Science Education:* Grade Band Endpoints • Ch. 5, PS3.C: Relationship Between Energy and Forces, p. 127 **IIB:** *Next Generation Science Standards:* Disciplinary Core Ideas Column • K: PS3.C: Relationship Between Energy and Forces, p.5 • Gr. 4: PS3.C: Relationship Between Energy and Forces, p. 23 • MS: PS3.C: Relationship Between Energy and Forces, p. 40 • HS: PS3.C: Relationship Between Energy and Forces, p. 74 **IIC:** *NSTA Quick Reference Guide to the NGSS K-12:* Disciplinary Core Ideas Column • K: PS3.C: Relationship Between Energy and Forces, p. 93 • Gr. 4: PS3.C: Relationship Between Energy and Forces, p.109 • MS: PS3.C: Relationship Between Energy and Forces p. 125 • HS: PS3.C: Relationship Between Energy and Forces p. 152
III. Curriculum, Instruction, and Formative Assessment	**IIIA:** *Disciplinary Core Ideas: Reshaping Teaching and Learning* • By the End of Grade 5, p. 66 • By the End of Grade 8, p. 66 • By the End of Grade 12, p. 67 **IIIB:** *Uncovering Student Ideas:* Assessment Probe and Suggestions for Instruction • USI.PS1: Gravity Rocks, pp. 171, 174; Rolling to a Stop, pp. 91, 93–94
IV. Research on Commonly Held Ideas	**IVA:** *Benchmarks for Science Literacy:* Chapter 15 Research • 4E: Heat Transfer, p. 337; Energy Forms and Energy Transformation, p. 338 **IVB:** *Making Sense of Secondary Science- Research into Children's Ideas* • Ch. 19: Energy, Movement, and Force, pp. 144–145 **IVC:** *Uncovering Student Ideas:* Related Research • USI.PS1: Gravity Rocks, pp. 173–174; Rolling to a Stop, p. 93 **IVD:** *Disciplinary Core Ideas: Reshaping Teaching and Learning* • Ch. 4: PS3.C: Relationship Between Energy and Forces: Common Prior Conceptions, p. 64

Section and Outcome	Selected Sources and Readings for Study and Reflection Read and examine **related** parts of
V. **K–12** **Articulation** **and** **Connections**	**VA:** *Next Generation Science Standards:* Appendices: Progression • Appendix E: PS3.C: Relationship Between Energy and Forces p. 8 **VB:** *NSTA Quick Reference Guide to the NGSS K-12:* Progression • PS3.C: Relationship Between Energy and Forces, p. 65 **VC:** *The NSTA Atlas of the Three Dimensions:* Map Page • 3.6: Conservation of Energy and Energy Transfer (PS3.C)
VI. **Assessment** **Expectation**	**VIA:** *State Standards* • Examine your state's standards **VIB:** *Next Generation Science Standards:* Performance Expectations • K: K-PS2-1, p. 5 • Gr. 4: 4-PS3-3, p. 23 • MS: MS-PS3-2, p. 40 • HS: HS-PS3-5, p. 74 **VIC:** *NSTA Quick Reference Guide to the NGSS K-12:* Performance Expectations Column • K: K-PS2-1, p. 93 • Grade 4: 4-PS3-3, p.109 • MS: MS-PS3-2, p. 125 • HS: HS-PS3–5, p. 152

Review Chapter 2 instructions on how to use this guide.

Visit curriculumtopicstudy2.org for more information about CTS and additional resources.

NOTE: Atlas page numbers have not been provided because *The NSTA Atlas of the Three Dimensions* was produced concurrently with this edition. Titles and map codes are accurate.

Additional Readings:

Available for download at www.curriculumtopicstudy2.org

Copyright © 2020 by Corwin Press, Inc. All rights reserved. Reprinted from *Science Curriculum Topic Study: Bridging the Gap Between Three-Dimensional Standards, Research, and Practice* (2nd ed.) by Page Keeley and Joyce Tugel. Thousand Oaks, CA: Corwin, www.corwin.com. Reproduction authorized for educational use by educators, local school sites, and/or noncommercial or nonprofit entities that have purchased the book.

Sound
Grades K–12 Standards- and Research-Based Study of a Curricular Topic

Section and Outcome	Selected Sources and Readings for Study and Reflection Read and examine **related** parts of
I. **Content Knowledge**	**IA:** *Science for All Americans* • Ch. 4: Motion, pp. 52–55 (paras. 5–7) **IB:** *Science Matters* (2009 edition) • Ch. 2: Other Kinds of Energy, p. 30 **IC:** *Framework for K–12 Science Education:* Narrative Section • Ch. 5: PS4.A: Wave Properties, pp. 131–132 **ID:** *Disciplinary Core Ideas: Reshaping Teaching and Learning* • Ch. 5: PS4.A: Wave Properties, pp. 76–83 **IE:** *The NSTA Atlas of the Three Dimensions:* Narrative Page • 3.8: Wave Properties (PS4.A)
II. **Concepts, Core Ideas, or Practices**	**IIA:** *Framework for K–12 Science Education:* Grade Band Endpoints • Ch. 5: PS4.A: Wave Properties, pp. 132–133 **IIB:** *Next Generation Science Standards:* Disciplinary Core Ideas Column • Grade 1: PS.4.A: Wave Properties, p. 9 • Grade 4: PS.4.A: Wave Properties, p. 24 • MS: PS.4.A: Wave Properties, p. 42 • HS: PS.4.A: Wave Properties, pp. 76 **IIC:** *NSTA Quick Reference Guide to the NGSS K–12:* Disciplinary Core Ideas Column • Grade 1: PS.4.A: Wave Properties, p. 95 • Grade 4: PS.4.A: Wave Properties, p. 109 • MS: PS.4.A: Wave Properties, p. 127 • HS: PS.4.A: Wave Properties, p. 152
III. **Curriculum, Instruction, and Formative Assessment**	**IIIA:** *Disciplinary Core Ideas: Reshaping Teaching and Learning* • Ch. 4: By End of Grade 2, PS.4.A: p. 88; PS4.C: pp. 89–90 • Ch. 4: By End of Grade 5, PS.4.A: p. 90 • Ch. 4: By End of Grade 8, PS.4.A: p. 92 • Ch. 4: By End of Grade 12, PS.4.A: p. 83 **IIIB:** *Uncovering Student Ideas:* Assessment Probe and Suggestions for Instruction • USI.1: Making Sound, pp. 37, 42 • USI.K-2: Rubber Band Box, pp. 83, 85–86
IV. **Research on Commonly Held Ideas**	**IVA:** *Making Sense of Secondary Science: Research Into Children's Ideas* • Ch. 4: Hearing, pp. 45–46 • Ch. 18: Sound, pp. 133–137 **IVB:** *Uncovering Student Ideas:* Related Research • USI.1: Making Sound, p. 41 • USI.K-2: Rubber Band Box, p. 85

Section and Outcome	Selected Sources and Readings for Study and Reflection Read and examine **related** parts of
V. K–12 Articulation and Connections	**VA:** *Next Generation Science Standards:* Appendices: Progression • Appendix E: PS.4.A: Wave Properties, p. 8 **VB:** *NSTA Quick Reference Guide to the NGSS K–12:* Progression • PS.4.A: Wave Properties, p. 66 **VC:** *The NSTA Atlas of the Three Dimensions:* Map Page • 3.8: Wave Properties (PS4.A)
VI. Assessment Expectation	**VIA:** *State Standards* • Examine your state's standards **VIB:** *Next Generation Science Standards:* Performance Expectations • Grade 1: 1-PS4-1, p. 9 • Grade 4: 4-PS4-1, p. 24 • MS: MS-PS4-1, MS-PS4-2, p. 42 • HS: HS-PS4-1, HS-PS4-2, HS-PS4-5, p. 76 **VIC:** *NSTA Quick Reference Guide to the NGSS K–12:* Performance Expectations • Grade 1: 1-PS4-1, p. 95 • Grade 4: 4-PS4-1, p. 109 • MS: MS-PS4-1, MS-PS4-2, p. 127 • HS: HS-PS4-1, HS-PS4-2, HS-PS4-5, pp. 152–153

Review Chapter 2 instructions on how to use this guide.

Visit curriculumtopicstudy2.org for more information about CTS and additional resources.

NOTE: *Atlas* page numbers have not been provided because *The NSTA Atlas of the Three Dimensions* was produced concurrently with this edition. Titles and map codes are accurate.

Additional Readings:

Available for download at www.curriculumtopicstudy2.org

Copyright © 2020 by Corwin Press, Inc. All rights reserved. Reprinted from *Science Curriculum Topic Study: Bridging the Gap Between Three-Dimensional Standards, Research, and Practice* (2nd ed.) by Page Keeley and Joyce Tugel. Thousand Oaks, CA: Corwin, www.corwin.com. Reproduction authorized for educational use by educators, local school sites, and/or noncommercial or nonprofit entities that have purchased the book.

Transfer of Energy
Grades 3–12 Standards- and Research-Based Study of a Curricular Topic

Section and Outcome	Selected Sources and Readings for Study and Reflection Read and examine **related** parts of
I. Content Knowledge	**IA:** *Science for All Americans* • Ch. 4: Energy Transformation, pp. 49–52 **IB:** *Science Matters* (2009 edition) • Ch. 2: Good News: The First Law, pp. 31–33; Moving Heat, pp. 33–34; Bad News: The Second Law, pp. 36–41 **IC:** *Framework for K–12 Science Education:* Narrative Section • Ch. 5: PS3.B: Conservation of Energy and Energy Transfer, pp. 124–125 **ID:** *Disciplinary Core Ideas: Reshaping Teaching and Learning* • Ch. 4: PS3.B: Conservation of Energy and Energy Transfer, pp. 62–63 **IE:** *The NSTA Atlas of the Three Dimensions:* Narrative Page • 3.6: Conservation of Energy and Energy Transfer (PS3.B)
II. Concepts, Core Ideas, or Practices	**IIA:** *Framework for K–12 Science Education:* Grade Band Endpoints • Ch. 5: PS3.B: Conservation of Energy and Energy Transfer, pp. 125–126 **IIB:** *Next Generation Science Standards:* Disciplinary Core Ideas Column • Grade 4: PS3.B: Conservation of Energy and Energy Transfer, p. 23 • MS: PS3.B: Conservation of Energy and Energy Transfer, p. 40 • HS: PS3.B: Conservation of Energy and Energy Transfer, p. 74 **IIC:** *NSTA Quick Reference Guide to the NGSS K–12:* Disciplinary Core Ideas Column • Grade 4: PS3.B: Conservation of Energy and Energy Transfer, pp. 108–109 • MS: PS3.B: Conservation of Energy and Energy Transfer, p. 126 • HS: PS3.B: Conservation of Energy and Energy Transfer, pp. 150–151
III. Curriculum, Instruction, and Formative Assessment	**IIIA:** *Uncovering Student Ideas:* Assessment Probe and Suggestions for Instruction • USI.1: Mitten Problem, pp. 107, 112–113; Objects and Temperature, pp. 115, 120–121 • USI.2: Ice Cold Lemonade, pp. 77, 81–82; Mixing Water, pp. 83, 88–89 • USI.PS3: Hot Soup, pp. 195, 199–200; Cold Spoons, p. 201, 205; How Can I Keep It Cold? pp. 207, 211–212 **IIIB:** *Disciplinary Core Ideas: Reshaping Teaching and Learning* • Ch. 4: By the End of Grade 5, PS3.B, p. 64 • Ch. 4: By the End of Grade 8, PS3.B, p. 64 • Ch. 4: By the End of Grade 12, PS3.B, p. 64
IV. Research on Commonly Held Ideas	**IVA:** *Benchmarks for Science Literacy:* Chapter 15 Research • 4E: Heat Transfer, p. 337; Energy Forms and Energy Transformation, p. 338 **IVB:** *Making Sense of Secondary Science: Research Into Children's Ideas* • Ch. 19: Heat and Temperature, pp. 138–141; Energy Transfer Processes, pp. 141–142 **IVC:** *Uncovering Student Ideas:* Related Research • USI.1: Mitten Problem, p. 111; Objects and Temperature, p. 119 • USI.2: Ice Cold Lemonade, pp. 80–81; Mixing Water, pp. 87–88 • USI.PS3: Hot Soup, p. 199; Cold Spoons, pp. 201, 204–205; How Can I Keep It Cold? pp. 210–211

Section and Outcome	Selected Sources and Readings for Study and Reflection Read and examine **related** parts of
	IVD: *Disciplinary Core Ideas: Reshaping Teaching and Learning* • Ch. 4: PS3.B: Conservation of Energy and Energy Transfer: Common Prior Conceptions, pp. 61–62
V. **K–12 Articulation and Connections**	**VA:** *Next Generation Science Standards:* Appendices: Progression • Appendix E: PS3.B: Conservation of Energy and Energy Transfer, p. 7 **VB:** *NSTA Quick Reference Guide to the NGSS K–12:* Progression • PS3.B: Conservation of Energy and Energy Transfer, p. 65 **VC:** *The NSTA Atlas of the Three Dimensions:* Map Page • 3.6: Conservation of Energy and Energy Transfer (PS3.B & PS3.C)
VI. **Assessment Expectation**	**VIA:** *State Standards* • Examine your state's standards **VIB:** *Next Generation Science Standards:* Performance Expectations • Grade 4: 4-PS3-2, 4-PS3-3, 4-PS3-4, p. 23 • MS: MS-PS3-3, MS-PS3-4, MS-PS3-5, p. 40 • HS: HS-PS3-3, HS-PS3-4, p. 74 **VIC:** *NSTA Quick Reference Guide to the NGSS K–12:* Performance Expectations Column • Grade 4: 4-PS3-2, 4-PS3-3, 4-PS3-4, pp. 108–109 • MS: MS-PS3-3, MS-PS3-4, MS-PS3-5, p. 126 • HS: HS-PS3-1, HS-PS3-4, pp. 150–151

Review Chapter 2 instructions on how to use this guide.

Visit curriculumtopicstudy2.org for more information about CTS and additional resources.

NOTE: This study can be combined with the following crosscutting concept guide: Energy and Matter: Flows, Cycles, and Conservation

NOTE: *Atlas* page numbers have not been provided because *The NSTA Atlas of the Three Dimensions* was produced concurrently with this edition. Titles and map codes are accurate.

Additional Readings:

 Available for download at www.curriculumtopicstudy2.org

Copyright © 2020 by Corwin Press, Inc. All rights reserved. Reprinted from *Science Curriculum Topic Study: Bridging the Gap Between Three-Dimensional Standards, Research, and Practice* (2nd ed.) by Page Keeley and Joyce Tugel. Thousand Oaks, CA: Corwin, www.corwin.com. Reproduction authorized for educational use by educators, local school sites, and/or noncommercial or nonprofit entities that have purchased the book.

Visible Light and Electromagnetic Radiation
Grades K–12 Standards- and Research-Based Study of a Curricular Topic

Section and Outcome	Selected Sources and Readings for Study and Reflection Read and examine **related** parts of
I. Content Knowledge	**IA:** *Science for All Americans* • Ch. 4: Motion, pp. 52–55 (start at fifth paragraph) **IB:** *Science Matters* (2009 edition) • Ch. 3: Electromagnetic Radiation, pp. 55–65 **IC:** *Framework for K–12 Science Education:* Narrative Section • Ch. 5: PS4.B: Electromagnetic Radiation, pp. 133–134 **ID:** *Disciplinary Core Ideas: Reshaping Teaching and Learning* • Ch. 5: PS4.B: Electromagnetic (EM) Radiation, pp. 83–85 **IE:** *The NSTA Atlas of the Three Dimensions:* Narrative Page • 3.9: Electromagnetic Radiation and Information Technologies (PS4.B)
II. Concepts, Core Ideas, or Practices	**IIA:** *Framework for K–12 Science Education:* Grade Band Endpoints • Ch. 5: PS4.B: Electromagnetic Radiation, pp. 134–136 **IIB:** *Next Generation Science Standards:* Disciplinary Core Ideas Column • Grade 1: PS4.B: Electromagnetic Radiation, p. 9 • Grade 4: PS4.B: Electromagnetic Radiation, p. 25 • MS: PS4.B: Electromagnetic Radiation, p. 42 • HS: PS4.B: Electromagnetic Radiation, p. 76 **IIC:** *NSTA Quick Reference Guide to the NGSS K–12:* Disciplinary Core Ideas Column • Grade 1: PS4.B: Electromagnetic Radiation, p. 95 • Grade 4: PS4.B: Electromagnetic Radiation, p. 110 • MS: PS4.B: Electromagnetic Radiation, p. 127 • HS: PS4.B: Electromagnetic Radiation, pp. 152–153
III. Curriculum, Instruction, and Formative Assessment	**IIIA:** *Disciplinary Core Ideas: Reshaping Teaching and Learning* • Ch. 5: PS4.B: By the End of Grade 2, pp. 88–89 • Ch. 5: PS4.B: By the End of Grade 5, pp. 90–91 • Ch. 5: PS4.B: By the End of Grade 8, pp. 92–93 • Ch. 5: PS4.B: By the End of Grade 12, pp. 93–94 **IIIB:** *Uncovering Student Ideas:* Assessment Probe and Suggestions for Instruction • USI.1: Can It Reflect Light? pp. 17, 19–20, 22–23; Apple in the Dark, pp. 25, 29–30; Birthday Candles, pp. 31, 35–36
IV. Research on Commonly Held Ideas	**IVA:** *Benchmarks for Science Literacy:* Chapter 15 Research • 4F: Light, pp. 338–339 **IVB:** *Making Sense of Secondary Science: Research Into Children's Ideas* • Ch. 4: Vision, pp. 41–45 • Ch. 17: Light, pp. 128–132 **IVC:** *Uncovering Student Ideas:* Related Research • USI.1: Can It Reflect Light? pp. 20–21; Apple in the Dark, pp. 28–29; Birthday Candles, pp. 34–35

Section and Outcome	Selected Sources and Readings for Study and Reflection Read and examine **related** parts of
V. **K–12 Articulation and Connections**	**VA:** *Next Generation Science Standards:* Appendices: Progression • Appendix E: PS4.B: Electromagnetic Radiation, p. 8 **VB:** *NSTA Quick Reference Guide to the NGSS K–12:* Progression • PS4.B: Electromagnetic Radiation, p. 67 **VC:** *The NSTA Atlas of the Three Dimensions:* Map Page • 3.9: Electromagnetic Radiation and Information Technologies (PS4.B & PS4.C)
VI. **Assessment Expectation**	**VIA:** *State Standards* • Examine your state's standards **VIB:** *Next Generation Science Standards:* Performance Expectations • Grade 1: 1-PS4-2, 1-PS4-3, p. 9 • Grade 4: 4-PS4-2, p. 25 • MS: MS-PS4-2, p. 42 • HS: HS-PS4-3, HS-PS4-4, HS-PS4-5, p. 76 **VIC:** *NSTA Quick Reference Guide to the NGSS K–12:* Performance Expectations • Grade 1: 1-PS4-2, 1-PS4-3, p. 95 • Grade 4: 4-PS4-2, p. 110 • MS: MS-PS4-2, p. 127 • HS: HS-PS4-3, HS-PS4-4, HS-PS4-5, pp. 152–153

Review Chapter 2 instructions on how to use this guide.

Visit curriculumtopicstudy2.org for more information about CTS and additional resources.

NOTE: *Atlas* page numbers have not been provided because *The NSTA Atlas of the Three Dimensions* was produced concurrently with this edition. Titles and map codes are accurate.

Additional Readings:

Available for download at www.curriculumtopicstudy2.org

Copyright © 2020 by Corwin Press, Inc. All rights reserved. Reprinted from *Science Curriculum Topic Study: Bridging the Gap Between Three-Dimensional Standards, Research, and Practice* (2nd ed.) by Page Keeley and Joyce Tugel. Thousand Oaks, CA: Corwin, www.corwin.com. Reproduction authorized for educational use by educators, local school sites, and/or noncommercial or nonprofit entities that have purchased the book.

Waves and Information Technologies
Grades K–12 Standards- and Research-Based Study of a Curricular Topic

Section and Outcome	Selected Sources and Readings for Study and Reflection Read and examine **related** parts of
I. Content Knowledge	**IA:** *Science for All Americans* • Ch. 8: Communication, pp. 118–120; Information Processing, pp. 120–123 **IB:** *Science Matters* (2009 edition) • Ch. 3: The Wireless World, pp. 65–66 **IC:** *Framework for K–12 Science Education:* Narrative Section • Ch. 5: PS.4.C: Information Technologies and Instrumentation, pp. 136–137 **ID:** *Disciplinary Core Ideas: Reshaping Teaching and Learning* • Ch. 5: PS4.C: Information Technologies and Instrumentation, pp. 85–87 **IE:** *The NSTA Atlas of the Three Dimensions:* Narrative Page • 3.9: Electromagnetic Radiation and Information Technologies (PS4.C)
II. Concepts, Core Ideas, or Practices	**IIA:** *Framework for K–12 Science Education:* Grade Band Endpoints • Ch. 5: PS.4.C: Information Technologies and Instrumentation, p. 137 **IIB:** *Next Generation Science Standards:* Disciplinary Core Ideas Column • Grade 1: PS.4.C: Information Technologies and Instrumentation, p. 9 • Grade 4: PS.4.C: Information Technologies and Instrumentation, p. 24 • MS: PS.4.C: Information Technologies and Instrumentation, p. 42 • HS: PS.4.C: Information Technologies and Instrumentation, p. 76 **IIC:** *NSTA Quick Reference Guide to the NGSS K–12:* Disciplinary Core Ideas Column • Grade 1: PS.4.C: Information Technologies and Instrumentation, p. 95 • Grade 4: PS.4.C: Information Technologies and Instrumentation, p. 110 • MS: PS.4.C: Information Technologies and Instrumentation, p. 127 • HS: PS.4.C: Information Technologies and Instrumentation, p. 153
III. Curriculum, Instruction, and Formative Assessment	**IIIA:** *Disciplinary Core Ideas: Reshaping Teaching and Learning* • Ch. 5: PS.4.C: By the End of Grade 2, pp. 89–90 • Ch. 5: PS.4.C: By the End of Grade 5, pp. 91–92 • Ch. 5: PS.4.C: By the End of Grade 8, p. 93 • Ch. 5: PS.4.C: By the End of Grade 12, p. 94
IV. Research on Commonly Held Ideas	No research available for this section
V. K–12 Articulation and Connections	**VA:** *Next Generation Science Standards:* Appendices: Progression • Appendix E: PS.4.C: Information Technologies and Instrumentation, p. 8 **VB:** *NSTA Quick Reference Guide to the NGSS K–12:* Progression • PS.4.C: Information Technologies and Instrumentation, p. 67 **VC:** *The NSTA Atlas of the Three Dimensions:* Map Page • 3.9: Electromagnetic Radiation and Information Technologies (PS4B & PS4.C)

Section and Outcome	Selected Sources and Readings for Study and Reflection Read and examine **related** parts of
VI. **Assessment Expectation**	**VIA:** *State Standards* • Examine your state's standards **VIB:** *Next Generation Science Standards:* Performance Expectations • Grade 1: 1-PS4-4, p. 9 • Grade 4: 4-PS4-3, p. 24 • MS: MS-PS4-3, p. 42 • HS: HS-PS4-5, p. 76 **VIC:** *NSTA Quick Reference Guide to the NGSS K–12:* Performance Expectations • Grade 1: 1-PS4-4, p. 95 • Grade 4: 4-PS4-3, p. 110 • MS: MS-PS4-3, p. 127 • HS: HS-PS4-5, p. 153

Review Chapter 2 instructions on how to use this guide.

Visit curriculumtopicstudy2.org for more information about CTS and additional resources.

NOTE: *Atlas* page numbers have not been provided because *The NSTA Atlas of the Three Dimensions* was produced concurrently with this edition. Titles and map codes are accurate.

Additional Readings:

Available for download at www.curriculumtopicstudy2.org

Copyright © 2020 by Corwin Press, Inc. All rights reserved. Reprinted from *Science Curriculum Topic Study: Bridging the Gap Between Three-Dimensional Standards, Research, and Practice* (2nd ed.) by Page Keeley and Joyce Tugel. Thousand Oaks, CA: Corwin, www.corwin.com. Reproduction authorized for educational use by educators, local school sites, and/or noncommercial or nonprofit entities that have purchased the book.

Waves and Wave Properties
Grades K–12 Standards- and Research-Based Study of a Curricular Topic

Section and Outcome	Selected Sources and Readings for Study and Reflection Read and examine **related** parts of
I. Content Knowledge	**IA:** *Science for All Americans* • Ch. 4: Motion, pp. 52–55 (start at fifth paragraph) **IB:** *Science Matters* (2009 edition) • Ch. 3: Characteristics of Waves, p. 56 (second paragraph) **IC:** *Framework for K–12 Science Education:* Narrative Section • Ch. 5: PS4.A: Wave Properties, pp. 131–132 (stop at Grade Band Endpoints) **ID:** *Disciplinary Core Ideas: Reshaping Teaching and Learning* • Ch. 5: PS4.A: Wave Properties, pp. 76–83 **IE:** *The NSTA Atlas of the Three Dimensions:* Narrative Page • 3.8: Wave Properties (PS4.A)
II. Concepts, Core Ideas, or Practices	**IIA:** *Framework for K–12 Science Education:* Grade Band Endpoints • Ch. 5: PS4.A: Wave Properties, pp. 132–133 **IIB:** *Next Generation Science Standards:* Disciplinary Core Ideas Column • Grade 1: PS.4.A: Wave Properties, p. 9 • Grade 4: PS.4.A: Wave Properties, p. 24 • MS: PS.4.A: Wave Properties, p. 42 • HS: PS.4.A: Wave Properties, p. 76 **IIC:** *NSTA Quick Reference Guide to the NGSS K–12:* Disciplinary Core Ideas Column • Grade 1: PS.4.A: Wave Properties, p. 95 • Grade 4: PS.4.A: Wave Properties, p. 109 • MS: PS.4.A: Wave Properties, p. 127 • HS: PS.4.A: Wave Properties, pp. 152–153
III. Curriculum, Instruction, and Formative Assessment	**IIIA:** *Uncovering Student Ideas:* Assessment Probe and Suggestions for Instruction • USI.1: Making Sound, pp. 37, 42 • USI.K–2: Do the Waves Move the Boat? pp. 75, 78; Rubber Band Box, pp. 83, 85–86 **IIIB:** *Disciplinary Core Ideas: Reshaping Teaching and Learning* • Ch. 5: PS4.A: By the End of Grade 2, pp. 87–88 • Ch. 5: PS4.A: By the End of Grade 5, p. 90 • Ch. 5: PS4.A: By the End of Grade 8, p. 92 • Ch. 5: PS4.A: By the End of Grade 12, p. 93
IV. Research on Commonly Held Ideas	**IVA:** *Making Sense of Secondary Science: Research Into Children's Ideas* • Ch. 18: Sound Transmission, pp. 135–136 **IVB:** *Uncovering Student Ideas:* Related Research • USI.1: Making Sound, p. 41 • USI.K–2: Do the Waves Move the Boat? pp. 77–78; Rubber Band Box, p. 85

Section and Outcome	Selected Sources and Readings for Study and Reflection Read and examine **related** parts of
V. **K–12** **Articulation** **and** **Connections**	**VA:** ***Next Generation Science Standards:*** Appendices: Progression • Appendix E: PS.4.A: Wave Properties, p. 8 **VB:** ***NSTA Quick Reference Guide to the NGSS K–12:*** Progression • PS.4.A: Wave Properties, p. 66 **VC:** ***The NSTA Atlas of the Three Dimensions:*** Map Page • 3.8: Wave Properties (PS4.A)
VI. **Assessment** **Expectation**	**VIA:** ***State Standards*** • Examine your state's standards **VIB:** ***Next Generation Science Standards:*** Performance Expectations • Grade 1: 1-PS4-1, p. 9 • Grade 4: 4-PS4-1, p. 24 • MS: MS-PS4-1, MS-PS4-2, p. 42 • HS: HS-PS4-1, HS-PS4-2, HS-PS4-3, HS-PS4-5, p. 76 **VIC:** ***NSTA Quick Reference Guide to the NGSS K–12:*** Performance Expectations • Grade 1: 1-PS4-1, p. 95 • Grade 4: 4-PS4-1, p. 109 • MS: MS-PS4-1, MS-PS4-2, p. 127 • HS: HS-PS4-1, HS-PS4-2, HS-PS4-3, HS-PS4-5, pp. 152–153

Review Chapter 2 instructions on how to use this guide.

Visit curriculumtopicstudy2.org for more information about CTS and additional resources.

NOTE: *Atlas* page numbers have not been provided because *The NSTA Atlas of the Three Dimensions* was produced concurrently with this edition. Titles and map codes are accurate.

Additional Readings:

Available for download at www.curriculumtopicstudy2.org

Copyright © 2020 by Corwin Press, Inc. All rights reserved. Reprinted from *Science Curriculum Topic Study: Bridging the Gap Between Three-Dimensional Standards, Research, and Practice* (2nd ed.) by Page Keeley and Joyce Tugel. Thousand Oaks, CA: Corwin, www.corwin.com. Reproduction authorized for educational use by educators, local school sites, and/or noncommercial or nonprofit entities that have purchased the book.

Category C: Earth and Space Science Guides

The twenty-one CTS guides in this section are divided into three subsections. The guides focus on materials, structures, processes, and systems on Earth and beyond, ranging in scale from local areas on Earth to the vast universe and from observable processes to processes that take billions of years.

Earth Structure, Materials, and Systems Guides focus on the small and large-scale composition of Earth and the systems, including human-designed, that drive Earth's conditions. The alphabetically arranged guides in this section include

- Earth's Materials and Systems
- Earth's Natural Resources
- Global Climate Change
- Human Impact on Earth Systems
- Structure of the Solid Earth
- Water Cycle and Distribution
- Water in the Earth System
- Weather and Climate

Earth History and Processes That Change the Earth Guides focus on Earth's continual evolution on both small and large-time scales. The alphabetically arranged guides in this section include

- Biogeology
- Earthquakes and Volcanoes
- Earth's History
- Natural Hazards
- Plate Tectonics
- Weathering, Erosion, and Deposition

Earth in Space, Solar System, and the Universe Guides focus on Earth and its place in the Universe, at varying scales of time and space. The alphabetically arranged guides in this section include

- Earth and Our Solar System
- Earth, Moon, Sun System
- Earth-Sun System
- Formation of the Earth, Solar System, and the Universe
- Phases of the Moon

- Seasons and Seasonal Patterns in the Sky
- Stars and Galaxies

NOTES FOR USING CATEGORY C GUIDES

Overall

- *Atlas* page numbers have not been provided because *The NSTA Atlas of the Three Dimensions* was produced concurrently with this edition. Titles and map codes are accurate.

- The same resources are not always included for each section. For example, some guides include readings from *Science Matters* for section I, others do not.

- Eliminate redundancy. Some readings include the exact same information. However, even when the information is the same, there may be an advantage in how the information is presented. Select the reading based on the resources you have available to use for CTS and/or the advantage of using one over the other.

Section I

- Readings from the *Framework* and *Atlas* narrative are exactly the same. Choose one of these resources for this section.

- When reading the *Framework*, stop at Grade Band Endpoints.

- Some *Atlas* maps combine topics. When using the *Atlas* narrative for this section, if there is more than one core idea included in the narrative, focus on the one that is listed on your CTS guide.

- When reading *Disciplinary Core Ideas* for this section, focus on the content that helps you understand this topic. There are also suggestions for instruction embedded in this reading that can be added to CTS section III.

Section II

- Readings from the *Framework, NGSS,* and *NSTA Quick Reference Guide* are practically the same. In a few cases, a *Framework* idea was not included in the *NGSS* or was moved to a different grade span. In the *Framework*, goals are described in grade bands K–2, 3–5, 6–8, and 9–12. In the *NGSS* and *NSTA Quick Reference Guide*, they are phrased as disciplinary core ideas and designate a specific grade.

- The readings from the *NGSS* and the *NSTA Quick Reference Guide* are exactly the same. Choose one of these resources. An advantage to using the *NSTA Quick Reference Guide* is that the disciplinary core idea is matched to the performance expectation (section VI) on the chart.

Section III

- Readings listed for *Disciplinary Core Ideas* are the longest readings. In some guides these have been broken down into subsections. You can read all the designated pages or focus on the subsections that are of interest to you. If doing CTS with a group, you may consider assigning subsections so one person does not have to read all the designated pages.

- Sometimes readings about instructional implications from *Disciplinary Core Ideas* may mention and cite a research study on students' commonly held ideas or difficulties. This can be combined with section IV.

- Several readings may be listed from the *Uncovering Student Ideas in Science* series. This does not mean you need to read them all. Choose ones from the books you have available. The page numbers list the probe that can be used to elicit students' ideas, and the teacher notes contain suggestions for instruction that are designed to address the ideas elicited by the probe. Some of these books also have a section on curricular considerations. This can be added to your reading.

- There are always new books released in the *Uncovering Student Ideas in Science* series that may include probes that are not listed on the topic guides. Check the CTS website for a list of new probes published after 2019 that can be used with CTS.

Section IV

- There are fewer available research studies on commonly held ideas for earth science in the resources used for this section. If you have access to professional journals that include research on students' commonly held ideas, you can add articles to the additional readings listed at the end of each guide.

- If you are using the research summaries from the *Uncovering Student Ideas* series, they usually include references to the commonly held ideas identified in *Benchmarks* Chapter 15 Research and *Making Sense of Secondary Science* so there is no need to use all three of the resources listed. In addition, the *Uncovering Student Ideas* series includes research published after *Benchmarks* and *Making Sense of Secondary science*.

Section V

- Readings from the *NGSS* Appendix E and the *NSTA Quick Reference Guide* both describe a progression. The difference is the progression is summarized for each grade span in the *NGSS* Appendix E. In the *NSTA Quick Reference Guide*, they are listed by the disciplinary core ideas along with the code for the performance expectation. You can choose one or both of these resources, depending on the level of specificity you desire.

- The *Atlas* includes the same information as the *NSTA Quick Reference Guide* but the visual mapping of the ideas allows you to see precursor ideas and connections, as well as connections to other ideas. Be sure to read the front matter of the *Atlas* before using the maps. The information will help you use the maps effectively.

- The *Atlas* maps often combine two or more topics. Follow the connections that match the topic you are studying.

Section VI

- The *NGSS* and the *NSTA Quick Reference Guide* provide the exact same information. Clarifications and assessment boundaries are also included in both. Choose one of these resources based on the advantages listed below or the resources you have available.

- An advantage to using the *NSTA Quick Reference Guide* is that the performance expectation is listed next to the disciplinary core idea included in that performance expectation.

- An advantage to using the *NGSS* that is not evident in the *NSTA Quick Reference Guide* is that the scientific and engineering practice and crosscutting concept chart, included below the performance expectations, shows the other two dimensions that are part of the performance expectation.

Earth's Materials and Systems

Grades K–12 Standards- and Research-Based Study of a Curricular Topic

Section and Outcome	Selected Sources and Readings for Study and Reflection Read and examine **related** parts of
I. Content Knowledge	**IA:** *Science for All Americans* • Ch. 4: The Earth, pp. 42–44; Processes That Shape the Earth, pp. 44–46 **IB:** *Science Matters* (2009 edition) • Ch. 13: A Window Into the Solid Earth, pp. 228–230 • Ch. 14: Earth Cycles, pp. 233–250 **IC:** *Framework for K–12 Science Education:* Narrative Section • Ch. 7: Core Idea ESS2: Earth's Systems, p. 179 • Ch. 7: ESS2.A: Earth's Materials and Systems, pp. 179–180 **ID:** *Disciplinary Core Ideas: Reshaping Teaching and Learning* • Ch. 11: What Is This Disciplinary Core Idea and Why Is It Important? pp. 205–206 • ESS2.A: Earth's Materials and Systems, pp. 206–207 **IE:** *The NSTA Atlas of the Three Dimensions:* Narrative Page • 5.2: Earth's Systems (ESS2.A)
II. Concepts, Core Ideas, or Practices	**IIA:** *Framework for K–12 Science Education:* Grade Band Endpoints • Ch. 7: ESS2.A: Earth's Materials and Systems, pp. 180–182 **IIB:** *Next Generation Science Standards:* Disciplinary Core Ideas Column • Grade 2: ESS2.A: Earth's Materials and Systems, p. 15 • Grade 4: ESS2.A: Earth's Materials and Systems, p. 26 • Grade 5: ESS2.A: Earth's Materials and Systems, p. 30 • MS: ESS2.A: Earth's Materials and Systems, p. 58 • HS: ESS2.A: Earth's Materials and Systems, pp. 96, 98 **IIC:** *NSTA Quick Reference Guide to the NGSS K–12:* Disciplinary Core Ideas • Grade 2: ESS2.A: Earth's Materials and Systems, p. 96 • Grade 4: ESS2.A: Earth's Materials and Systems, p. 107 • Grade 5: ESS2.A: Earth's Materials and Systems, p. 112 • MS: ESS2.A: Earth's Materials and Systems, p. 134 • HS: ESS2.A: Earth's Materials and Systems, pp. 163–164
III. Curriculum, Instruction, and Formative Assessment	**IIIA:** *Disciplinary Core Ideas: Reshaping Teaching and Learning* • Ch. 11: Lower Elementary, p. 215 • Ch. 11: Upper Elementary, p. 216 • Ch. 11: Middle School, pp. 217–218 • Ch. 11: High School, pp. 218–220 • Ch. 11: What Approaches Can We Use to Teach About This Disciplinary Core Idea, pp. 220–222 **IIIB:** *Uncovering Student Ideas:* Assessment Probe and Suggestions for Instruction • USI.E&ES: What Do You Know About Soil? pp. 19, 22; Land or Water? p. 25, 27; Does the Ocean Influence Our Weather or Climate? pp. 65, 68; Can a Plant Break Rocks? pp. 103, 105–106; Grand Canyon, pp. 107, 110; What Is the Inside of the Earth Like? pp. 121, 123–124
IV. Research on Commonly Held Ideas	**IVA:** *Benchmarks for Science Literacy:* Chapter 15 Research • 4C: Processes That Shape the Earth, p. 336 **IVB:** *Making Sense of Secondary Science: Research Into Children's Ideas* • Ch. 12: Water Cycle, pp. 101–102

Section and Outcome	Selected Sources and Readings for Study and Reflection Read and examine **related** parts of
	• Ch. 13: Wind, p. 111 • Ch. 14: Rocks, pp. 112–114; Soil, p. 114 **IVC: *Uncovering Student Ideas:*** Related Research • USI.E&ES: What Do You Know About Soil? pp. 21–22; Land or Water? p. 27; Does the Ocean Influence Our Climate or Weather? p. 67; Can a Plant Break Rocks? p. 105; Grand Canyon, pp. 109–110; What Is the Inside of the Earth Like? p. 123 **IVD: *Disciplinary Core Ideas:*** Reshaping Teaching and Learning • ESS2.A: Earth Materials and Systems, p. 207 (second column)
V. **K–12 Articulation and Connections**	**VA: *Next Generation Science Standards:*** Appendices: Progression • Appendix E: ESS2.A: Earth's Materials and Systems, p. 2 **VB: *NSTA Quick Reference Guide to the NGSS K–12:*** Progression • ESS2.A: Earth's Materials and Systems, p. 76 **VC: *The NSTA Atlas of the Three Dimensions:*** Map Page • 5.2: Earth's Systems (ESS2.A & ESS2.E)
VI. **Assessment Expectation**	**VIA: *State Standards*** • Examine your state's standards **VIB: *Next Generation Science Standards:*** Performance Expectations • Grade 2: 2-ESS2-1, p. 15 • Grade 4: 4-ESS2-1, p. 26 • Grade 5: 5-ESS2-1, p. 30 • MS: MS-ESS2-1, p. 58 • HS: ESS2-2, HS-ESS2-3, p. 96; HS-ESS2-4, p. 98 **VIC: *NSTA Quick Reference Guide to the NGSS K–12:*** Performance Expectations Column • Grade 2: 2-ESS2-1, p. 96 • Grade 4: 4-ESS2-1, p. 107 • Grade 5: 5-ESS2-1, p. 112 • MS: MS-ESS2-1; MS-ESS2-2, p. 134 • HS: ESS2-2, HS-ESS2-3, HS-ESS2-4, pp. 163–164

Review Chapter 2 instructions on how to use this guide.
Visit curriculumtopicstudy2.org for more information about CTS and additional resources.

NOTE: *Atlas* page numbers have not been provided because *The NSTA Atlas of the Three Dimensions* was produced concurrently with this edition. Titles and map codes are accurate.

Additional Readings:

 Available for download at www.curriculumtopicstudy2.org

Copyright © 2020 by Corwin Press, Inc. All rights reserved. Reprinted from *Science Curriculum Topic Study: Bridging the Gap Between Three-Dimensional Standards, Research, and Practice* (2nd ed.) by Page Keeley and Joyce Tugel. Thousand Oaks, CA: Corwin, www.corwin.com. Reproduction authorized for educational use by educators, local school sites, and/or noncommercial or nonprofit entities that have purchased the book.

Earth's Natural Resources
Grades K–12 Standards- and Research-Based Study of a Curricular Topic

Section and Outcome	Selected Sources and Readings for Study and Reflection Read and examine **related** parts of
I. Content Knowledge	**IA:** *Science for All Americans* • Ch. 4: The Earth, p. 44 (eighth paragraph) • Ch. 8: Energy Sources, pp. 114–115 **IB:** *Science Matters* (2009 edition) • Ch. 13: Searching for Buried Treasure, pp. 230–231 **IC:** *Framework for K–12 Science Education:* Narrative Section • Ch. 7: ESS3.A: Natural Resources, pp. 191–192 **ID:** *Disciplinary Core Ideas: Reshaping Teaching and Learning* • Ch. 12: What Is This Disciplinary Core Idea and Why Is It Important? p. 225 • Ch. 12: ESS3.A: Natural Resources, p. 227 **IE:** *The NSTA Atlas of the Three Dimensions:* Narrative Page • 5.5: Natural Resources and Natural Hazards (ESS3.A & ESS3.B)
II. Concepts, Core Ideas, or Practices	**IIA:** *Framework for K–12 Science Education:* Grade Band Endpoints • Ch. 7: ESS3.A: Natural Resources, p. 192 **IIB:** *Next Generation Science Standards:* Disciplinary Core Ideas Column • K: ESS3.A: Natural Resources, p. 6 • Grade 4: ESS3.A: Natural Resources, p. 23 • MS: ESS3.A: Natural Resources, p. 58 • HS: ESS3.A: Natural Resources, p. 99 **IIC:** *NSTA Quick Reference Guide to the NGSS K–12:* Disciplinary Core Ideas Column • K: ESS3.A: Natural Resources, p. 92 • Grade 4: ESS3.A: Natural Resources, p. 108 • MS: ESS3.A: Natural Resources, p. 136 • HS: ESS3.A: Natural Resources, p. 165
III. Curriculum, Instruction, and Formative Assessment	**IIIA:** *Disciplinary Core Ideas: Reshaping Teaching and Learning* • Ch. 12: K–2: Lower Elementary, p. 230 (first paragraph) • Ch. 12: 3–5: Upper Elementary, p. 231 • Ch. 12: 6–8: Middle School, p. 231 (first paragraph) • Ch. 12: 9–12: High School, pp. 231–232 **IIIB:** *Uncovering Student Ideas:* Assessment Probe and Suggestions for Instruction • USI.2: Where Does Oil Come From? pp. 151, 155–156 • USI.E&ES: Renewable or Nonrenewable? pp. 145, 148
IV. Research on Commonly Held Ideas	**IVA:** *Disciplinary Core Ideas: Reshaping Teaching and Learning* • ESS3.A: Natural Resources, p. 227 (second paragraph) **IVB:** *Uncovering Student Ideas:* Related Research • USI.2: Where Does Oil Come From? p. 155 • USI.E&ES: Renewable or Nonrenewable? p. 148

Section and Outcome	Selected Sources and Readings for Study and Reflection Read and examine **related** parts of
V. **K–12** **Articulation** **and** **Connections**	**VA:** *Next Generation Science Standards:* Appendices: Progression • Appendix E: ESS3.A: Natural Resources, p. 3 **VB:** *NSTA Quick Reference Guide to the NGSS K–12:* Progression • ESS3.A: Natural Resources, p. 78 **VC:** *The NSTA Atlas of the Three Dimensions:* Map Page • 5.5: Natural Resources and Natural Hazards (ESS3.A & ESS3.B)
VI. **Assessment** **Expectation**	**VIA:** *State Standards* • Examine your state's standards **VIB:** *Next Generation Science Standards:* Performance Expectations • K: K-ESS3-1, p. 6 • Grade 4: 4-ESS3-1, p. 23 • MS: MS-ESS3-1, p. 58 • HS: HS-ESS3-1, HS-ESS3-2, p. 99 **VIC:** *NSTA Quick Reference Guide to the NGSS K–12:* Performance Expectations • K: K-ESS3-1, p. 92 • Grade 4: 4-ESS3-1, p. 108 • MS: MS-ESS3-1, p. 136 • HS: HS-ESS3-1, HS-ESS3-2, p. 165

Review Chapter 2 instructions on how to use this guide.

Visit curriculumtopicstudy2.org for more information about CTS and additional resources.

NOTE: *Atlas* page numbers have not been provided because *The NSTA Atlas of the Three Dimensions* was produced concurrently with this edition. Titles and map codes are accurate.

Additional Readings:

Available for download at www.curriculumtopicstudy2.org

Copyright © 2020 by Corwin Press, Inc. All rights reserved. Reprinted from *Science Curriculum Topic Study: Bridging the Gap Between Three-Dimensional Standards, Research, and Practice* (2nd ed.) by Page Keeley and Joyce Tugel. Thousand Oaks, CA: Corwin, www.corwin.com. Reproduction authorized for educational use by educators, local school sites, and/or noncommercial or nonprofit entities that have purchased the book.

Global Climate Change
Grades 6–12 Standards- and Research-Based Study of a Curricular Topic

Section and Outcome	Selected Sources and Readings for Study and Reflection Read and examine **related** parts of
I. Content Knowledge	**IA:** *Framework for K–12 Science Education:* Narrative Section • Ch. 7: ESS3.D: Global Climate Change, pp. 196–198 **IB:** *Science Matters* (2009 Edition) • Ch. 19: The Greenhouse Effect and Global Climate Change, pp. 338–343 **IC:** *Disciplinary Core Ideas: Reshaping Teaching and Learning* • Ch. 12: What Is This Disciplinary Core Idea and Why Is It Important? p. 226 • Ch. 12: ESS3.D: Global Climate Change, pp. 229–230 **ID:** *The NSTA Atlas of the Three Dimensions:* Narrative Page • 5.6: Humans Impacts on Earth Systems (ESS3.D)
II. Concepts, Core Ideas, or Practices	**IIA:** *Framework for K–12 Science Education:* Grade Band Endpoints • Ch. 7: ESS3.D: Global Climate Change, p. 198 **IIB:** *Next Generation Science Standards:* Disciplinary Core Ideas Column • MS: ESS3.D: Global Climate Change, p. 59 • HS: ESS3.D: Global Climate Change, pp. 98, 99 **IIC:** *NSTA Quick Reference Guide to the NGSS K–12:* Disciplinary Core Ideas Column • MS: ESS3.D: Global Climate Change, p. 137 • HS: ESS3.D: Global Climate Change, p. 166
III. Curriculum, Instruction, and Formative Assessment	**IIIA:** *Disciplinary Core Ideas: Reshaping Teaching and Learning* • Elementary School: Foundations for Understanding Global Climate Change, pp. 232–233 • Middle School: An Explicit Focus on Learning the Drivers of Global Climate Change, pp. 233–234 • High School: Defining and Broadening an Understanding of Global Climate Change, p. 234 **IIIB:** *Uncovering Student Ideas:* Assessment Probe and Suggestions for Instruction • USI.4: Global Warming; pp. 143, 148–149 • USI.E&ES: Coldest Winter Ever, pp. 69, 71–72; Are They Talking About Weather or Climate? pp. 73, 75–76; What Are the Signs of Global Warming? pp. 77, 80
IV. Research on Commonly Held Ideas	**IVA:** *Disciplinary Core Ideas: Reshaping Teaching and Learning* • ESS3.D: Global Climate Change, pp. 229 (last paragraph)–230 **IVB:** *Uncovering Student Ideas:* Related Research • USI.4: Global Warming; p. 148 • USI.E&ES: Coldest Winter Ever, p. 71; Are They Talking About Weather or Climate? p. 75; What Are the Signs of Global Warming? p. 80
V. K–12 Articulation and Connections	**VA:** *Next Generation Science Standards:* Appendices: Progression • Appendix E: ESS3.D: Global Climate Change, p. 3 **VB:** *NSTA Quick Reference Guide to the NGSS K–12:* Progression • ESS3.D: Global Climate Change, p. 79 **VC:** *The NSTA Atlas of the Three Dimensions:* Map Page • 5.6: Humans Impacts on Earth Systems (ESS3.C & ESS3.D)

Section and Outcome	Selected Sources and Readings for Study and Reflection Read and examine **related** parts of
VI. **Assessment Expectation**	**VIA:** *State Standards* • Examine your state's standards **VIB:** *Next Generation Science Standards:* Performance Expectations • MS: MS-ESS3-5, p. 59 • HS: HS-ESS3-5, p. 98; HS-ESS3-6, p. 99 **VIC:** *NSTA Quick Reference Guide to the NGSS K–12:* Performance Expectations • MS: MS-ESS3-5, p. 137 • HS: HS-ESS3-5, HS-ESS3-6, p. 166

Review Chapter 2 instructions on how to use this guide.
Visit curriculumtopicstudy2.org for more information about CTS and additional resources.

NOTE: This study can be combined with the CTS guide Human Impacts on Earth's Systems.

NOTE: *Atlas* page numbers have not been provided because *The NSTA Atlas of the Three Dimensions* was produced concurrently with this edition. Titles and map codes are accurate.

Additional Readings:

Available for download at www.curriculumtopicstudy2.org

Copyright © 2020 by Corwin Press, Inc. All rights reserved. Reprinted from *Science Curriculum Topic Study: Bridging the Gap Between Three-Dimensional Standards, Research, and Practice* (2nd ed.) by Page Keeley and Joyce Tugel. Thousand Oaks, CA: Corwin, www.corwin.com. Reproduction authorized for educational use by educators, local school sites, and/or noncommercial or nonprofit entities that have purchased the book.

Human Impact on Earth Systems
Grades K–12 Standards- and Research-Based Study of a Curricular Topic

Section and Outcome	Selected Sources and Readings for Study and Reflection Read and examine **related** parts of
I. **Content Knowledge**	**IA:** *Science for All Americans* • Ch. 4: The Earth, p. 42 (second paragraph) • Ch. 4: Processes That Shape the Earth, p. 46 **IB:** *Science Matters* (2009 Edition) • Ch. 19: Humans and the Environment, pp. 334–343 **IC:** *Framework for K–12 Science Education:* Narrative Section • Ch. 7: ESS3.C: Human Impacts on Earth Systems, pp. 194–195 **ID:** *The NSTA Atlas of the Three Dimensions:* Narrative Page • 5.6: Humans Impacts on Earth Systems (ESS3.C)
II. **Concepts, Core Ideas, or Practices**	**IIA:** *Framework for K–12 Science Education:* Grade Band Endpoints • Ch. 7: ESS3.C: Human Impacts on Earth Systems, pp. 195–196 **IIB:** *Next Generation Science Standards:* Disciplinary Core Ideas Column • K: ESS3.C: Human Impacts on Earth Systems, p. 6 • Grade 5: ESS3.C: Human Impacts on Earth Systems, p. 30 • MS: ESS3.C: Human Impacts on Earth Systems, p. 60 • HS: ESS3.C: Human Impacts on Earth Systems, p. 99 **IIC:** *NSTA Quick Reference Guide to the NGSS K–12:* Disciplinary Core Ideas Column • K: ESS3.C: Human Impacts on Earth Systems, p. 93 • Grade 5: ESS3.C: Human Impacts on Earth Systems, p. 112 • MS: ESS3.C: Human Impacts on Earth Systems, pp. 136–137 • HS: ESS3.C: Human Impacts on Earth Systems, p. 166
III. **Curriculum, Instruction, and Formative Assessment**	**IIIA:** *Disciplinary Core Ideas: Reshaping Teaching and Learning* • Ch. 12: K–2: Lower Elementary, pp. 230–231 • Ch. 12: 3–5: Upper Elementary, p. 231 • Ch. 12: 6–8: Middle School, p. 231 • Ch. 12: 9–12: High School, pp. 231–232 • What Approaches Can We Use to Teach About This Disciplinary Core Idea? pp. 234–237 **IIIB:** *Uncovering Student Ideas:* Assessment Probe and Suggestions for Instruction • USI.E&ES: Acid Rain, pp. 151, 154; Is Natural Better? pp. 161, 164; The Greenhouse Effect, pp. 165, 168–169
IV. **Research on Commonly Held Ideas**	**IVA:** *Disciplinary Core Ideas: Reshaping Teaching and Learning* • ESS3.C: Human Impacts on Earth Systems, pp. 228 (last paragraph)–229 **IVB:** *Uncovering Student Ideas:* Related Research • USI.E&ES: Acid Rain, p. 153; Is Natural Better? pp. 163–164; The Greenhouse Effect, p. 168

Section and Outcome	Selected Sources and Readings for Study and Reflection Read and examine **related** parts of
V. K–12 Articulation and Connections	**VA:** *Next Generation Science Standards:* Appendices: Progression • Appendix E: ESS3.C: Human Impacts on Earth Systems, p. 3 **VB:** *NSTA Quick Reference Guide to the NGSS K–12:* Progression • ESS3.C: Human Impacts on Earth Systems, p. 79 **VC:** *The NSTA Atlas of the Three Dimensions:* Map Page • 5.6: Humans Impacts on Earth Systems (ESS3.C & ESS3.D)
VI. Assessment Expectation	**VIA:** *State Standards* • Examine your state's standards **VIB:** *Next Generation Science Standards:* Performance Expectations • K: K-ESS3-2, K-ESS3-3, p. 6 • Grade 5: 5-ESS3-1, p. 30 • MS: MS-ESS3-3, MS-ESS3-4, p. 60 • HS: HS-ESS3-3, HS-ESS3-4, p. 99 **VIC:** *NSTA Quick Reference Guide to the NGSS K–12:* Performance Expectations Column • K: K-ESS3-2, p. 92; K-ESS3-3, p. 93 • Grade 5: 5-ESS3-1, p. 112 • MS: p. 136; MS-ESS3-4, p. 137 • HS: HS-ESS3-3, HS-ESS3-4, p. 166

Review Chapter 2 instructions on how to use this guide.

Visit curriculumtopicstudy2.org for more information about CTS and additional resources.

NOTE: This study can be combined with CTS guide Global Climate Change.

NOTE: *Atlas* page numbers have not been provided because *The NSTA Atlas of the Three Dimensions* was produced concurrently with this edition. Titles and map codes are accurate.

Additional Readings:

 Available for download at **www.curriculumtopicstudy2.org**

Copyright © 2020 by Corwin Press, Inc. All rights reserved. Reprinted from *Science Curriculum Topic Study: Bridging the Gap Between Three-Dimensional Standards, Research, and Practice* (2nd ed.) by Page Keeley and Joyce Tugel. Thousand Oaks, CA: Corwin, www.corwin.com. Reproduction authorized for educational use by educators, local school sites, and/or noncommercial or nonprofit entities that have purchased the book.

Structure of the Solid Earth
Grades 6–12 Standards- and Research-Based Study of a Curricular Topic

Section and Outcome	Selected Sources and Readings for Study and Reflection Read and examine **related** parts of
I. Content Knowledge	**IA:** *Science for All Americans* • Ch. 4: The Earth, p. 42 **IB:** *Science Matters* (2009 edition) • Ch. 13: A Window Into the Solid Earth, pp. 228–232 **IC:** *Framework for K–12 Science Education:* Narrative Section • Ch. 7: ESS2.A: Earth's Materials and Systems, pp. 179–180 (first three paragraphs) **ID:** *The NSTA Atlas of the Three Dimensions:* Narrative Page • 5.2: Earth's Systems (ESS2.A)
II. Concepts, Core Ideas, or Practices	**IIA:** *Framework for K–12 Science Education: Grade* Band Endpoints • Ch. 7: ESS2.A: Earth's Materials and Systems, pp. 180–182 **IIB:** *Next Generation Science Standards:* Disciplinary Core Ideas Column • MS: ESS2.A: Earth's Materials and Systems, p. 58 • HS: ESS2.A: Earth's Materials and Systems, p. 96; PS4.A: Wave Properties, p. 97 **IIC:** *NSTA Quick Reference Guide to the NGSS K–12:* Disciplinary Core Ideas Column • MS: ESS2.A: Earth's Materials and Systems, p. 134 • HS: ESS2.A: Earth's Materials and Systems, pp. 163–164
III. Curriculum, Instruction, and Formative Assessment	**IIIA:** *Disciplinary Core Ideas: Reshaping Teaching and Learning* • Ch. 10: 6–8: Middle School, p. 217 • Ch. 10: 9–12: High School, p. 219 (second paragraph) **IIIB:** *Uncovering Student Ideas:* Assessment Probe and Suggestions for Instruction • USI.E&ES: What's Beneath Us? pp. 15, 17; What Is the Inside of the Earth Like? pp. 121, 123–124
IV. Research on Commonly Held Ideas	**IVA:** *Uncovering Student Ideas:* Related Research • USI.E&ES: What's Beneath Us? p. 17; What Is the Inside of the Earth Like? p. 123
V. K–12 Articulation and Connections	**VA:** *Next Generation Science Standards:* Appendices: Progression • Appendix E: ESS2.A: Earth's Materials and Systems, p. 2 **VB:** *NSTA Quick Reference Guide to the NGSS K–12:* Progression • ESS2.A: Earth's Materials and Systems, p. 76 **VC:** *The NSTA Atlas of the Three Dimensions:* Map Page • 5.2: Earth's Systems (ESS2.A & ESS2.E)

Section and Outcome	Selected Sources and Readings for Study and Reflection Read and examine **related** parts of
VI. Assessment Expectation	**VIA: *State Standards*** • Examine your state's standards **VIB: *Next Generation Science Standards:*** Performance Expectations • MS: MS-ESS2-1, p. 58 • HS: HS-ESS2-3, p. 96 **VIC: *NSTA Quick Reference Guide to the NGSS K–12:*** Performance Expectations Column • MS: MS-ESS2-1, p. 134 • HS: HS-ESS2-3, p. 163

Review Chapter 2 instructions on how to use this guide.

Visit curriculumtopicstudy2.org for more information about CTS and additional resources.

NOTE: *Atlas* page numbers have not been provided because *The NSTA Atlas of the Three Dimensions* was produced concurrently with this edition. Titles and map codes are accurate.

Additional Readings:

Available for download at www.curriculumtopicstudy2.org

Copyright © 2020 by Corwin Press, Inc. All rights reserved. Reprinted from *Science Curriculum Topic Study: Bridging the Gap Between Three-Dimensional Standards, Research, and Practice* (2nd ed.) by Page Keeley and Joyce Tugel. Thousand Oaks, CA: Corwin, www.corwin.com. Reproduction authorized for educational use by educators, local school sites, and/or noncommercial or nonprofit entities that have purchased the book.

Water Cycle and Distribution
Grades K–8 Standards- and Research-Based Study of a Curricular Topic

Section and Outcome	Selected Sources and Readings for Study and Reflection Read and examine **related** parts of
I. Content Knowledge	**IA:** *Science for All Americans* • Ch. 4: The Earth, pp. 43–44 **IB:** *Science Matters* (2009 edition) • Ch. 14: The Water Cycle, pp. 239–245 **IC:** *Framework for K–12 Science Education:* Narrative Section • Ch. 7: ESS2.C: The Roles of Water in Earth's Surface Processes, p. 184 **ID: Disciplinary Core Ideas: Reshaping Teaching and Learning** • ESS2.C: The Roles of Water in Earth's Surface Properties, p. 210 **IE:** *The NSTA Atlas of the Three Dimensions:* Narrative Page • 5.4: Weather and Climate (ESS2.C)
II. Concepts, Core Ideas, or Practices	**IIA:** *Framework for K–12 Science Education:* Grade Band Endpoints • Ch. 7: ESS2.C: The Roles of Water in Earth's Surface Processes, pp. 184–186 **IIB:** *Next Generation Science Standards:* Disciplinary Core Ideas Column • Grade 2: ESS2.C: The Roles of Water in Earth's Surface Processes, p. 15 • Grade 5: ESS2.C: The Roles of Water in Earth's Surface Processes, p. 50 • MS: ESS2.C: The Roles of Water in Earth's Surface Processes, p. 81 **IIC:** *NSTA Quick Reference Guide to the NGSS K–12:* Disciplinary Core Ideas • Grade 2: ESS2.C: The Roles of Water in Earth's Surface Processes, p. 96 • Grade 5: ESS2.C: The Roles of Water in Earth's Surface Processes, p. 112 • MS: ESS2.C: The Roles of Water in Earth's Surface Processes, pp. 134–135
III. Curriculum, Instruction, and Formative Assessment	**IIIA:** *Disciplinary Core Ideas: Reshaping Teaching and Learning* • Ch. 11: ESS2.A: Earth Materials and Systems, p. 207 (second paragraph) • Ch. 11: Lower Elementary, pp. 215–216 • Ch. 11: Upper Elementary, p. 216 • Ch. 11: Middle School, pp. 217–218 **IIIA:** *Uncovering Student Ideas:* Assessment Probe and Suggestions for Instruction • USI.1: Wet Jeans, pp. 165, 168, 171–172 • USI.3: Where Did the Water Come From? pp. 163, 165–166, 168 • USI.4: Where Would It Fall? pp. 157, 160 • USI.E&ES: Land or Water? pp. 25, 27; Where Is Most of the Fresh Water? pp. 29, 31; Groundwater, pp. 33, 35; How Many Oceans and Seas? pp. 37, 39; Water Cycle Diagram, pp. 49, 51; Where Did the Water in the Puddle Go? pp. 53, 55–56
IV. Research on Commonly Held Ideas	**IVA:** *Benchmarks for Science Literacy:* Chapter 15 Research • 4B: Water Cycle, p. 336 **IVB:** *Making Sense of Secondary Science: Research Into Children's Ideas* • Ch. 12: Water Cycle, pp. 101–102 **IVC:** *Uncovering Student Ideas:* Related Research • USI.1: Wet Jeans, pp. 169–170 • USI.3: Where Did the Water Come From? p. 167

Section and Outcome	Selected Sources and Readings for Study and Reflection Read and examine **related** parts of
	• USI.4: Where Would It Fall? pp. 157–158 • USI.E&ES: Land or Water? p. 27; Where Is Most of the Fresh Water? p. 31; Groundwater, p. 35; How Many Oceans and Seas? p. 39; Water Cycle Diagram, p. 51; Where Did the Water in the Puddle Go? p. 55
V. K–12 Articulation and Connections	**VA:** *Next Generation Science Standards:* Appendices: Progression • Appendix E: ESS2.C: The Roles of Water in Earth's Surface Processes, p. 3 **VB:** *NSTA Quick Reference Guide to the NGSS K–12:* Progression Chart • ESS2.C: The Roles of Water in Earth's Surface Processes, p. 77 **VC:** *The NSTA Atlas of the Three Dimensions:* Map Page • 5.4: Weather and Climate (ESS2.C & ESS2.D)
VI. Assessment Expectation	**VIA:** *State Standards* • Examine your state's standards **VIB:** *Next Generation Science Standards:* Performance Expectations • Grade 2: 2-ESS2-3, p. 15 • Grade 5: 5-ESS2-2, p. 30 • MS: MS-ESS2-4, p. 58 **VIC:** *NSTA Quick Reference Guide to the NGSS K–12:* Performance Expectations • Grade 2: 2-ESS2-3, p. 96 • Grade 5: 5-ESS2-2, p. 112 • MS: MS-ESS2-4, p. 135

Review Chapter 2 instructions on how to use this guide.

Visit curriculumtopicstudy2.org for more information about CTS and additional resources.

NOTE: *Atlas* page numbers have not been provided because *The NSTA Atlas of the Three Dimensions* was produced concurrently with this edition. Titles and map codes are accurate.

Additional Readings:

Available for download at www.curriculumtopicstudy2.org

Copyright © 2020 by Corwin Press, Inc. All rights reserved. Reprinted from *Science Curriculum Topic Study: Bridging the Gap Between Three-Dimensional Standards, Research, and Practice* (2nd ed.) by Page Keeley and Joyce Tugel. Thousand Oaks, CA: Corwin, www.corwin.com. Reproduction authorized for educational use by educators, local school sites, and/or noncommercial or nonprofit entities that have purchased the book.

Water in the Earth System
Grades K–12 Standards- and Research-Based Study of a Curricular Topic

Section and Outcome	Selected Sources and Readings for Study and Reflection Read and examine **related** parts of
I. Content Knowledge	**IA:** *Science for All Americans* • Ch. 4: The Earth, pp. 42–44 **IB:** *Science Matters* (2009 edition) • Ch. 14: The Water Cycle, pp. 239–245 **IC:** *Framework for K–12 Science Education:* Narrative Section • Ch. 7: ESS2.A: Earth Materials and Systems, pp. 179–180 • Ch. 7: ESS2.C: The Roles of Water in Earth's Surface Processes, p. 184 **ID:** *Disciplinary Core Ideas: Reshaping Teaching and Learning* • ESS2.A: Earth Materials and Systems, pp. 206–207 • ESS2.C: The Roles of Water in Earth's Surface Properties, pp. 210–211 **IE:** *The NSTA Atlas of the Three Dimensions:* Narrative Page • 5.2: Earth Systems (ESS2.A) • 5.4: Weather and Climate (ESS2.C)
II. Concepts, Core Ideas, or Practices	**IIA:** *Framework for K–12 Science Education:* Grade Band Endpoints • Ch. 7: ESS2.A: Earth Materials and Systems, pp. 180–181 • Ch. 7: ESS2.C: The Roles of Water in Earth's Surface Processes, pp. 184–186 **IIB:** *Next Generation Science Standards:* Disciplinary Core Ideas Column • Grade 2: ESS2.A: Earth's Materials and Systems; ESS2.C: The Roles of Water in Earth's Surface Processes, p. 15 • Grade 4: ESS2.A: Earth's Materials and Systems, p. 26 • Grade 5: ESS2.A: Earth's Materials and Systems, p. 50; ESS2.C: The Roles of Water in Earth's Surface Processes, p. 30 • MS: ESS2.C: The Roles of Water in Earth's Surface Processes, pp. 58, 59 • HS: ESS2.C: The Roles of Water in Earth's Surface Processes, p. 96 **IIC:** *NSTA Quick Reference Guide to the NGSS K–12:* Disciplinary Core Ideas Column • Grade 2: ESS2.A: Earth's Materials and Systems, p. 96; ESS2.C: The Roles of Water in Earth's Surface Processes, p. 96 • Grade 4: ESS2.A: Earth's Materials and Systems, p. 107 • Grade 5: ESS2.A: Earth's Materials and Systems, p. 112; ESS2.C: The Roles of Water in Earth's Surface Processes, p. 112 • MS: ESS2.C: The Roles of Water in Earth's Surface Processes, p. 135 • HS: ESS2.C: The Roles of Water in Earth's Surface Processes, p. 164
III. Curriculum, Instruction, and Formative Assessment	**IIIA:** *Disciplinary Core Ideas: Reshaping Teaching and Learning* • Ch. 11: Lower Elementary, p. 215 • Ch. 11: Upper Elementary, p. 216 • Ch. 11: Middle School, pp. 217–218 • Ch. 11: High School, pp. 218–220 **IIIB:** *Uncovering Student Ideas:* Assessment Probe and Suggestions for Instruction • USI.1: Wet Jeans, pp. 165, 168, 171–172 • USI.E&ES: Where Is Most of the Fresh Water? pp. 29, 31; Groundwater, pp. 33, 35; How Many Oceans and Seas? pp. 37, 39; In Which Direction Does the Water Swirl? pp. 61, 63; Does the Ocean Influence Our Weather or Climate? pp. 65, 68; Grand Canyon, 107, 110; What Is a Watershed? pp. 155, 158

Section and Outcome	Selected Sources and Readings for Study and Reflection Read and examine **related** parts of
IV. **Research on Commonly Held Ideas**	**IVA: *Benchmarks for Science Literacy:*** Chapter 15 Research<ul><li>4B: Water Cycle, p. 336</li></ul>**IVB: *Making Sense of Secondary Science: Research Into Children's Ideas***<ul><li>Ch. 12: Water, pp. 98–103</li></ul>**IVC: *Uncovering Student Ideas:*** Related Research<ul><li>USI.1: Wet Jeans, pp. 169–170</li><li>USI.3: Where Did the Water Come From? p. 167</li><li>USI.E&ES: Where Is Most of the Fresh Water? p. 31; Groundwater, p. 35; How Many Oceans and Seas? p. 39; Where Did the Water in the Puddle Go? p. 55; In Which Direction Does the Water Swirl? p. 63; Does the Ocean Influence Our Weather or Climate? p. 67; Grand Canyon, pp. 109–110; What Is a Watershed? pp. 157–158</li></ul>
V. **K–12 Articulation and Connections**	**VA: *Next Generation Science Standards:*** Appendices: Progression<ul><li>Appendix E: ESS2.A: Earth Materials and Systems, p. 2</li><li>Appendix E: ESS2.C: The Roles of Water in Earth's Surface Processes, p. 3</li></ul>**VB: *NSTA Quick Reference Guide to the NGSS K–12:*** Progression Chart<ul><li>ESS2.A: Earth Materials and Systems, p. 76</li><li>ESS2.C: The Roles of Water in Earth's Surface Processes, p. 77</li></ul>**VC: *The NSTA Atlas of the Three Dimensions:*** Map Page<ul><li>5.2: Earth Systems (ESS2.A & ESS2.E)</li><li>5.4: Weather and Climate (ESS2.C & ESS2.D)</li></ul>
VI. **Assessment Expectation**	**VIA: *State Standards***<ul><li>Examine your state's standards</li></ul>**VIB: *Next Generation Science Standards:*** Performance Expectations<ul><li>Grade 2: 2-ESS2-1, 2-ESS2-3, p. 15</li><li>Grade 4: 4-ESS2-1, p. 26</li><li>Grade 5: 5-ESS2-1, 5-ESS2-2, p. 30</li><li>MS: MS-ESS2-4, p. 58; MS-ESS2-5, ESS2-6, p. 59</li><li>HS-ESS2-5, p. 96</li></ul>**VIC: *NSTA Quick Reference Guide to the NGSS K–12:*** Performance Expectations<ul><li>Grade 2: 2-ESS2-1, ESS2-2, 2-ESS2-3, p. 96</li><li>Grade 4: 4-ESS2-1, p. 107</li><li>Grade 5: 5-ESS2-1, 5-ESS2-2, p. 112</li><li>MS: MS-ESS2-4, ESS2-6, p. 135</li><li>HS: HS-ESS2-2, p. 163; HS-ESS2-4, HS-ESS2-5, p. 164</li></ul>

Review Chapter 2 instructions on how to use this guide.
Visit curriculumtopicstudy2.org for more information about CTS and additional resources.

SEE ALSO: Weathering and Erosion or Weather and Climate

NOTE: *Atlas* page numbers have not been provided because *The NSTA Atlas of the Three Dimensions* was produced concurrently with this edition. Titles and map codes are accurate.

Additional Readings:

 Available for download at **www.curriculumtopicstudy2.org**

Copyright © 2020 by Corwin Press, Inc. All rights reserved. Reprinted from *Science Curriculum Topic Study: Bridging the Gap Between Three-Dimensional Standards, Research, and Practice* (2nd ed.) by Page Keeley and Joyce Tugel. Thousand Oaks, CA: Corwin, www.corwin.com. Reproduction authorized for educational use by educators, local school sites, and/or noncommercial or nonprofit entities that have purchased the book.

Weather and Climate
Grades K–12 Standards- and Research-Based Study of a Curricular Topic

Section and Outcome	Selected Sources and Readings for Study and Reflection Read and examine **related** parts of
I. Content Knowledge	**IA:** *Science for All Americans* • Ch. 4: The Earth, pp. 43–44 **IB:** *Science Matters* (2009 edition) • Ch. 14: The Atmospheric Cycle, pp. 245–249 **IC:** *Framework for K–12 Science Education:* Narrative Section • Ch. 7: ESS2.D: Weather and Climate, pp. 186–187 **ID:** *Disciplinary Core Ideas: Reshaping Teaching and Learning* • Ch. 11: ESS2.D: Weather and Climate, pp. 211–214 **IE:** *The NSTA Atlas of the Three Dimensions:* Narrative Page • 5.4: Weather and Climate (ESS2.D)
II. Concepts, Core Ideas, or Practices	**IIA:** *Framework for K–12 Science Education:* Grade Band Endpoints • Ch. 7: ESS2.D: Weather and Climate, pp. 188–189 **IIB:** *Next Generation Science Standards:* Disciplinary Core Ideas Column • K: ESS2.D: Weather and Climate, p. 7 • Grade 3: ESS2.D: Weather and Climate, p. 21 • MS: ESS2.D: Weather and Climate, p. 82 • HS: ESS2.D: Weather and Climate, pp. 96, 98 **IIC:** *NSTA Quick Reference Guide to the NGSS K–12:* Disciplinary Core Ideas Column • K: ESS2.D: Weather and Climate, p. 92 • Grade 3: ESS2.D: Weather and Climate, p. 105 • MS: ESS2.D: Weather and Climate, p. 135 • HS: ESS2.D: Weather and Climate, p. 163, 164
III. Curriculum, Instruction, and Formative Assessment	**IIIA:** *Uncovering Student Ideas:* Assessment Probe and Suggestions for Instruction • USI.3: What Are Clouds Made Of? pp. 155, 159–160; Rainfall, pp. 171, 175 • USI.E&ES: Weather Predictors, pp. 57, 59–60; Does the Ocean Influence Our Weather or Climate? pp. 65, 68; Coldest Winter Ever! pp. 69, 71–72; Are They Talking About Weather or Climate? pp. 73, 75–76 **IIIB:** *Disciplinary Core Ideas: Reshaping Teaching and Learning* • Ch. 11: K–2: Lower Elementary, p. 216 • Ch. 11: 3–5: Upper Elementary, p. 216 (last paragraph) • Ch. 11: 6–8: Middle School, pp. 217 (last paragraph)–218 • Ch. 11: 9–12: High School, p. 220
IV. Research on Commonly Held Ideas	**IVA:** *Making Sense of Secondary Science: Research Into Children's Ideas* • Ch. 13: Existence of Air, pp. 104–105; Wind, p. 111 **IVB:** *Uncovering Student Ideas:* Related Research • USI.3: What Are Clouds Made Of? pp. 158–159; Rainfall, pp. 174–175 • USI.E&ES: Weather Predictors, p. 59; Does the Ocean Influence Our Weather or Climate? p. 67; Coldest Winter Ever! p. 71; Are They Talking About Weather or Climate? p. 75 **IVC:** *Disciplinary Core Ideas: Reshaping Teaching and Learning* • Ch. 11: ESS2.D: Weather and Climate, p. 214 (first column)

Section and Outcome	Selected Sources and Readings for Study and Reflection Read and examine **related** parts of
V. **K–12 Articulation and Connections**	**VA:** *Next Generation Science Standards:* Appendices: Progression • Appendix E: ESS2.D: Weather and Climate, p. 3 **VB:** *NSTA Quick Reference Guide to the NGSS K–12:* Progression • ESS2.D: Weather and Climate, p. 78 **VC:** *The NSTA Atlas of the Three Dimensions:* Map Page • 5.4: Weather and Climate (ESS2.C & ESS2.D)
VI. **Assessment Expectation**	**VIA:** *State Standards* • Examine your state's standards **VIB:** *Next Generation Science Standards:* Performance Expectations • K: K-ESS2-1, p. 7 • Grade 3: 2-ESS2-1, 3-ESS2-2, p. 21 • MS: MS-ESS2-5, MS-ESS2-6, p. 59 • HS: HS-ESS2-2, p. 96; HS-ESS2-4, p. 98; ESS2-6, HS-ESS2-7, p. 96 **VIC:** *NSTA Quick Reference Guide to the NGSS K–12:* Performance Expectations • K: K-ESS2-1, p. 92 • Grade 3: 2-ESS2-1, 3-ESS2-2, p. 105 • MS: MS-ESS2-5, MS-ESS2-6, p. 135 • HS: HS-ESS2-2, p. 163; HS-ESS2-4, ESS2-6, p. 164; HS-ESS2-7, p. 165

Review Chapter 2 instructions on how to use this guide.

Visit curriculumtopicstudy2.org for more information about CTS and additional resources.

NOTE: This study can be combined with Global Climate Change.

NOTE: *Atlas* page numbers have not been provided because *The NSTA Atlas of the Three Dimensions* was produced concurrently with this edition. Titles and map codes are accurate.

Additional Readings:

Available for download at www.curriculumtopicstudy2.org

Biogeology
Grades K–12 Standards- and Research-Based Study of a Curricular Topic

Section and Outcome	Selected Sources and Readings for Study and Reflection Read and examine **related** parts of
I. Content Knowledge	**IA:** *Science for All Americans* • Ch. 4: Processes That Shape the Earth, p. 45 (last paragraph) **IB:** *Science Matters* (2009 edition) • Ch. 14: Earth's Cycles: Frontiers, pp. 249–250 • Ch. 19: Nutrients and the Carbon Cycle, pp. 332–333 **IC:** *Framework for K–12 Science Education:* Narrative Section • Ch. 7: ESS2.E: Biogeology, pp. 189–190 **ID:** *Disciplinary Core Ideas: Reshaping Teaching and Learning* • Ch. 11: ESS2.E: Biogeology, pp. 214–215 **VC:** *The NSTA Atlas of the Three Dimensions:* Narrative Page • 5.2: Earth Systems (ESS2.E)
II. Concepts, Core Ideas, or Practices	**IIA:** *Framework for K–12 Science Education:* Grade Band Endpoints • Ch. 7: ESS2.E: Biogeology, p. 190 **IIB:** *Next Generation Science Standards:* Disciplinary Core Ideas Column • K: ESS2.E: Biogeology, p. 6 • Grade 4: ESS2.E: Biogeology, p. 26 • HS: ESS2.E: Biogeology, pp. 96–97 **IIC:** *NSTA Quick Reference Guide to the NGSS K–12:* Disciplinary Core Ideas Column • K: ESS2.E: Biogeology, p. 92 • Grade 4: ESS2.E: Biogeology, p. 107 • HS: ESS2.E: Biogeology, p. 165
III. Curriculum, Instruction, and Formative Assessment	**IIIA:** *Disciplinary Core Ideas: Reshaping Teaching and Learning* • Ch. 11: K–2: Lower Elementary, pp. 215–216 • Ch. 11: 3–5: Upper Elementary, p. 216 • Ch. 11: 6–8: Middle School, pp. 217–218 • Ch. 11: 9–12: High School, p. 218–220 **IIIB:** *Uncovering Student Ideas:* Assessment Probe and Suggestions for Instruction • USI.E&ES: Can a Plant Break Rocks? pp. 103, 105–106
IV. Research on Commonly Held Ideas	**IVA:** *Uncovering Student Ideas:* Related Research • USI.E&ES: Can a Plant Break Rocks? p. 105 **IVB:** *Disciplinary Core Ideas: Reshaping Teaching and Learning* • Biogeology, p. 215 (first column)
V. K–12 Articulation and Connections	**VA:** *Next Generation Science Standards:* Appendices: Progression • Appendix E: ESS2.E: Biogeology, p. 3 **VB:** *NSTA Quick Reference Guide to the NGSS K–12:* Progression • ESS2.E: Biogeology, p. 78 **VC:** *The NSTA Atlas of the Three Dimensions:* Map Page • 5.2: Earth Systems (ESS2.A & ESS2.E)

Section and Outcome	Selected Sources and Readings for Study and Reflection Read and examine **related** parts of
VI. Assessment Expectation	**VIA:** *State Standards* • Examine your state's standards **VIB:** *Next Generation Science Standards:* Performance Expectations • K: K-ESS2-2, p. 6 • Grade 4: 4-ESS2-1, p. 26 • HS: HS-ESS2-7, p. 96 **VIC:** *NSTA Quick Reference Guide to the NGSS K–12:* Performance Expectations • K: K-ESS2-2, p. 96 • Grade 4: 4-ESS2-1, p. 107 • HS: HS-ESS2-7, p. 165

Review Chapter 2 instructions on how to use this guide.

Visit curriculumtopicstudy2.org for more information about CTS and additional resources.

NOTE: *Atlas* page numbers have not been provided because *The NSTA Atlas of the Three Dimensions* was produced concurrently with this edition. Titles and map codes are accurate.

Additional Readings:

 Available for download at **www.curriculumtopicstudy2.org**

Copyright © 2020 by Corwin Press, Inc. All rights reserved. Reprinted from *Science Curriculum Topic Study: Bridging the Gap Between Three-Dimensional Standards, Research, and Practice* (2nd ed.) by Page Keeley and Joyce Tugel. Thousand Oaks, CA: Corwin, www.corwin.com. Reproduction authorized for educational use by educators, local school sites, and/or noncommercial or nonprofit entities that have purchased the book.

Earthquakes and Volcanoes
Grades 3–12 Standards- and Research-Based Study of a Curricular Topic

Section and Outcome	Selected Sources and Readings for Study and Reflection Read and examine **related** parts of
I. **Content Knowledge**	**IA:** *Science for All Americans* • Ch. 4: Processes That Shape the Earth, pp. 44–45 **IB:** *Science Matters* (2009 edition) • Ch. 13: The Restless Earth, pp. 215–218; Earthquakes, pp. 226–227 • Ch. 14: Igneous Rocks, pp. 235–236 **IC:** *Framework for K–12 Science Education:* Narrative Section • Ch. 7: ESS2.B: Plate Tectonics and Large-Scale System Interactions, p. 182 • Ch. 7: ESS3.B: Natural Hazards, pp. 192–193 **ID:** *Disciplinary Core Ideas: Reshaping Teaching and Learning* • Ch. 11: ESS2.B: Plate Tectonics and Large-Scale System Interactions, pp. 207–210 **IE:** *The NSTA Atlas of the Three Dimensions:* Narrative Page • 5.3: Plate Tectonics and the History of Planet Earth (ESS2.B) • 5.5: Natural Resources and Natural Hazards (ESS3.B)
II. **Concepts, Core Ideas, or Practices**	**IIA:** *Framework for K–12 Science Education:* Grade Band Endpoints • Ch. 7: ESS2.B: Plate Tectonics and Large-Scale System Interactions, p. 183 • Ch. 7: ESS3.B: Natural Hazards, pp. 193–194 **IIB:** *Next Generation Science Standards:* Disciplinary Core Ideas Column • Grade 4: ESS2.B: Plate Tectonics and Large-Scale System Interactions; ESS3.B: Natural Hazards, p. 26 • MS: ESS2.B: Plate Tectonics and Large-Scale System Interactions, p. 57; ESS3.B: Natural Hazards, p. 60 • HS: ESS2.B: Plate Tectonics and Large-Scale System Interactions, p. 94 **IIC:** *NSTA Quick Reference Guide to the NGSS K–12:* Disciplinary Core Ideas Column • Grade 4: ESS2.B: Plate Tectonics and Large-Scale System Interactions, p. 107; ESS3.B: Natural Hazards, p. 108 • MS: ESS2.A: Earth's Materials and Systems, p. 134; ESS3.B: Natural Hazards, p. 136 • HS: ESS2.B: Plate Tectonics and Large-Scale System Interactions, p. 163
III. **Curriculum, Instruction, and Formative Assessment**	**IIIA:** *Disciplinary Core Ideas: Reshaping Teaching and Learning* • Ch. 11: Upper Elementary, p. 216 (second column, third paragraph) • Ch. 12: Upper Elementary, p. 231 • Ch. 11: Middle School, p. 217 (first two paragraphs) • Ch. 12: Middle School, p. 231 • Ch. 12: High School, p. 219 (first column, second paragraph) • Ch. 12: What Approaches Can We Use to Teach About This Disciplinary Core Idea, pp. 220–222 **IIIB:** *Uncovering Student Ideas:* Assessment Probe and Suggestions for Instruction • USI.E&ES: What Do You Know About Volcanoes and Earthquakes? pp. 135, 138–139

Section and Outcome	Selected Sources and Readings for Study and Reflection Read and examine **related** parts of
IV. **Research on Commonly Held Ideas**	**IVA:** *Benchmarks for Science Literacy:* Chapter 15 Research • 4C: Processes That Shape the Earth, p. 336 **IVB:** *Making Sense of Secondary Science: Research Into Children's Ideas* • Ch. 14: Mountains and Volcanoes, pp. 113–114 **IVC:** *Disciplinary Core Ideas: Reshaping Teaching and Learning* • Ch. 11: ESS2.B: Plate Tectonics and Large-Scale System Interactions, pp. 209–210 **IVD:** *Uncovering Student Ideas:* Related Research • USI.E&ES: What Do You Know About Volcanoes and Earthquakes? p. 138
V. **K–12 Articulation and Connections**	**VA:** *Next Generation Science Standards:* Appendices: Progression • Appendix E: ESS2.B: Plate Tectonics and Large-Scale System Interactions, p. 2 • Appendix E: ESS3.B: Natural Hazards, p. 3 **VB:** *NSTA Quick Reference Guide to the NGSS K–12:* Progression • ESS2.B: Plate Tectonics and Large-Scale System Interactions, p. 77 • ESS3.B: Natural Hazards, p. 79 **VC:** *The NSTA Atlas of the Three Dimensions:* Map Page • 5.3: Plate Tectonics and the History of Planet Earth (ESS1.C & ESS2.B) • 5.5: Natural Resources and Natural Hazards (ESS3.A & ESS3.B)
VI. **Assessment Expectation**	**VIA:** *State Standards* • Examine your state's standards **VIB:** *Next Generation Science Standards:* Performance Expectations • Grade 4: 4-ESS2-2, 4-ESS3-2, p. 26 • MS: MS-ESS2-2, p. 57; MS-ESS3-2, p. 60 • HS: HS-ESS1-5, p. 94; HS-ESS2-2, p. 96 **VIC:** *NSTA Quick Reference Guide to the NGSS K–12:* Performance Expectations • Grade 4: 4-ESS2-2, p. 107; 4-ESS3-2, p. 108 • MS: MS-ESS2-2, p. 134; MS-ESS3-3, p. 136 • HS: HS-ESS1-5, p. 162; HS-ESS2-2, p. 163

Review Chapter 2 instructions on how to use this guide.

Visit curriculumtopicstudy2.org for more information about CTS and additional resources.

NOTE: *Atlas* page numbers have not been provided because *The NSTA Atlas of the Three Dimensions* was produced concurrently with this edition. Titles and map codes are accurate.

Additional Readings:

 Available for download at www.curriculumtopicstudy2.org

Copyright © 2020 by Corwin Press, Inc. All rights reserved. Reprinted from *Science Curriculum Topic Study: Bridging the Gap Between Three-Dimensional Standards, Research, and Practice* (2nd ed.) by Page Keeley and Joyce Tugel. Thousand Oaks, CA: Corwin, www.corwin.com. Reproduction authorized for educational use by educators, local school sites, and/or noncommercial or nonprofit entities that have purchased the book.

Earth's History
Grades K–12 Standards- and Research-Based Study of a Curricular Topic

Section and Outcome	Selected Sources and Readings for Study and Reflection Read and examine **related** parts of
I. Content Knowledge	**IA: *Science for All Americans*** • Ch. 4: Processes That Shape the Earth, pp. 44–46 • Ch. 10: Extending Time, pp. 151–152 • Ch. 10: Moving the Continents, pp. 152–153 **IB: *Science Matters*** (2009 edition) • Ch. 13: The Restless Earth, pp. 215–223 • Ch. 14: The Rock Cycle, pp. 235–239 • Ch. 18: Fossils and Evolution, pp. 314–315 **IC: *Framework for K–12 Science Education:*** Narrative Section • Ch. 7: ESS1.C: The History of Planet Earth, p. 177–178 **ID: *The NSTA Atlas of the Three Dimensions:*** Narrative Page • 5.3: Plate Tectonics and the History of Planet Earth (ESS1.C)
II. Concepts, Core Ideas, or Practices	**IIA: *Framework for K–12 Science Education:*** Grade Band Endpoints • Ch. 7: ESS1.C: The History of Planet Earth, p. 178–179 **IIB: *Next Generation Science Standards:*** Disciplinary Core Ideas Column • Grade 2: ESS1.C: The History of Planet Earth, p. 15 • Grade 4: ESS1.C: The History of Planet Earth, p. 26 • MS: ESS1.C: The History of Planet Earth, p. 57 • HS: ESS1.C: The History of Planet Earth, p. 94 **IIC: *NSTA Quick Reference Guide to the NGSS K–12:*** Disciplinary Core Ideas Column • Grade 2: ESS1.C: The History of Planet Earth, p. 96 • Grade 4: ESS1.C: The History of Planet Earth, p. 107 • MS: ESS1.C: The History of Planet Earth, p. 134 • HS: ESS1.C: The History of Planet Earth, p. 162
III. Curriculum, Instruction, and Formative Assessment	**IIIA: *Disciplinary Core Ideas: Reshaping Teaching and Learning*** • Ch. 10: ESS1.C: The History of Planet Earth, p. 188 • Ch. 10: K–2: Lower Elementary, p. 189 (last paragraph) • Ch. 10: 3–5: Upper Elementary, p. 190 (last paragraph) • Ch. 10: 6–8: Middle School, p. 193 • Ch. 10: 9–12: High School, p. 196 • Ch. 10: What Approaches Can We Use to Teach About This Disciplinary Core Idea? pp. 200–201 **IIIB: *Uncovering Student Ideas:*** Assessment Probe and Suggestions for Instruction • USI.2: Mountaintop Fossil, pp. 165, 169 • USI.E&ES: How Old Is Earth? pp. 87, 89–90; Sedimentary Rock Layers, pp. 95, 97–98
IV. Research on Commonly Held Ideas	**IVA: *Benchmarks for Science Literacy:*** Chapter 15 Research • 4C: Processes That Shape the Earth, p. 336 **IVB: *Uncovering Student Ideas:*** Related Research • USI.2: Mountaintop Fossil, pp. 168–169 • USI.E&ES: How Old Is Earth? p. 89; Sedimentary Rock Layers, p. 97

Section and Outcome	Selected Sources and Readings for Study and Reflection Read and examine **related** parts of
V. K–12 Articulation and Connections	**VA:** *Next Generation Science Standards:* Appendices: Progression<ul><li>Appendix E: ESS1.C: The History of Planet Earth, p. 2</li></ul>**VB:** *NSTA Quick Reference Guide to the NGSS K–12:* Progression<ul><li>ESS1.C: The History of Planet Earth, p. 76</li></ul>**VC:** *The NSTA Atlas of the Three Dimensions:* Map Page<ul><li>5.3: Plate Tectonics and the History of Planet Earth (ESS1.C & ESS2.B)</li></ul>
VI. Assessment Expectation	**VIA:** *State Standards*<ul><li>Examine your state's standards</li></ul>**VIB:** *Next Generation Science Standards:* Performance Expectations<ul><li>Grade 2: 2-ESS1-1, p. 15</li><li>Grade 4: 4-ESS1-1, p. 26</li><li>MS: MS-ESS1-4, MS-ESS2-4, MS-ESS2-3, p. 57</li><li>HS: HS-ESS1-5, ESS1-6, p. 94</li></ul>**VIC:** *NSTA Quick Reference Guide to the NGSS K–12:* Performance Expectations<ul><li>Grade 2: 2-ESS1-1, p. 96</li><li>Grade 4: 4-ESS1-1, p. 107</li><li>MS: MS-ESS1-4; MS-ESS2-3, p. 134</li><li>HS: HS-ESS1-5, ESS1-6, p. 162</li></ul>

Review Chapter 2 instructions on how to use this guide.

Visit curriculumtopicstudy2.org for more information about CTS and additional resources.

SEE ALSO: Plate Tectonics

NOTE: *Atlas* page numbers have not been provided because *The NSTA Atlas of the Three Dimensions* was produced concurrently with this edition. Titles and map codes are accurate.

Additional Readings:

Available for download at www.curriculumtopicstudy2.org

Copyright © 2020 by Corwin Press, Inc. All rights reserved. Reprinted from *Science Curriculum Topic Study: Bridging the Gap Between Three-Dimensional Standards, Research, and Practice* (2nd ed.) by Page Keeley and Joyce Tugel. Thousand Oaks, CA: Corwin, www.corwin.com. Reproduction authorized for educational use by educators, local school sites, and/or noncommercial or nonprofit entities that have purchased the book.

Natural Hazards
Grades K–12 Standards- and Research-Based Study of a Curricular Topic

Section and Outcome	Selected Sources and Readings for Study and Reflection Read and examine **related** parts of
I. Content Knowledge	**IA:** *Framework for K–12 Science Education:* Narrative Section • Ch. 7: ESS3.B: Natural Hazards, pp. 192–193 **IB:** *Disciplinary Core Ideas: Reshaping Teaching and Learning* • Ch. 12: ESS3.B: Natural Hazards, pp. 227–228 **IB:** *The NSTA Atlas of the Three Dimensions:* Narrative Page • 5.5: Natural Resources and Natural Hazards (ESS3.B)
II. Concepts, Core Ideas, or Practices	**IIA:** *Framework for K–12 Science Education:* Grade Band Endpoints • Ch. 7: ESS3.B: Natural Hazards, pp. 193–194 **IIB:** *Next Generation Science Standards:* Disciplinary Core Ideas Column • K: ESS3.B: Natural Hazards, p. 7 • Grade 3: ESS3.B: Natural Hazards, p. 21 • Grade 4: ESS3.B: Natural Hazards, p. 26 • MS: ESS3.B: Natural Hazards, p. 60 • HS: ESS3.B: Natural Hazards, p. 99 **IIC:** *NSTA Quick Reference Guide to the NGSS K–12:* Disciplinary Core Ideas Column • K: ESS3.B: Natural Hazards, p. 92 • Grade 3: ESS3.B: Natural Hazards, p. 105 • Grade 4: ESS3.B: Natural Hazards, p. 108 • MS: ESS3.B: Natural Hazards, p. 136 • HS: ESS3.B: Natural Hazards, p. 165
III. Curriculum, Instruction, and Formative Assessment	**IIIA:** *Disciplinary Core Ideas: Reshaping Teaching and Learning* • Ch. 12: K–2: Lower Elementary, pp. 230–231 • Ch. 12: 3–5: Upper Elementary, p. 231 • Ch. 12: 6–8: Middle School, p. 231 • Ch. 12: 9–12: High School, pp. 231–232
IV. Research on Commonly Held Ideas	**IVA:** *Disciplinary Core Ideas: Reshaping Teaching and Learning* • ESS3.A: Natural Resources, p. 228 (last paragraph before ESS3.C)
V. K–12 Articulation and Connections	**VA:** *Next Generation Science Standards:* Appendices: Progression • Appendix E: ESS3.B: Natural Hazards, p. 3 **VB:** *NSTA Quick Reference Guide to the NGSS K–12:* Progression • ESS3.B: Natural Hazards, p. 79 **VC:** *The NSTA Atlas of the Three Dimensions:* Map Page • 5.5: Natural Resources and Natural Hazards (ESS3.A & ESS3.B)

Section and Outcome	Selected Sources and Readings for Study and Reflection Read and examine **related** parts of
VI. **Assessment** **Expectation**	**VIA:** *State Standards* • Examine your state's standards **VIB:** *Next Generation Science Standards:* Performance Expectations • K: K-ESS3-2, p. 7 • Grade 3: 3-ESS3-1, p. 21 • Grade 4: 4-ESS3-2, p. 26 • MS: MS-ESS3-2, p. 60 • HS: HS-ESS3-1, p. 99 **VIC:** *NSTA Quick Reference Guide to the NGSS K–12:* Performance Expectations Column • K: K-ESS3-2, p. 92 • Grade 3: 3-ESS3-1, p. 105 • Grade 4: 4-ESS3-2, p. 108 • MS: MS-ESS3-2, p. 136 • HS: HS-ESS3-1, p. 165

Review Chapter 2 instructions on how to use this guide.

Visit curriculumtopicstudy2.org for more information about CTS and additional resources.

NOTE: *Atlas* page numbers have not been provided because *The NSTA Atlas of the Three Dimensions* was produced concurrently with this edition. Titles and map codes are accurate.

Additional Readings:

Available for download at www.curriculumtopicstudy2.org

Copyright © 2020 by Corwin Press, Inc. All rights reserved. Reprinted from *Science Curriculum Topic Study: Bridging the Gap Between Three-Dimensional Standards, Research, and Practice* (2nd ed.) by Page Keeley and Joyce Tugel. Thousand Oaks, CA: Corwin, www.corwin.com. Reproduction authorized for educational use by educators, local school sites, and/or noncommercial or nonprofit entities that have purchased the book.

Plate Tectonics
Grades K–12 Standards- and Research-Based Study of a Curricular Topic

Section and Outcome	Selected Sources and Readings for Study and Reflection Read and examine **related** parts of
I. **Content Knowledge**	**IA:** *Science for All Americans* • Ch. 4: Processes That Shape the Earth, pp. 44–45 • Ch. 10: Moving the Continents, pp. 152–153 **IB:** *Science Matters* (2009 edition) • Ch. 13: Plate Tectonics, pp. 218–226 **IC:** *Framework for K–12 Science Education:* Narrative Section • Ch. 7: ESS2.B: Plate Tectonics and Large-Scale System Interactions, p. 182 **ID:** *Disciplinary Core Ideas: Reshaping Teaching and Learning* • Ch. 11: ESS2.B: Plate Tectonics and Large-Scale System Interactions, pp. 207–210 **IE:** *The NSTA Atlas of the Three Dimensions:* Narrative Page • 5.3: Plate Tectonics and the History of Planet Earth (ESS2.B)
II. **Concepts, Core Ideas, or Practices**	**IIA:** *Framework for K–12 Science Education:* Grade Band Endpoints • Ch. 7: ESS2.B: Plate Tectonics and Large-Scale System Interactions, p. 183 **IIB:** *Next Generation Science Standards:* Disciplinary Core Ideas Column • Grade 2: ESS2.B: Plate Tectonics and Large-Scale System Interactions, p. 15 • Grade 4: ESS2.B: Plate Tectonics and Large-Scale System Interactions, p. 26 • MS: ESS2.B: Plate Tectonics and Large-Scale System Interactions, p. 57 • HS: ESS2.B: Plate Tectonics and Large-Scale System Interactions, p. 96 **IIC:** *NSTA Quick Reference Guide to the NGSS K–12:* Disciplinary Core Ideas • Grade 2: ESS2.B: Plate Tectonics and Large-Scale System Interactions, p. 96 • Grade 4: ESS2.B: Plate Tectonics and Large-Scale System Interactions, p. 107 • MS: ESS2.B: Plate Tectonics and Large-Scale System Interactions, p. 134 • HS: ESS2.B: Plate Tectonics and Large-Scale System Interactions, p. 162
III. **Curriculum, Instruction, and Formative Assessment**	**IIIA:** *Disciplinary Core Ideas: Reshaping Teaching and Learning* • Ch. 11: Upper Elementary, p. 216 (second column) • Ch. 11: Middle School, p. 217 • Ch. 11: High School, p. 219 • What Approaches Can We Use to Teach About This Disciplinary Core Idea, pp. 220–222 **IIIB:** *Uncovering Student Ideas:* Assessment Probe and Suggestions for Instruction • USI.E&ES: Describing Earth's Plates, pp. 125, 128; Where Do You Find Earth's Plates? pp. 131, 133–134
IV. **Research on Commonly Held Ideas**	**IVA:** *Benchmarks for Science Literacy:* Chapter 15 Research • 4C: Processes That Shape the Earth, p. 336 **IVB:** *Making Sense of Secondary Science: Research Into Children's Ideas* • Ch. 14: Mountains and Volcanoes, pp. 113–114 **IVC:** *Disciplinary Core Ideas: Reshaping Teaching and Learning* • Ch. 11: ESS2.B: Plate Tectonics and Large-Scale System Interactions, pp. 209–210

Section and Outcome	Selected Sources and Readings for Study and Reflection Read and examine **related** parts of
	IVD: *Uncovering Student Ideas:* Related Research • USI.E&ES: Describing Earth's Plates, pp. 127–128; Where Do You Find Earth's Plates? p. 133
V. **K–12 Articulation and Connections**	**VA:** *Next Generation Science Standards:* Appendices: Progression • Appendix E: ESS2.B: Plate Tectonics and Large-Scale System Interactions, p. 2 **VB:** *NSTA Quick Reference Guide to the NGSS K–12:* Progression • ESS2.B: Plate Tectonics and Large-Scale System Interactions, p. 77 **VC:** *The NSTA Atlas of the Three Dimensions:* Map Page • 5.3: Plate Tectonics and the History of Planet Earth (ESS1.C & ESS2.B)
VI. **Assessment Expectation**	**VIA:** *State Standards* • Examine your state's standards **VIB:** *Next Generation Science Standards:* Performance Expectations • Grade 2: 2-ESS2-2, p. 15 • Grade 4: 4-ESS2-2, p. 26 • MS: MS-ESS2-3, p. 57 • HS: HS-ESS1-5, p. 96 **VIC:** *NSTA Quick Reference Guide to the NGSS K–12:* Performance Expectations (Left Column) • Grade 2: 2-ESS2-2, p. 96 • Grade 4: 4-ESS2-2, p. 107 • MS: MS-ESS2-3, p. 134 • HS: HS-ESS1-5, p. 162

Review Chapter 2 instructions on how to use this guide.

Visit curriculumtopicstudy2.org for more information about CTS and additional resources.

NOTE: *Atlas* page numbers have not been provided because *The NSTA Atlas of the Three Dimensions* was produced concurrently with this edition. Titles and map codes are accurate.

Additional Readings:

Available for download at www.curriculumtopicstudy2.org

Copyright © 2020 by Corwin Press, Inc. All rights reserved. Reprinted from *Science Curriculum Topic Study: Bridging the Gap Between Three-Dimensional Standards, Research, and Practice* (2nd ed.) by Page Keeley and Joyce Tugel. Thousand Oaks, CA: Corwin, www.corwin.com. Reproduction authorized for educational use by educators, local school sites, and/or noncommercial or nonprofit entities that have purchased the book.

Weathering, Erosion, and Deposition
Grades K–12 Standards- and Research-Based Study of a Curricular Topic

Section and Outcome	Selected Sources and Readings for Study and Reflection Read and examine **related** parts of
I. Content Knowledge	**IA:** *Science for All Americans* • Ch. 4: Processes That Shape the Earth, pp. 45–46 **IB:** *Science Matters* (2009 edition) • Ch. 14: Earth Cycles, pp. 233–239 **IC**: *Framework for K–12 Science Education:* Narrative Section • Ch. 7: ESS2.A: Earth's Materials and Systems, pp. 179–180 • Ch. 7: ESS2.C: The Roles of Water in Earth's Surface Properties, p. 184 **ID**: *Disciplinary Core Ideas: Reshaping Teaching and Learning* • Ch. 11: ESS2.A: Earth Materials and Systems, pp. 206–207 • Ch. 11: ESS2.C: The Roles of Water in Earth's Surface Properties, pp. 210–211 **IE**: *The NSTA Atlas of the Three Dimensions:* Narrative Page • 5.2: Earths Systems (ESS2.A) • 5.4: Weather and Climate (ESS2.C)
II. Concepts, Core Ideas, or Practices	**IIA**: *Framework for K–12 Science Education:* Grade Band Endpoints • Ch. 7: ESS2.A: Earth's Materials and Systems, pp. 180–181 • Ch. 7: ESS2.C: The Roles of Water in Earth's Surface Properties, pp. 184–186 **IIB**: *Next Generation Science Standards:* Disciplinary Core Ideas Column • Grade 2: ESS1.C: The History of Planet Earth; ESS2.A: Earth's Materials and Systems, p. 15 • Grade 4: ESS2.A: Earth's Materials and Systems, p. 26 • MS: ESS2.A: Earth's Materials and Systems, pp. 57, 58; ESS2.C: The Roles of Water in Earth's Surface Properties, p. 57 • HS: ESS2.A: Earth's Materials and Systems; ESS2.C: The Roles of Water in Earth's Surface Properties, p. 96 **IIC**: *NSTA Quick Reference Guide to the NGSS K–12:* Disciplinary Core Ideas Column • Grade 2: ESS1.C: The History of Planet Earth, p. 96; ESS2.A: Earth's Materials and Systems, p. 96 • Grade 4: ESS2.A: Earth's Materials and Systems, p. 107 • MS: ESS2.A: Earth's Materials and Systems, p. 134; ESS2.C: The Roles of Water in Earth's Surface Properties, p. 134 • HS: ESS2.A: Earth's Materials and Systems, p. 164; ESS2.C: The Roles of Water in Earth's Surface Properties, p. 164
III. Curriculum, Instruction, and Formative Assessment	**IIIA**: *Disciplinary Core Ideas: Reshaping Teaching and Learning* • Ch. 10: Lower Elementary, pp. 215–216 (second paragraph) • Ch. 10: Upper Elementary, p. 216 (first paragraph) • Ch. 10: Middle School, pp. 217–218 (third paragraph) • Ch. 10: High School, pp. 219 (last paragraph)–220 **IIIB**: *Uncovering Student Ideas:* Assessment Probe and Suggestions for Instruction • USI.1: Mountain Age, pp. 181, 186–187 • USI.E&ES: Is It Erosion, pp. 99, 101–102; Can A Plant Break Rocks? pp. 103, 105–106; Grand Canyon, pp. 107, 110; Mountains and Beaches, p. 111, 114; How Do Rivers Form? p. 117; 119–120

Section and Outcome	Selected Sources and Readings for Study and Reflection Read and examine **related** parts of
IV. Research on Commonly Held Ideas	**IVA:** *Benchmarks for Science Literacy:* Chapter 15 Research • 4C: Processes That Shape the Earth, p. 336 **IVB:** *Making Sense of Secondary Science: Research Into Children's Ideas* • Ch. 14: Rocks, pp. 112–114 **IVC:** *Uncovering Student Ideas:* Related Research • USI.1: Mountain Age, p. 185 • USI.E&ES: Is It Erosion, p. 101; Can A Plant Break Rocks? p. 105; Grand Canyon, pp. 109–110; Mountains and Beaches, pp. 113–114; How Do Rivers Form? p. 119
V. K–12 Articulation and Connections	**VA:** *Next Generation Science Standards:* Appendices: Progression • Appendix E: ESS2.A: Earth's Materials and Systems, p. 2 • Appendix E: ESS2.C: The Roles of Water in Earth's Surface Properties, p. 3 **VB:** *NSTA Quick Reference Guide to the NGSS K–12:* Progression • ESS2.A: Earth's Materials and Systems, p. 76; ESS2.C: The Roles of Water in Earth's Surface Properties, p. 77 **VC:** *The NSTA Atlas of the Three Dimensions:* Map Page • 5.2: Earths Systems (ESS2.A & ESS2.E) • 5.4: Weather and Climate (ESS2.C & ESS2.D)
VI. Assessment Expectation	**VIA:** *State Standards* • Examine your state's standards **VIB:** *Next Generation Science Standards:* Performance Expectations • Grade 2: 2-ESS1-1, 2-ESS2-1, p. 15 • Grade 4: 4-ESS2-1, p. 26 • MS: MS-ESS2-1, p. 58; MS-ESS2-2, p. 57 • HS: HS-ESS2-2, ESS2-5, p. 96 **VIC:** *NSTA Quick Reference Guide to the NGSS K–12:* Performance Expectations Column • Grade 2: 2-ESS1-1, 2-ESS2-1, p. 96 • Grade 4: 4-ESS2-1, p. 107 • MS: MS-ESS2-1, MS-ESS2-2, p. 134 • HS: HS-ESS2-2, p. 163; ESS2-5, p. 164

Review Chapter 2 instructions on how to use this guide.
Visit curriculumtopicstudy2.org for more information about CTS and additional resources.

NOTE: *Atlas* page numbers have not been provided because *The NSTA Atlas of the Three Dimensions* was produced concurrently with this edition. Titles and map codes are accurate.

Additional Readings:

 Available for download at www.curriculumtopicstudy2.org

Copyright © 2020 by Corwin Press, Inc. All rights reserved. Reprinted from *Science Curriculum Topic Study: Bridging the Gap Between Three-Dimensional Standards, Research, and Practice* (2nd ed.) by Page Keeley and Joyce Tugel. Thousand Oaks, CA: Corwin, www.corwin.com. Reproduction authorized for educational use by educators, local school sites, and/or noncommercial or nonprofit entities that have purchased the book.

Earth and Our Solar System
Grades K–12 Standards- and Research-Based Study of a Curricular Topic

Section and Outcome	Selected Sources and Readings for Study and Reflection Read and examine **related** parts of
I. Adult Content Knowledge	**IA:** *Science for All Americans* • Ch. 4: The Universe, pp. 40–42 • Ch. 10: Displacing Earth from the Center of the Universe, pp. 147–149 **IB:** *Science Matters* (2009 edition) • Ch. 10: The Solar System, pp. 172–176 **IC:** *Framework for K–12 Science Education:* Narrative Section • Ch. 7: ESS1.B: Earth and the Solar System, pp. 175–176 **ID:** *Disciplinary Core Ideas: Reshaping Teaching and Learning* • Ch. 10: ESS1.B: Earth and the Solar System, pp. 187–188 **IE:** *The NSTA Atlas of the Three Dimensions:* Narrative Page • 5.1: The Earth, the Solar System, and the Universe (ESS1.B)
II. Concepts, Core Ideas, or Practices	**IIA:** *Framework for K–12 Science Education:* Grade Band Endpoints • Ch. 7: ESS1.B: Earth and the Solar System, p. 176 **IIB:** *Next Generation Science Standards:* Disciplinary Core Ideas Column • Grade 1: ESS1.B: Earth and the Solar System, p. 11 • Grade 5: ESS1.B: Earth and the Solar System, p. 31 • MS: ESS1.B: Earth and the Solar System, pp. 56 • HS: ESS1.B: Earth and the Solar System, p. 92 **IIC:** *NSTA Quick Reference Guide to the NGSS K–12:* Disciplinary Core Ideas Column • Grade 1: ESS1.B: Earth and the Solar System, p. 95 • Grade 5: ESS1.B: Earth and the Solar System, p. 111 • MS: ESS1.B: Earth and the Solar System, p. 133 • HS: ESS1.B: Earth and the Solar System, p. 162
III. Curriculum, Instruction, and Formative Assessment	**IIIA:** *Disciplinary Core Ideas: Reshaping Teaching and Learning* • Ch. 10: K–2: Lower Elementary, pp. 188–189 • Ch. 10: 3–5: Upper Elementary, pp. 189–190 • Ch. 10: 6–8: Middle School, pp. 190–193 • Ch. 10: 9–12: High School, pp. 193–196 • Ch. 10: What Approaches Can We Use to Teach About This Disciplinary Core Idea? pp. 197–199 **IIIB:** *Uncovering Student Ideas:* Assessment Probe and Suggestions for Instruction • USI.1: Gazing at the Moon, pp. 189, 194–195; Going Through a Phase, pp. 197, 202–203 • USI.2: Darkness at Night, pp. 171, 174–175; Objects in the Sky, pp. 185, 188–189 • USI.4: Moonlight, pp. 161, 164; Lunar Eclipse, pp. 167, 170; Solar Eclipse, pp. 173, 176 • USI.K–2: What Lights Up the Moon? pp. 109, 111; When Is the Next Full Moon? pp. 113, 115–116 • USI.A: Sunrise to Sunset, pp. 43, 45–46; What's Moving, pp. 51, 53–54; How Far Away Is the Sun? pp. 61, 63–64; What's Inside Our Solar System? pp. 147, 151; How Do Planets Orbit the Sun? pp. 153, 156–157

Section and Outcome	Selected Sources and Readings for Study and Reflection Read and examine **related** parts of
IV. **Research on Commonly Held Ideas**	**IVA: Benchmarks for Science Literacy:** Chapter 15 Research • 4A: The Universe, p. 335 **IVB: Making Sense of Secondary Science: Research Into Children's Ideas** • Ch. 24: The Earth in Space, pp. 168–175 **IVC: Uncovering Student Ideas:** Related Research • USI.1: Gazing at the Moon, pp. 192–193; Going Through a Phase, p. 201 • USI.2: Darkness at Night, p. 174; Objects in the Sky, p. 188 • USI.4: Moonlight, p. 164; Lunar Eclipse, p. 170; Solar Eclipse, p. 176 • USI.K-2: What Lights Up the Moon? p. 111; When Is the Next Full Moon? p. 115 • USI.A: Sunrise to Sunset, p. 45; What's Moving, p. 53; How Far Away Is the Sun? p. 63; What's Inside Our Solar System? pp. 149–150; How Do Planets Orbit the Sun? pp. 155–156
V. **K–12 Articulation and Connections**	**VA: Next Generation Science Standards:** Appendices: Progression • Appendix E: ESS1.B: Earth and the Solar System, p. 2 **VB: NSTA Quick Reference Guide to the NGSS K–12:** Progression • ESS1.B: Earth and the Solar System, p. 75 **VC: The NSTA Atlas of the Three Dimensions:** Map Page • 5.1: The Earth, the Solar System, and the Universe (ESS1.A & ESS1.B)
VI. **Assessment Expectation**	**VIA: State Standards** • Examine your state's standards **VIB: Next Generation Science Standards:** Performance Expectations • Grade 1: 1-ESS1-1, 1-ESS1-2, p. 11 • Grade 5: 5-ESS1-2, p. 31 • MS: MS-ESS1-2, MS-ESS1-3, p. 56 • HS: HS-ESS1-4, p. 92 **VIC: NSTA Quick Reference Guide to the NGSS K–12:** Performance Expectations • Grade 1: 1-ESS1-1, 1-ESS1-2, pp. 94–95 • Grade 5: 5-ESS1-2, p. 111 • MS: MS-ESS1-2, MS-ESS1-3, p. 133 • HS: HS-ESS1-4, p. 162

Review Chapter 2 instructions on how to use this guide.
Visit curriculumtopicstudy2.org for more information about CTS and additional resources.

NOTE: *Atlas* page numbers have not been provided because *The NSTA Atlas of the Three Dimensions* was produced concurrently with this edition. Titles and map codes are accurate.

Additional Readings:

 Available for download at www.curriculumtopicstudy2.org

Copyright © 2020 by Corwin Press, Inc. All rights reserved. Reprinted from *Science Curriculum Topic Study: Bridging the Gap Between Three-Dimensional Standards, Research, and Practice* (2nd ed.) by Page Keeley and Joyce Tugel. Thousand Oaks, CA: Corwin, www.corwin.com. Reproduction authorized for educational use by educators, local school sites, and/or noncommercial or nonprofit entities that have purchased the book.

Earth, Moon, Sun System
Grades K–12 Standards- and Research-Based Study of a Curricular Topic

Section and Outcome	Selected Sources and Readings for Study and Reflection Read and examine **related** parts of
I. **Content Knowledge**	**IA: *Science for All Americans*** • Ch. 4: The Earth, p. 43 **IB: *Framework for K–12 Science Education:*** Narrative Section • Ch. 7: ESS1.B: Earth and the Solar System, pp. 175–176 **IC: *The NSTA Atlas of the Three Dimensions:*** Narrative Page • 5.1: The Earth, the Solar System, and the Universe (ESS1.B)
II. **Concepts, Core Ideas, or Practices**	**IIA: *Framework for K–12 Science Education:*** Grade Band Endpoints • Ch. 7: ESS1.B: Earth and the Solar System, p. 176 **IIB: *Next Generation Science Standards:*** Disciplinary Core Ideas Column • Grade 1: ESS1.A: The Universe and Its Stars; p. 11 • Grade 5: ESS1.B: Earth and the Solar System, p. 31 • MS: ESS1.A: The Universe and Its Stars; ESS1.B: Earth and the Solar System, p. 56 • HS: ESS1.B: Earth and the Solar System, p. 92 **IIC: *NSTA Quick Reference Guide to the NGSS K–12:*** Disciplinary Core Ideas Column • Grade 1: ESS1.A: The Universe and Its Stars, p. 94 • Grade 5: ESS1.B: Earth and the Solar System, p. 111 • MS: ESS1.A: The Universe and Its Stars; ESS1.B: Earth and the Solar System, p. 133 • HS: ESS1.B: Earth and the Solar System, p. 162
III. **Curriculum, Instruction, and Formative Assessment**	**IIIA: *Disciplinary Core Ideas: Reshaping Teaching and Learning*** • Ch. 10: K–2: Lower Elementary, pp. 188–189 • Ch. 10: 3–5: Upper Elementary, p. 190 • Ch. 10: 6–8: Middle School, pp. 190–192 • Ch. 10: 9–12: High School, pp. 193–194 • Ch. 10: What Approaches Can We Use to Teach About This Disciplinary Core Idea? pp. 197–198 **IIIB: *Uncovering Student Ideas:*** Assessment Probe and Suggestions for Instruction • USI.1: Gazing at the Moon, pp. 189, 194–195; Going Through a Phase, pp. 197, 202–203 • USI.2: Objects in the Sky, pp. 185, 188–189 • USI.4: Moonlight, pp. 161, 164; Lunar Eclipse, pp. 176, 170; Solar Eclipse, pp. 173, 176 • USI.K–2: What Lights Up the Moon? pp. 109, 111; When Is the Next Full Moon? pp. 113, 115–116 • USI.A: Section 3: Modeling the Moon (twelve probes), pp. 86–142
IV. **Research on Commonly Held Ideas**	**IVA: *Benchmarks for Science Literacy:*** Chapter 15 Research • 4B: Explanations of Astronomical Phenomena, pp. 335–336 **IVB: *Making Sense of Secondary Science: Research Into Children's Ideas*** • Ch. 24: The Earth, Moon, and Sun, p. 171; The Phases of the Moon and Eclipses, pp. 171–173

Section and Outcome	Selected Sources and Readings for Study and Reflection Read and examine **related** parts of
	IVC: *Uncovering Student Ideas:* Related Research • USI.1: Gazing at the Moon, pp. 192–193; Going Through a Phase, p. 201 • USI.2: Objects in the Sky, p. 188 • USI.4: Moonlight, p. 164; Lunar Eclipse, p. 170; Solar Eclipse, p. 176 • USI.K-2: What Lights Up the Moon? p. 111; When Is the Next Full Moon? p. 115 • USI.A: Section 3: Modeling the Moon (twelve probes), pp. 86–142
V. **K–12 Articulation and Connections**	**VA: *Next Generation Science Standards:*** Appendices: Progression • Appendix E: ESS1.B: Earth and the Solar System, p. 2 **VB: *NSTA Quick Reference Guide to the NGSS K–12:*** Progression • ESS1.B: Earth and the Solar System, p. 75 **VC: *The NSTA Atlas of the Three Dimensions:*** Map Page • 5.1: The Earth, the Solar System, and the Universe (ESS1.A & ESS1.B)
VI. **Assessment Expectation**	**VIA: *State Standards*** • Examine your state's standards **VIB: *Next Generation Science Standards:*** Performance Expectations • Grade 1: 1-ESS1-1, p. 11 • Grade 5: 5-ESS1-2, p. 31 • MS: MS-ESS1-1, MS-ESS1-2, p. 56 • HS: HS-ESS1-4, p. 92 **VIC: *NSTA Quick Reference Guide to the NGSS K–12:*** Performance Expectations • Grade 1: 1-ESS1-1, p. 94 • Grade 5: 5-ESS1-2, p. 111 • MS: MS-ESS1-1, MS-ESS1-2, p. 133 • HS: HS-ESS1-4, p. 162

Review Chapter 2 instructions on how to use this guide.

Visit curriculumtopicstudy2.org for more information about CTS and additional resources.

NOTE: *Atlas* page numbers have not been provided because *The NSTA Atlas of the Three Dimensions* was produced concurrently with this edition. Titles and map codes are accurate.

Additional Readings:

Available for download at www.curriculumtopicstudy2.org

Copyright © 2020 by Corwin Press, Inc. All rights reserved. Reprinted from *Science Curriculum Topic Study: Bridging the Gap Between Three-Dimensional Standards, Research, and Practice* (2nd ed.) by Page Keeley and Joyce Tugel. Thousand Oaks, CA: Corwin, www.corwin.com. Reproduction authorized for educational use by educators, local school sites, and/or noncommercial or nonprofit entities that have purchased the book.

Earth-Sun System
Grades K–8 Standards- and Research-Based Study of a Curricular Topic

Section and Outcome	Selected Sources and Readings for Study and Reflection Read and examine **related** parts of
I. Content Knowledge	**IA:** *Science for All Americans* • Ch. 4: The Earth, p. 43 **IB:** *Framework for K–12 Science Education:* Narrative Section • Ch. 7: ESS1.B: Earth and the Solar System, pp. 175–176 **IC:** *Disciplinary Core Ideas: Reshaping Teaching and Learning* • Ch. 10: ESS1.B: Earth and the Solar System, pp. 187–188 **ID:** *The NSTA Atlas of the Three Dimensions:* Narrative Page • 5.1: The Earth, the Solar System, and the Universe (ESS1.B)
II. Concepts, Core Ideas, or Practices	**IIA:** *Framework for K–12 Science Education:* Grade Band Endpoints • Ch. 7: ESS1.B: Earth and the Solar System, p. 176 **IIB:** *Next Generation Science Standards:* Disciplinary Core Ideas Column • Grade 1: ESS1.B: Earth and the Solar System, p. 11 • Grade 5: ESS1.B: Earth and the Solar System, p. 31 • MS: ESS1.B: Earth and the Solar System, p. 56 **IIC:** *NSTA Quick Reference Guide to the NGSS K–12:* Disciplinary Core Ideas Column • Grade 1: ESS1.B: Earth and the Solar System, p. 95 • Grade 5: ESS1.B: Earth and the Solar System, p. 111 • MS: ESS1.B: Earth and the Solar System, p. 133
III. Curriculum, Instruction, and Formative Assessment	**IIIA:** *Disciplinary Core Ideas: Reshaping Teaching and Learning* • Ch. 10: Lower Elementary, pp. 188–189 • Ch. 10: Upper Elementary, p. 190 • Ch. 10: Middle School, pp. 190–191 • Ch. 10: What Approaches Can We Use to Teach About This Disciplinary Core Idea? pp. 197–198 **IIIB:** *Uncovering Student Ideas:* Assessment Probe and Suggestions for Instruction • USI.2: Darkness at Night, pp. 171, 174–175 • USI.3: Summer Talk, pp. 177, 182–183; Me and My Shadow, pp. 185, 188–189 • USI.K–2: When Is My Shadow the Longest? pp. 105, 107–108 • USI.A: Section 2: The Sun Earth System, pp. 43–83 (nine probes)
IV. Research on Commonly Held Ideas	**IVA:** *Benchmarks for Science Literacy:* Chapter 15 Research • 4B: Explanations of Astronomical Phenomena, pp. 335–336 **IVB:** *Making Sense of Secondary Science: Research Into Children's Ideas* • Ch. 24: Day and Night, pp. 168–171; The Changing Year, pp. 173–174 **IVC:** *Uncovering Student Ideas:* Related Research • USI.2: Darkness at Night, pp. 174 • USI.3: Summer Talk, pp. 181–182; Me and My Shadow, p. 188 • USI.K–2: When Is My Shadow the Longest? p. 107 • USI.A: Section 2: The Sun Earth System, pp. 43–83 (nine probes)

Section and Outcome	Selected Sources and Readings for Study and Reflection Read and examine **related** parts of
V. **K–12 Articulation and Connections**	**VA:** *Next Generation Science Standards:* Appendices: Progression • Appendix E: ESS1.B: Earth and the Solar System, p. 2 **VB:** *NSTA Quick Reference Guide to the NGSS K–12:* Progression • ESS1.B: Earth and the Solar System, p. 75 **VC:** *The NSTA Atlas of the Three Dimensions:* Map Page • 5.1: The Earth, the Solar System, and the Universe (ESS1.A & ESS1.B)
VI. **Assessment Expectation**	**VIA:** *State Standards* • Examine your state's standards **VIB:** *Next Generation Science Standards:* Performance Expectations • Grade 1: 1-ESS1-2, p. 11 • Grade 5: 5-ESS1-2, p. 31 • MS: MS-ESS1-1, p. 56 **VIC:** *NSTA Quick Reference Guide to the NGSS K–12:* Performance Expectations • Grade 1: 1-ESS1-2, p. 95 • Grade 5: 5-ESS1-2, p. 111 • MS: MS-ESS1-1, p. 133

Review Chapter 2 instructions on how to use this guide.

Visit curriculumtopicstudy2.org for more information about CTS and additional resources.

NOTE: *Atlas* page numbers have not been provided because *The NSTA Atlas of the Three Dimensions* was produced concurrently with this edition. Titles and map codes are accurate.

Additional Readings:

Available for download at www.curriculumtopicstudy2.org

Copyright © 2020 by Corwin Press, Inc. All rights reserved. Reprinted from *Science Curriculum Topic Study: Bridging the Gap Between Three-Dimensional Standards, Research, and Practice* (2nd ed.) by Page Keeley and Joyce Tugel. Thousand Oaks, CA: Corwin, www.corwin.com. Reproduction authorized for educational use by educators, local school sites, and/or noncommercial or nonprofit entities that have purchased the book.

Formation of Earth, the Solar System, and the Universe
Grades 6–12 Standards- and Research-Based Study of a Curricular Topic

Section and Outcome	Selected Sources and Readings for Study and Reflection Read and examine **related** parts of
I. Content Knowledge	**IA:** *Science for All Americans* • Ch. 4: The Universe, pp. 40–42 **IB:** *Science Matters* (2009 edition) • Ch. 10: The Solar System and Formation of the Moon, pp. 172–174 • Ch. 11: The Big Bang, pp. 185–189 **IC:** *Framework for K–12 Science Education:* Narrative Section • Ch. 7: ESS1.A: The Universe and Its Stars, pp. 173–174 • Ch. 7: ESS1.B: Earth and the Solar System, p. 175 (first paragraph) **ID:** *The NSTA Atlas of the Three Dimensions:* Narrative Page • 5.1: The Earth, the Solar System, and the Universe (ESS1.A & ESS1.B)
II. Concepts, Core Ideas, or Practices	**IIA:** *Framework for K–12 Science Education:* Grade Band Endpoints • Ch. 7: ESS1.A: The Universe and Its Stars, Grades 8–12, p. 174 **IIB:** *Next Generation Science Standards:* Disciplinary Core Ideas Column • MS: ESS1.B: Earth and the Solar System, p. 56 • HS: ESS1.A: The Universe and Its Stars, p. 92 **IIC:** *NSTA Quick Reference Guide to the NGSS K–12:* Disciplinary Core Ideas Column • MS: ESS1.B: Earth and the Solar System, p. 133 • HS: ESS1.A: The Universe and Its Stars, p. 161
III. Curriculum, Instruction, and Formative Assessment	**IIIA:** *Disciplinary Core Ideas: Reshaping Teaching and Learning* • Ch. 10: Middle School, pp. 192–193 • Ch. 10: High School, pp. 194–196 • Ch. 10: What Approaches Can We Use to Teach About This Disciplinary Core Idea? pp. 198 (last paragraph)–200 **IIIB:** *Uncovering Student Ideas:* Assessment Probe and Suggestions for Instruction • USI.A: Teaching and Learning Considerations, p. 183 (last paragraph); Expanding Universe, pp. 233, 236–237; Is the Big Bang "Just a Theory?" pp. 239, 242–243 • USI.E&SS: How Old Is Earth? pp. 87, 89–90
IV. Research on Commonly Held Ideas	**IVA:** *Making Sense of Secondary Science: Research Into Children's Ideas* • Ch. 24: The Solar System and Beyond, pp. 174–175 **IVB:** *Uncovering Student Ideas:* Related Research • USI.A: Expanding Universe, pp. 235–236; Is the Big Bang "Just a Theory?" pp. 241–242 • USI.E&SS: How Old Is Earth? p. 89
V. K–12 Articulation and Connections	**VA:** *Next Generation Science Standards:* Appendices: Progression • Appendix E: ESS1.A: The Universe and Its Stars, p. 2 **VB:** *NSTA Quick Reference Guide to the NGSS K–12:* Progression • ESS1.A: The Universe and Its Stars, p. 75 • ESS1.B: Earth and the Solar System, p. 75 **VC:** *The NSTA Atlas of the Three Dimensions:* Map Page • 5.1: The Earth, the Solar System, and the Universe (ESS1.A & ESS1.B)

Section and Outcome	Selected Sources and Readings for Study and Reflection Read and examine **related** parts of
VI. Assessment Expectation	**VIA: *State Standards*** • Examine your state's standards **VIB: *Next Generation Science Standards:* Performance Expectations** • MS: MS-ESS1-2, p. 56 • HS: HS-ESS1-2, p. 92 **VIC: *NSTA Quick Reference Guide to the NGSS K–12:* Performance Expectations Column** • MS: MS-ESS1-2, p. 133 • HS: HS-ESS1-2, p. 161

Review Chapter 2 instructions on how to use this guide.

Visit curriculumtopicstudy2.org for more information about CTS and additional resources.

NOTE: *Atlas* page numbers have not been provided because *The NSTA Atlas of the Three Dimensions* was produced concurrently with this edition. Titles and map codes are accurate.

Additional Readings:

 Available for download at **www.curriculumtopicstudy2.org**

Copyright © 2020 by Corwin Press, Inc. All rights reserved. Reprinted from *Science Curriculum Topic Study: Bridging the Gap Between Three-Dimensional Standards, Research, and Practice* (2nd ed.) by Page Keeley and Joyce Tugel. Thousand Oaks, CA: Corwin, www.corwin.com. Reproduction authorized for educational use by educators, local school sites, and/or noncommercial or nonprofit entities that have purchased the book.

Phases of the Moon
Grades K–8 Standards- and Research-Based Study of a Curricular Topic

Section and Outcome	Selected Sources and Readings for Study and Reflection Read and examine **related** parts of
I. **Content Knowledge**	**IA:** *Science for All Americans* • Ch. 4: The Earth, p. 43 **IB:** *Framework for K–12 Science Education:* Narrative Section • Ch. 7: ESS1.B: Earth and the Solar System, pp. 175–176 **IC:** *Disciplinary Core Ideas: Reshaping Teaching and Learning* • Ch. 10: ESS1.B: Earth and the Solar System, p. 187 **ID:** *The NSTA Atlas of the Three Dimensions:* Narrative Page • 5.1: The Earth, the Solar System, and the Universe (ESS1.B)
II. **Concepts, Core Ideas, or Practices**	**IIA:** *Framework for K–12 Science Education:* Grade Band Endpoints • Ch. 7: ESS1.B: Earth and the Solar System, p. 176 **IIB:** *Next Generation Science Standards:* Disciplinary Core Ideas Column • Grade 1: ESS1.A: The Universe and Its Stars, p. 11 • Grade 5: ESS1.B: Earth and the Solar System, p. 31 • MS: ESS1.B: Earth and the Solar System, p. 56 **IIC:** *NSTA Quick Reference Guide to the NGSS K–12:* Disciplinary Core Ideas Column • Grade 1: ESS1.A: The Universe and Its Stars, p. 94 • Grade 5: ESS1.B: Earth and the Solar System, p. 111 • MS: ESS1.B: Earth and the Solar System, p. 133
III. **Curriculum, Instruction, and Formative Assessment**	**IIIA:** *Disciplinary Core Ideas: Reshaping Teaching and Learning* • Ch. 10: Lower Elementary, pp. 188–189 • Ch. 10: Upper Elementary, p. 190 • Ch. 10: Middle School, pp. 190–191 • Ch. 10: What Approaches Can We Use to Teach About This Disciplinary Core Idea? pp. 197–198 **IIIB:** *Uncovering Student Ideas:* Assessment Probe and Suggestions for Instruction • USI.1: Gazing at the Moon, pp. 189, 194–195; Going Through a Phase, pp. 197, 202–203 • USI.K-2: When Is the Next Full Moon? pp. 113, 115–116 • USI.A: Earth or Moon Shadow? pp. 103, 106; Moon Phase and Solar Eclipse, pp. 109, 111–112; Chinese Moon, pp. 123, 125–126; Crescent Moon, pp. 127, 129
IV. **Research on Commonly Held Ideas**	**IVA:** *Benchmarks for Science Literacy:* Chapter 15 Research • 4B: Explanations of Astronomical Phenomena, pp. 335–336 **IVB:** *Making Sense of Secondary Science: Research Into Children's Ideas* • Ch. 24: The Earth, Moon, and Sun, p. 171; The Phases of the Moon and Eclipses, pp. 171–173 **IVC:** *Uncovering Student Ideas:* Related Research • USI.1: Gazing at the Moon, pp. 192–193; Going Through a Phase, p. 201 • USI.K-2: When Is the Next Full Moon? p. 115 • USI.A: USI.A: Earth or Moon Shadow? p. 105; Moon Phase and Solar Eclipse, p. 111; Chinese Moon, p. 125; Crescent Moon, p. 129

Section and Outcome	Selected Sources and Readings for Study and Reflection Read and examine **related** parts of
V. **K–12 Articulation and Connections**	**VA:** *Next Generation Science Standards:* Appendices: Progression<ul><li>Appendix E: ESS1.B: Earth and the Solar System, p. 2</li></ul>**VB:** *NSTA Quick Reference Guide to the NGSS K–12:* Progression<ul><li>ESS1.B: Earth and the Solar System, p. 75</li></ul>**VC:** *The NSTA Atlas of the Three Dimensions:* Map Page<ul><li>5.1: The Earth, the Solar System, and the Universe (ESS1.B)</li></ul>
VI. **Assessment Expectation**	**VIA:** *State Standards*<ul><li>Examine your state's standards</li></ul>**VIB:** *Next Generation Science Standards:* Performance Expectations<ul><li>Grade 1: 1-ESS1-1, p. 11</li><li>Grade 5: 5-ESS1-2, p. 31</li><li>MS: MS-ESS1-1, p. 56</li></ul>**VIC:** *NSTA Quick Reference Guide to the NGSS K–12:* Performance Expectations<ul><li>Grade 1: 1-ESS1-1, p. 94</li><li>Grade 5: 5-ESS1-2, p. 111</li><li>MS: MS-ESS1-1, p. 133</li></ul>

Review Chapter 2 instructions on how to use this guide.

Visit curriculumtopicstudy2.org for more information about CTS and additional resources.

NOTE: *Atlas* page numbers have not been provided because *The NSTA Atlas of the Three Dimensions* was produced concurrently with this edition. Titles and map codes are accurate.

Additional Readings:

 Available for download at www.curriculumtopicstudy2.org

Copyright © 2020 by Corwin Press, Inc. All rights reserved. Reprinted from *Science Curriculum Topic Study: Bridging the Gap Between Three-Dimensional Standards, Research, and Practice* (2nd ed.) by Page Keeley and Joyce Tugel. Thousand Oaks, CA: Corwin, www.corwin.com. Reproduction authorized for educational use by educators, local school sites, and/or noncommercial or nonprofit entities that have purchased the book.

Seasons and Seasonal Patterns in the Sky
Grades K–8 Standards- and Research-Based Study of a Curricular Topic

Section and Outcome	Selected Sources and Readings for Study and Reflection Read and examine **related** parts of
I. Content Knowledge	**IA:** *Science for All Americans* • Ch. 4: The Earth, p. 43 **IB:** *Science Matters* (2009 edition) • Ch. 14: Earth Cycles, pp. 242–243 (third paragraph) **IC:** *Framework for K–12 Science Education:* Narrative Section • Ch. 7: ESS1.B: Earth and the Solar System, pp. 175–176 (last paragraph) **ID:** *The NSTA Atlas of the Three Dimensions:* Narrative Page • 5.1: The Earth, the Solar System, and the Universe (ESS1.B)
II. Concepts, Core Ideas, or Practices	**IIA:** *Framework for K–12 Science Education:* Grade Band Endpoints • Ch. 7: ESS1.B: Earth and the Solar System, p. 176 **IIB:** *Next Generation Science Standards:* Disciplinary Core Ideas Column • Grade 1: ESS1.B: Earth and the Solar System, p. 11 • Grade 5: ESS1.B: Earth and the Solar System, p. 31 • MS: ESS1.A: The Universe and Its Stars; ESS1.B: Earth and the Solar System, p. 56 **IIC:** *NSTA Quick Reference Guide to the NGSS K–12:* Disciplinary Core Ideas Column • Grade 1: ESS1.B: Earth and the Solar System, p. 95 • Grade 5: ESS1.B: Earth and the Solar System, p. 111 • MS: ESS1.B: Earth and the Solar System, p. 133
III. Curriculum, Instruction, and Formative Assessment	**IIIA:** *Disciplinary Core Ideas: Reshaping Teaching and Learning* • Ch. 10: Lower Elementary, p. 189 (first column) • Ch. 10: Upper Elementary, p. 190 (first column) • Ch. 10: Middle School, pp. 190–191 • Ch. 10: What Approaches Can We Use to Teach About This Disciplinary Core Idea? pp. 197–198 **IIIB:** *Uncovering Student Ideas:* Assessment Probe and Suggestions for Instruction • USI.3: Summer Talk, pp. 177, 182–183 • USI.A: Changing Constellations, pp. 75, 77–78; Shorter Days in Winter, pp. 69, 74; Why Is It Warmer in Summer? pp. 79, 82–83
IV. Research on Commonly Held Ideas	**IVA:** *Benchmarks for Science Literacy:* Chapter 15 Research • 4B: Explanations of Astronomical Phenomena, pp. 335–336 **IVB:** *Making Sense of Secondary Science: Research Into Children's Ideas* • Ch. 24: The Changing Year, pp. 173–174 **IVC:** *Uncovering Student Ideas:* Related Research • USI.3: Summer Talk, pp. 181–182 • USI.A: Changing Constellations, p. 77; Shorter Days in Winter, p. 77; Why Is It Warmer in Summer? pp. 81–82

Section and Outcome	Selected Sources and Readings for Study and Reflection Read and examine **related** parts of
V. K–12 Articulation and Connections	**VA:** *Next Generation Science Standards:* Appendices: Progression • Appendix E: ESS1.B: Earth and the Solar System, p. 2 **VB:** *NSTA Quick Reference Guide to the NGSS K–12:* Progression • ESS1.B: Earth and the Solar System, p. 75 **VC:** *The NSTA Atlas of the Three Dimensions:* Map Page • 5.1: The Earth, the Solar System, and the Universe (ESS1.A & ESS1.B)
VI. Assessment Expectation	**VIA:** *State Standards* • Examine your state's standards **VIB:** *Next Generation Science Standards:* Performance Expectations • Grade 1: 1-ESS1-2, p. 11 • Grade 5: 5-ESS1-2, p. 31 • MS: MS-ESS1-1, p. 56 **VIC:** *NSTA Quick Reference Guide to the NGSS K–12:* Performance Expectations • Grade 1: 1-ESS1-2, p. 95 • Grade 5: 5-ESS1-2, p. 111 • MS: MS-ESS1-1, p. 133

Review Chapter 2 instructions on how to use this guide.

Visit curriculumtopicstudy2.org for more information about CTS and additional resources.

NOTE: *Atlas* page numbers have not been provided because *The NSTA Atlas of the Three Dimensions* was produced concurrently with this edition. Titles and map codes are accurate.

Additional Readings:

Available for download at www.curriculumtopicstudy2.org

Copyright © 2020 by Corwin Press, Inc. All rights reserved. Reprinted from *Science Curriculum Topic Study: Bridging the Gap Between Three-Dimensional Standards, Research, and Practice* (2nd ed.) by Page Keeley and Joyce Tugel. Thousand Oaks, CA: Corwin, www.corwin.com. Reproduction authorized for educational use by educators, local school sites, and/or noncommercial or nonprofit entities that have purchased the book.

Stars and Galaxies
Grades K–12 Standards- and Research-Based Study of a Curricular Topic

Section and Outcome	Selected Sources and Readings for Study and Reflection Read and examine **related** parts of
I. Content Knowledge	**IA:** *Science for All Americans* • Ch. 4: The Universe, pp. 40–42 **IB:** *Science Matters* (2009 edition) • Ch. 10: Astronomy, pp. 165–172; Galaxies, pp. 176–178 **IC:** *Framework for K–12 Science Education:* Narrative Section • Ch. 7: ESS1.A: The Universe and Its Stars, pp. 173–174 **ID:** *The NSTA Atlas of the Three Dimensions:* Narrative Page • 5.1: The Earth, the Solar System, and the Universe (ESS1.A)
II. Concepts, Core Ideas, or Practices	**IIA:** *Framework for K–12 Science Education:* Grade Band Endpoints • Ch. 7: ESS1.A: The Universe and Its Stars, p. 174 **IIB:** *Next Generation Science Standards:* Disciplinary Core Ideas Column • Grade 1: ESS1.A: The Universe and Its Stars, p. 11 • Grade 5: ESS1.A: The Universe and Its Stars, p. 31 • MS: ESS1.A: The Universe and Its Stars, p. 56 • HS: ESS1.A: The Universe and Its Stars, p. 92 **IIC:** *NSTA Quick Reference Guide to the NGSS K–12:* Disciplinary Core Ideas Column • Grade 1: ESS1.A: The Universe and Its Stars, p. 94 • Grade 5: ESS1.A: The Universe and Its Stars, p. 111 • MS: ESS1.A: The Universe and Its Stars, p. 133 • HS: ESS1.A: The Universe and Its Stars, p. 161
III. Curriculum, Instruction, and Formative Assessment	**IIIA:** *Disciplinary Core Ideas: Reshaping Teaching and Learning* • Ch. 10: Lower Elementary, pp. 188–189 • Ch. 10: Upper Elementary, pp. 189–190 • Ch. 10: Middle School, p. 190 (last paragraph) • Ch. 10: High School, pp. 193–196 • Ch. 10: What Approaches Can We Use to Teach About This Disciplinary Core Idea? pp. 197–200 **IIIB:** *Uncovering Student Ideas:* Assessment Probe and Suggestions for Instruction • USI.2: Emmy's Moon and Stars, pp. 177, 181–182 • USI.3: Where Do Stars Go? pp. 191, 195 • USI.A: Section 5: Stars, Galaxies, and the Universe, Teaching and Learning Considerations, p. 183; Is the Sun a Star? pp. 189, 192; Where Are the Stars in Orion? pp. 193, 196; What Are Stars Made of? pp. 203, 206; What Happens to Stars When They Die? pp. 209, 212–213; Do Stars Change? pp. 215, 218; Seeing into the Past, pp. 223, 225–226; What Is the Milky Way? pp. 227, 231
IV. Research on Commonly Held Ideas	**IVA:** *Benchmarks for Science Literacy:* Chapter 15 Research • 4A: The Universe, p. 335 **IVB:** *Making Sense of Secondary Science: Research Into Children's Ideas* • Ch. 24: The Solar System and Beyond, pp. 174–175 **IVC:** *Uncovering Student Ideas:* Related Research • USI.2: Emmy's Moon and Stars, p. 181 • USI.3: Where Do Stars Go? pp. 194–195

Section and Outcome	Selected Sources and Readings for Study and Reflection Read and examine **related** parts of
	• USI.A: Is the Sun a Star? p. 191; Where Are the Stars in Orion? pp. 195–196; What Are Stars Made of? pp. 205–206; What Happens to Stars When They Die? pp. 211–212; Do Stars Change? pp. 217–218; Seeing Into the Past, p. 225; What Is the Milky Way? p. 230
V. K–12 Articulation and Connections	**VA:** *Next Generation Science Standards:* Appendices: Progression • Appendix E: ESS1.A: The Universe and Its Stars, p. 2 **VB:** *NSTA Quick Reference Guide to the NGSS K–12:* Progression • ESS1.A: The Universe and Its Stars, p. 75 **VC:** *The NSTA Atlas of the Three Dimensions:* Map Page • 5.1: The Earth, the Solar System, and the Universe (ESS1.A & ESS1.B)
VI. Assessment Expectation	**VIA:** *State Standards* • Examine your state's standards **VIB:** *Next Generation Science Standards:* Performance Expectations • Grade 1: 1-ESS1-1, p. 11 • Grade 5: 5-ESS-1-1, 5-ESS1-2, p. 31 • MS: MS-ESS1-2, p. 56 • HS: HS-ESS1-1, HS-ESS1-2, HS-ESS1-3, p. 92 **VIC:** *NSTA Quick Reference Guide to the NGSS K–12:* Performance Expectations • Grade 1: 1-ESS1-1, p. 94 • Grade 5: 5-ESS-1-1, 5-ESS1-2, p. 111 • MS: MS-ESS1-2, p. 133 • HS: HS-ESS1-1, HS-ESS1-2, HS-ESS1-3, p. 161

Review Chapter 2 instructions on how to use this guide.

Visit curriculumtopicstudy2.org for more information about CTS and additional resources.

NOTE: *Atlas* page numbers have not been provided because *The NSTA Atlas of the Three Dimensions* was produced concurrently with this edition. Titles and map codes are accurate.

Additional Readings:

 Available for download at www.curriculumtopicstudy2.org

Copyright © 2020 by Corwin Press, Inc. All rights reserved. Reprinted from *Science Curriculum Topic Study: Bridging the Gap Between Three-Dimensional Standards, Research, and Practice* (2nd ed.) by Page Keeley and Joyce Tugel. Thousand Oaks, CA: Corwin, www.corwin.com. Reproduction authorized for educational use by educators, local school sites, and/or noncommercial or nonprofit entities that have purchased the book.

NOTES

Category D: Scientific and Engineering Practices Guides

The ten CTS guides in this section focus on the practices that scientists and engineers engage in as part of their work. Some guides address a practice only in science or in engineering. Other guides address a practice in both science and engineering. The guides in this section are arranged in the order they appear in the *Framework for K–12 Science Education* (NRC, 2012) and include

- Asking Questions in Science
- Defining Problems in Engineering
- Developing and Using Models
- Planning and Carrying Out Investigations
- Analyzing and Interpreting Data
- Using Mathematics and Computational Thinking
- Constructing Scientific Investigations
- Designing Solutions in Engineering
- Argumentation
- Obtaining, Evaluating, and Communicating Information

NOTES FOR USING CATEGORY D GUIDES

Overall

- *Atlas* page numbers have not been provided because *The NSTA Atlas of the Three Dimensions* was produced concurrently with this edition. Titles and map codes are accurate.
- Some readings combine the use of the practice in both science and engineering. If a topic guide is specific to science (or to engineering), focus only on that discipline.
- Eliminate redundancy. Some readings include the exact same information. However, even when the information is the same, there may be an advantage in how the information is presented. Select the reading based on the resources you have available to use for CTS and/or the advantage of using one over the other.

Section I

- Readings from the *Framework*, *NSTA Quick Reference Guide*, and *Atlas* narrative are exactly the same. Choose one of these resources for this section. The reading in the *NGSS* Appendix F is similar to the *Framework*, but not identical. However, they are similar enough that you don't need to read them both.

- Read only the narrative sections in the *Framework* or the *NSTA Quick Reference Guide*. Stop at Goals.

- Read only the narrative section in NGSS Appendix F. Stop at the chart.

- When reading *Helping Students Make Sense of the World*, focus on the content that helps you understand this practice. There are also suggestions for instruction embedded in this section that can be added to CTS section III.

Section II

- Readings from the *Framework* and the first bulleted reading in the *NSTA Quick Reference Guide* goals (by end of grade 12) are exactly the same. Stop at Progression.

- The *Framework* goals were used to develop the descriptions of the elements of the practice in the *NGSS*. In the *Framework* they are written as broad statements of what students should be able to do by the end of grade 12. They are not broken down by grade span.

- The readings from the *NGSS* are taken from elements of the *Framework* and are designated by grade. These are repeated in the *NSTA Quick Reference Guide* in broader grade spans of K–2, 3–5, 6–8, and 9–12. The information is the same. Choose the specificity (grade or grade span) when deciding which of these resources to use.

- The *NSTA Quick Reference Guide* describes the elements of the practice at K–2, 3–5, 6–8, 9–12 levels. These same elements appear on the *Atlas* map used in section V.

Section III

- Readings from the *Framework* and the *NSTA Quick Reference Guide* are exactly the same. Choose one of these resources for this section.

- Readings from the *Framework* and the *NSTA Quick Reference Guide* are narrative descriptions of what the practice looks like at different grade spans, including the kinds of instructional experiences students should have. This is different from section V, which shows the progression of specific elements of what students are expected to do at different grade spans.

- Readings from *Helping Students Make Sense of Science* are the longest readings. In some guides these have been broken down into subsections. You can read all the designated pages or focus on the subsections that are of interest to you. If using this resource with a group, you may consider assigning subsections so one person does not have to read all the designated pages.

- The *Uncovering Student Ideas in Science* series is not used in this section. However, a new book, *Uncovering Students Ideas about Engineering and Technology* (Keeley, Sneider, and Ravel, in press), is due to be released in 2020 and can be used with the engineering practices in this section. Check the CTS website for information on including this resource.

Section IV

- Except for *Benchmarks for Science Literacy:* Chapter 15, research summaries for the practices are not included in the common resources used for this section. Most of the research on commonly held ideas is focused on disciplinary core ideas. If you have access to professional journals, you may search for research articles on difficulties students encounter when they use the practices and add these to the additional readings at the end of the guides.

Section V

- Readings from the *NGSS* Appendix and the *NSTA Quick Reference Guide* are the same. However, the advantage to using the *NSTA Quick Reference Guide* is that each element of a practice is shown as a progression. The *NGSS* Appendix progression chart lists all the elements for each grade span but does not indicate the progression of each individual element.

- The *Atlas* includes the same information but the visual mapping of the practices allows you to see precursors and connections, as well as connections to other maps. Be sure to read the front matter of the *Atlas* before using the maps. The information will help you use the maps effectively.

- The *Atlas* is also used in sections I, II, and V and thus provides a versatile resource to use for half of the sections of CTS guides in this category.

Section VI

- The *NGSS* and the *NSTA Quick Reference Guide* provide the exact same information. Clarifications and assessment boundaries are also included in both. Choose one of these resources based on the advantages listed below or the resources you have available.

- An advantage to using the *NSTA Quick Reference Guide* is that the performance expectation is listed next to the disciplinary core idea that uses that practice.

- An advantage to using the *NGSS* that is not evident in the *NSTA Quick Reference Guide* is that the scientific and engineering practices chart, included below the performance expectations, shows which element of the practice is part of the performance expectation. It also adds the third dimension by showing which crosscutting concept is included with the practice.

Asking Questions in Science
Grades K–12 Standards- and Research-Based Study of a Curricular Topic

Section and Outcome	Selected Sources and Readings for Study and Reflection Read and examine **related** parts of
I. Content Knowledge	**IA:** *Framework for K–12 Science Education:* Narrative Section • Ch. 3: Practice 1: Asking Questions and Defining Problems, pp. 54–55 **IB:** *NSTA Quick Reference Guide to the NGSS K–12:* Scientific and Engineering Practices • Practice 1: Asking Questions and Defining Problems, pp. 4–5 **IC:** *Helping Students Make Sense of the World Using NGSS Practices* • Ch. 5: The Importance of Asking Questions in Science, pp. 87–88 **ID:** *The NSTA Atlas of the Three Dimensions:* Narrative Page • 1.1: Asking Questions and Defining Problems (AQDP) **IE:** *Next Generation Science Standards:* Appendices • Appendix F: Practice 1: Asking Questions and Defining Problems, p. 4
II. Concepts, Core Ideas, or Practices	**IIA:** *Framework for K–12 Science Education:* Goals • Ch. 3: Asking Questions and Defining Problems, p. 55 **IIB:** *Next Generation Science Standards:* Scientific and Engineering Practices Column • Grade 3: Asking Questions and Defining Problems, p. 18 • Grade 4: Asking Questions and Defining Problems, p. 23 • MS: Asking Questions and Defining Problems, pp. 38, 59 • HS: Asking Questions and Defining Problems, pp. 76, 85 **IIC:** *NSTA Quick Reference Guide to the NGSS K–12:* Scientific and Engineering Practices • K–12: Goals by the End of Grade 12, p. 5 • K–2: Asking Questions and Defining Problems, p. 88 • 3–5: Asking Questions and Defining Problems, p. 100 • 6–8: Asking Questions and Defining Problems, p. 116 • 9–12: Asking Questions and Defining Problems, p. 140
III. Curriculum, Instruction, and Formative Assessment	**IIIA:** *Framework for K–12 Science Education:* Progression Narrative • Ch. 3: Asking Questions and Defining Problems, p. 56 **IIIB:** *NSTA Quick Reference Guide to the NGSS K–12:* Progression Narrative • Asking Questions and Defining Problems, p. 5 **IIIC:** *Helping Students Make Sense of the World Using NGSS Practices* • Ch. 5: The Role of Questioning in the Science and Engineering Practices, pp. 98–107
IV. Research on Commonly Held Ideas	*Students' commonly held ideas about questioning are not available in the current CTS resources
V. K–12 Articulation and Connections	**VA:** *Next Generation Science Standards:* Appendices: Progression Chart • Appendix F: Practice 1: Asking Questions and Defining Problems, pp. 4–5 **VB:** *NSTA Quick Reference Guide to the NGSS K–12:* Progression Chart • Asking Questions and Defining Problems, Condensed Practices, p. 50

Section and Outcome	Selected Sources and Readings for Study and Reflection Read and examine **related** parts of
	VC: *The NSTA Atlas of the Three Dimensions:* Map Page • 1.1: Asking Questions and Defining Problems (AQDP)
VI. **Assessment Expectation**	**VIA:** *State Standards* • Examine your state's standards **VIB:** *Next Generation Science Standards:* Performance Expectations • Grade 3: 3-PS2-3, p. 18 • Grade 4: 4-PS3-3, p. 23 • MS: MS-PS2-3, p. 38; MS-ESS3-5, p. 59 • HS: HS-PS4-2, p. 76; HS-LS3-1, p. 85 **VIC:** *NSTA Quick Reference Guide to the NGSS K–12:* Performance Expectations • Grade 3: 3-PS2-3, p. 106 • Grade 4: 4-PS3-3, p. 109 • MS: MS-PS2-3, p. 124; MS-ESS3-5, p. 137 • HS: HS-PS4-2, p. 152; HS-LS3-1, p. 158

Review Chapter 2 instructions on how to use this guide.

Visit curriculumtopicstudy2.org for more information about CTS and additional resources.

NOTE: *Atlas* page numbers have not been provided because *The NSTA Atlas of the Three Dimensions* was produced concurrently with this edition. Titles and map codes are accurate.

Additional Readings:

 Available for download at www.curriculumtopicstudy2.org

Copyright © 2020 by Corwin Press, Inc. All rights reserved. Reprinted from *Science Curriculum Topic Study: Bridging the Gap Between Three-Dimensional Standards, Research, and Practice* (2nd ed.) by Page Keeley and Joyce Tugel. Thousand Oaks, CA: Corwin, www.corwin.com. Reproduction authorized for educational use by educators, local school sites, and/or noncommercial or nonprofit entities that have purchased the book.

Defining Problems (Engineering)
Grades K–12 Standards- and Research-Based Study of a Curricular Topic

Section and Outcome	Selected Sources and Readings for Study and Reflection Read and examine **related** parts of
I. Content Knowledge	**IA:** *Framework for K–12 Science Education:* Narrative Section • Ch. 3: Practice 1: Asking Questions and Defining Problems, pp. 54–55 **IB:** *NSTA Quick Reference Guide to the NGSS K–12:* Scientific and Engineering Practices • Practice 1: Asking Questions and Defining Problems, pp. 4–5 (stop at Goals) **IC:** *The NSTA Atlas of the Three Dimensions:* Narrative Page • 1.1: Asking Questions and Defining Problems (AQDP) **ID:** *Next Generation Science Standards:* Appendices • Appendix F: Practice 1: Asking Questions and Defining Problems, p. 4
II. Concepts, Core Ideas, or Practices	**IIA:** *Framework for K–12 Science Education:* Goals • Ch. 3: Asking Questions and Defining Problems, p. 55 **IIB:** *Next Generation Science Standards:* Scientific and Engineering Practices Column • K: Asking Questions and Defining Problems, pp. 7, 16 • Grade 3: Asking Questions and Defining Problems, pp. 18, 32 • Grade 4: Asking Questions and Defining Problems, p. 32 • MS: Asking Questions and Defining Problems, p. 63 • HS: Asking Questions and Defining Problems, pp. 76, 102 **IIC:** *NSTA Quick Reference Guide to the NGSS K–12:* Scientific and Engineering Practices • K–12: Goals by the End of Grade 12, p. 5 • K–2: Asking Questions and Defining Problems, p. 88 • 3–5: Asking Questions and Defining Problems, p. 100 • 6–8: Asking Questions and Defining Problems, p. 116 • 9–12: Asking Questions and Defining Problems, p. 140
III. Curriculum, Instruction, and Formative Assessment	**IIIA:** *Framework for K–12 Science Education:* Progression Narrative • Ch. 3: Asking Questions and Defining Problems, p. 56 **IIIB:** *NSTA Quick Reference Guide to the NGSS K–12:* Progression Narrative • Asking Questions and Defining Problems, p. 5 **IIIC:** *Helping Students Make Sense of the World Using NGSS Practices* • Ch. 13: Asking Questions and Defining Problems, pp. 285–288
IV. Research on Commonly Held Ideas	*Research on students' commonly held ideas about asking questions and defining problems in engineering is not available in the current CTS resources
V. K–12 Articulation and Connections	**VA:** *Next Generation Science Standards:* Appendices: Progression Chart • Appendix F: Practice 1: Asking Questions and Defining Problems, pp. 4–5 **VB:** *NSTA Quick Reference Guide to the NGSS K–12:* Progression Chart • Asking Questions and Defining Problems, Condensed Practices, p. 50 **VC:** *The NSTA Atlas of the Three Dimensions:* Map Page • 1.1, Asking Questions and Defining Problems (AQDP)

Section and Outcome	Selected Sources and Readings for Study and Reflection Read and examine **related** parts of
VI. **Assessment** **Expectation**	**VIA: *State Standards*** • Examine your state's standards **VIB: *Next Generation Science Standards:* Performance Expectations** • K: K-ESS3-2, p. 7; K-2-ETS1-1, p. 16 • Grade 3: 3-PS2-4, p. 18; 3-5-ETS1, p. 32 • Grade 4: 3-5-ETS1, p. 32 • MS: MS-ETS1-1, p. 63 • HS: HS-PS4-2, p. 76; HS-ETS1-1, p. 102 **VIC: *NSTA Quick Reference Guide to the NGSS K–12:* Performance Expectations** • K: K-ESS3-2, p. 92; K-2-ETS-1, p. 98 • Grade 3: 3-PS2-4, p. 106; 3-5-ETS1-1, p. 114 • MS: MS-ETS1-1, p. 138 • HS: HS-PS4-2, p. 152; HS-ETS-1-1, p. 167

Review Chapter 2 instructions on how to use this guide.

Visit curriculumtopicstudy2.org for more information about CTS and additional resources.

NOTE: *Atlas* page numbers have not been provided because *The NSTA Atlas of the Three Dimensions* was produced concurrently with this edition. Titles and map codes are accurate.

Additional Readings:

online resources — Available for download at **www.curriculumtopicstudy2.org**

Copyright © 2020 by Corwin Press, Inc. All rights reserved. Reprinted from *Science Curriculum Topic Study: Bridging the Gap Between Three-Dimensional Standards, Research, and Practice* (2nd ed.) by Page Keeley and Joyce Tugel. Thousand Oaks, CA: Corwin, www.corwin.com. Reproduction authorized for educational use by educators, local school sites, and/or noncommercial or nonprofit entities that have purchased the book.

Developing and Using Models
Grades K–12 Standards- and Research-Based Study of a Curricular Topic

Section and Outcome	Selected Sources and Readings for Study and Reflection Read and examine **related** parts of
I. Content Knowledge	**IA: *Science for All Americans*** • Ch. 11B: Models, pp. 168–172 **IB: *Framework for K–12 Science Education:* Narrative Section** • Ch. 3: Practice 2: Developing and Using Models, pp. 56–58 **IC: *NSTA Quick Reference Guide to the NGSS K–12:*** Scientific and Engineering Practices • Practice 2: Developing and Using Models, pp. 6–7 **ID: *Helping Students Make Sense of the World Using NGSS Practices*** • Ch. 6: Developing and Using Models, pp. 111–122 **IE: *Next Generation Science Standards:* Appendices** • Appendix F: Practice 2: Developing and Using Models, p. 6 **IF: *The NSTA Atlas of the Three Dimensions:* Narrative Page** • 1.2: Developing and Using Models (MOD)
II. Concepts, Core Ideas, or Practices	**IIA: *Framework for K–12 Science Education:* Goals** • Ch. 3: Developing and Using Models, p. 58 **IIB: *Next Generation Science Standards:* Scientific and Engineering Practices Column** • K: Developing and Using Models, pp. 6, 16 • Grade 1: Developing and Using Models, p. 16 • Grade 2: Developing and Using Models, pp. 14, 15, 16 • Grade 3: Developing and Using Models, p. 20 • Grade 4: Developing and Using Models, pp. 24, 25 • Grade 5: Developing and Using Models, pp. 28, 29, 30, • MS: Developing and Using Models, pp. 35, 37, 40, 42, 45, 47, 50, 56, 58, 59, 63 • HS: Developing and Using Models, pp. 68, 70, 74, 80, 81, 85, 92, 94, 96, 98 **IIC: *NSTA Quick Reference Guide to the NGSS K–12:*** Scientific and Engineering Practices • K–12: Goals by the End of Grade 12, p. 7 • K–2: Developing and Using Models, p. 88 • 3–5: Developing and Using Models, p. 100 • 6–8: Developing and Using Models, p. 116 • 9–12: Developing and Using Models, p. 140
III. Curriculum, Instruction, and Formative Assessment	**IIIA: *Framework for K–12 Science Education:* Progression Narrative** • Ch. 3: Developing and Using Models, pp. 58–59 **IIIB: *NSTA Quick Reference Guide to the NGSS K–12:* Progression Narrative** • Developing and Using Models, pp. 7–8 **IIIC: *Helping Students Make Sense of the World Using NGSS Practices*** • Ch. 6: What Is Not Intended by the Modeling Practice, p. 116 • Ch. 6: Elementary Vignette, pp. 123–127; High School Vignette, pp. 127–130 • Ch. 6: How Can We Work Toward Equity with Regard to Modeling? p. 131 • Ch. 6: How Can I Support and Assess Developing and Using Models in the Classroom? How Do I Get Started With Modeling? pp. 131–134
IV. Research on Commonly Held Ideas	**IVA: *Benchmarks for Science Literacy:* Chapter 15 Research** • 11B: Models, p. 357
V. K–12 Articulation and Connections	**VA: *Next Generation Science Standards:* Appendices: Progression Chart** • Appendix F: Practice 2: Developing and Using Models, p. 6

Section and Outcome	Selected Sources and Readings for Study and Reflection Read and examine **related** parts of
	VB: *NSTA Quick Reference Guide to the NGSS K–12:* Progression Chart • Developing and Using Models, Condensed Practices, p. 51 **VC: *The NSTA Atlas of the Three Dimensions:*** Map Page • 1.2: Developing and Using Models (MOD)
VI. **Assessment Expectation**	**VIA: *State Standards*** • Examine your state's standards **VIB: *Next Generation Science Standards:*** Performance Expectations • K: K-ESS3-1, p. 6; K-2-ETS1-2, p. 16 • Grade 1: K-2-ETS1-2, p. 16 • Grade 2: 2-LS2-2, p. 14; 2-ESS2-2, p. 15; K-2-ETS1-2, p. 16 • Grade 3: 3-LS1-1, p. 20 • Grade 4: 4-PS4-1, p. 24; 4-PS4-2, p. 25 • Grade 5: 5-PS1-1, p. 28; 5-PS3-1, p. 29; 5-LS2-1, p. 29; 5-ESS2-1, p. 30 • MS: MS-PS1-1, MS-PS1-4, p. 35; MS-PS1-5, p. 37; MS-PS3-2, p. 40; MS-PS4-2, p. 42; MS-LS1-2, p. 45; MS-LS1-7, MS-LS2-3, p. 47; MS-LS3-1, MS-LS3-2, p. 50; MS-ESS1-1, MS-ESS1-2, p. 56; MS-ESS2-1, MS-ESS2-4, p. 58; MS-ESS2-6, p. 59; MS-ETS1-4, p. 63 • HS: HS-PS1-1, HS-PS1-8, p. 68; HS-PS1-4, p. 70; HS-PS1-8, P. 68; HS-PS3-2, HS-PS3-5, p. 74; HS-LS1-2, p. 80; HS-LS1-4, p. 85; HS-LS1-5, HS-LS1-7, HS-LS2-5, p. 81; HS-ESS1-1, p. 92; HS-ESS2-1, p. 94; HS-ESS2-3, HS-ESS2-6, p. 96; HS-ESS2-4, p. 98 **VIC: *NSTA Quick Reference Guide to the NGSS K–12:*** Performance Expectations • K: K-ESS3-1, p. 92; K-2-ETS1-2, p. 98 • Grade 1: K-2-ETS1-2, p. 98 • Grade 2: 2-LS2-2, 2-ESS2-2, p. 96; K-2-ETS1-2, p. 98 • Grade 3: 3-LS1-1, p. 104 • Grade 4: 4-PS4-1, p. 109, 4-PS4-2, p. 110 • Grade 5: 5-PS1-1, p. 112; 5-PS3-1, p. 113; 5-LS2-1, p. 111; 5-ESS2-1, p. 112 • MS: MS-PS1-1, MS-PS1-4, MS-PS1-5, pp. 122–123; MS-PS3-2, p. 125; MS-PS4-2, p. 127; MS-LS1-2, MS-LS1-7, pp. 128–129; MS-LS2-3, p. 130; MS-LS3-1, MS-LS3-2, p. 131; MS-ESS1-1, MS-ESS1-2, p. 133; MS-ESS2-1, MS-ESS2-4, MS-ESS2-6, pp. 134–135; MS-ETS1-4, p. 138 • HS: HS-PS1-1, HS-PS1-4, HS-PS1-8, pp. 147–148; HS-PS3-2, HS-PS3-5, pp. 151–152; HS-LS1-2, HS-LS1-4, HS-LS1-5, HS-LS1-7, pp. 154–155; HS-LS2-5, p. 157; HS-ESS1-1, p. 161; HS-ESS2-1, HS-ESS2-3, HS-ESS2-4, HS-ESS2-6, pp. 163–164

Review Chapter 2 instructions on how to use this guide.

Visit curriculumtopicstudy2.org for more information about CTS and additional resources.

NOTE: *Atlas* page numbers have not been provided because *The NSTA Atlas of the Three Dimensions* was produced concurrently with this edition. Titles and map codes are accurate.

Additional Readings:

 Available for download at www.curriculumtopicstudy2.org

Copyright © 2020 by Corwin Press, Inc. All rights reserved. Reprinted from *Science Curriculum Topic Study: Bridging the Gap Between Three-Dimensional Standards, Research, and Practice* (2nd ed.) by Page Keeley and Joyce Tugel. Thousand Oaks, CA: Corwin, www.corwin.com. Reproduction authorized for educational use by educators, local school sites, and/or noncommercial or nonprofit entities that have purchased the book.

Planning and Carrying Out Investigations
Grades K–12 Standards- and Research-Based Study of a Curricular Topic

Section and Outcome	Selected Sources and Readings for Study and Reflection Read and examine **related** parts of
I. Content Knowledge	**IA:** *Science for All Americans* • Ch. 1: Scientific Inquiry, pp. 3–7 **IB:** *Framework for K–12 Science Education:* Narrative Section • Ch. 3: Practice 3: Planning and Carrying Out Investigations, pp. 59–60 **IC:** *NSTA Quick Reference Guide to the NGSS K–12:* Scientific and Engineering Practices • Practice 3: Planning and Carrying Out Investigations, p. 9 **ID:** *Helping Students Make Sense of the World Using NGSS Practices* • Ch. 7: What Is the Planning and Carrying Out Investigations Practice All About? pp. 138–141 **IE:** *The NSTA Atlas of the Three Dimensions:* Narrative Page • 1.3: Planning and Carrying Out Investigations (INV) **IF:** *Next Generation Science Standards:* Appendices • Appendix F: Practice 3: Planning and Carrying Out Investigations, p. 7
II. Concepts, Core Ideas, or Practices	**IIA:** *Framework for K–12 Science Education:* Goals • Ch. 3: Planning and Carrying Out Investigations, p. 60 **IIB:** *Next Generation Science Standards:* Scientific and Engineering Practices Column • K: Planning and Carrying Out Investigations, pp. 5, 7 • Grade 1: Planning and Carrying Out Investigations, pp. 9, 11 • Grade 2: Planning and Carrying Out Investigations, pp. 13, 14 • Grade 3: Planning and Carrying Out Investigations, pp. 18, 32 • Grade 4: Planning and Carrying Out Investigations, pp. 23, 26, 32 • Grade 5: Planning and Carrying Out Investigations, pp. 28, 32 • MS: Planning and Carrying Out Investigations, pp. 38, 40, 45, 59 • HS: Planning and Carrying Out Investigations, pp. 68, 72, 74, 80, 96 **IIC:** *NSTA Quick Reference Guide to the NGSS K–12:* Scientific and Engineering Practices • K–12: Goals by the End of Grade 12, pp. 9–10 • K–2: Planning and Carrying Out Investigations, p. 88 • 3–5: Planning and Carrying Out Investigations, p. 100 • 6–8: Planning and Carrying Out Investigations, p. 116 • 9–12: Planning and Carrying Out Investigations, p. 140
III. Curriculum, Instruction, and Formative Assessment	**IIIA:** *Framework for K–12 Science Education:* Progression Narrative • Ch. 3: Planning and Carrying Out Investigations, pp. 60–61 **IIIB:** *NSTA Quick Reference Guide to the NGSS K–12:* Progression Narrative • Planning and Carrying Out Investigations, p. 10 **IIIC:** *Helping Students Make Sense of the World Using NGSS Practices* • Ch. 7: What Is Not Included in the Practice of Planning and Carrying Out an Investigation, p. 141 • Ch. 7: How Can I Support the Practice of Planning and Carrying Out Investigations in My Classroom? pp. 142–144 • Ch. 7: Vignettes: Elementary Example, pp. 145–146; Middle School Example, pp. 146–148; High School Example: pp. 149–154 • Ch. 7: Equity in Designing Investigations, pp. 154–155 • Ch. 7: How Do I Get Started With Students in Planning and Carrying Out Investigations? pp. 155–157
IV. Research on Commonly Held Ideas	**IVA:** *Benchmarks for Science Literacy:* Chapter 15 Research • 1B: Experimentation, p. 332 • 12C: Manipulation and Observation, p. 360 • 12E: Control of Variables, p. 360

Section and Outcome	Selected Sources and Readings for Study and Reflection Read and examine **related** parts of
V. K–12 Articulation and Connections	**VA:** *Next Generation Science Standards:* Appendices: Progression Chart • Appendix F: Practice 3: Planning and Carrying Out Investigations, pp. 7–8 **VB:** *NSTA Quick Reference Guide to the NGSS K–12:* Progression Chart • Planning and Carrying Out Investigations, Condensed Practices, p. 52 **VC:** *The NSTA Atlas of the Three Dimensions:* Map Page • 1.3: Planning and Carrying Out Investigations (INV)
VI. Assessment Expectation	**VIA:** *State Standards* • Examine your state's standards **VIB:** *Next Generation Science Standards:* Performance Expectations • K: K-PS2-1, p. 5; K-PS3-1, p. 7 • Grade 1: 1-PS4-1, 1-PS4-3, p. 9; 1-ESS1-2, p. 11 • Grade 2: 2-PS1-1, p. 13; 2-LS2-1, 2-LS4-1, p. 14 • Grade 3: 3-PS2-1, 3-PS2-2, p. 18; ETS-1-3, p. 32 • Grade 4: 4-PS3-2, p. 23; 4-ESS2-1, p. 26; ETS-1-3, p. 32 • Grade 5: 5-PS1-3, 5-PS1-4, p. 28; ETS-1-3, p. 32 • MS: MS-PS2-2, MS-PS2-5, p. 38; MS-PS3-4, p. 40; MS-LS1-1, p. 45; MS-ESS2-5, p. 59 • HS: HS-PS1-3, p. 68; HS-PS2-5, p. 72; HS-PS3-4, p. 74; HS-LS1-3, p. 80; HS-ESS2-5, p. 96 **VIC:** *NSTA Quick Reference Guide to the NGSS K–12:* Performance Expectations • K: K-PS2-1, K-PS3-1, p. 93 • Grade 1: 1-PS4-1, 1-PS4-3, 1-ESS1-2, p. 95 • Grade 2: 2-PS1-1, 2-LS2-1, 2-LS4-1, p. 96 • Grade 3: 3-PS2-1, 3-PS2-2, pp. 105–106 • Grade 4: 4-PS3-2, p. 108; 4-ESS2-1, p. 107 • Grade 5: 5-PS1-3, 5-PS1-4, p. 113 • MS: MS-PS2-2, p. 124; MS-PS2-5, p. 125; MS-PS3-4, p. 126; MS-LS1-1, p. 128; MS-ESS2-5, p. 135 • HS: HS-PS1-3, p. 147; HS-PS2-5, p. 150; HS-PS3-4, p. 151; HS-LS1-3, p. 154; HS-ESS2-5, p. 164

Review Chapter 2 instructions on how to use this guide.

Visit curriculumtopicstudy2.org for more information about CTS and additional resources.

NOTE: *Atlas* page numbers have not been provided because *The NSTA Atlas of the Three Dimensions* was produced concurrently with this edition. Titles and map codes are accurate.

Additional Readings:

Available for download at www.curriculumtopicstudy2.org

Copyright © 2020 by Corwin Press, Inc. All rights reserved. Reprinted from *Science Curriculum Topic Study: Bridging the Gap Between Three-Dimensional Standards, Research, and Practice* (2nd ed.) by Page Keeley and Joyce Tugel. Thousand Oaks, CA: Corwin, www.corwin.com. Reproduction authorized for educational use by educators, local school sites, and/or noncommercial or nonprofit entities that have purchased the book.

Analyzing and Interpreting Data
Grades K–12 Standards- and Research-Based Study of a Curricular Topic

Section and Outcome	Selected Sources and Readings for Study and Reflection Read and examine **related** parts of
I. Content Knowledge	**IA:** *Science for All Americans* • Ch. 9: Summarizing Data, pp. 137–139; Sampling, pp. 139–140 **IB:** *Framework for K–12 Science Education:* Narrative Section • Ch. 3: Practice 4: Analyzing and Interpreting Data, pp. 61–62 **IC:** *NSTA Quick Reference Guide to the NGSS K–12:* Scientific and Engineering Practices • Practice 4: Analyzing and Interpreting Data, p. 11 **ID:** *Helping Students Make Sense of the World Using NGSS Practices* • Ch. 8: What Is the Practice of Analyzing and Interpreting Data? pp. 159–163 **IE:** *Next Generation Science Standards:* Appendices • Appendix F: Practice 4: Analyzing and Interpreting Data, p. 9 **IF:** *The NSTA Atlas of the Three Dimensions:* Narrative Page • 1.4: Analyzing and Interpreting Data (DATA)
II. Concepts, Core Ideas, or Practices	**IIA:** *Framework for K–12 Science Education:* Goals • Ch. 3: Practice 4: Analyzing and Interpreting Data, pp. 62–63 **IIB:** *Next Generation Science Standards:* Scientific and Engineering Practices Column • K: Analyzing and Interpreting Data, pp. 5, 6, 7, 16 • Grade 1: Analyzing and Interpreting Data, pp. 11, 16 • Grade 2: Analyzing and Interpreting Data, pp. 13, 16 • Grade 3: Analyzing and Interpreting Data, pp. 19, 20, 21 • Grade 4: Analyzing and Interpreting Data, p. 26 • Grade 5: Analyzing and Interpreting Data, p. 31 • MS: Analyzing and Interpreting Data, pp. 37, 40, 47, 52, 56, 57, 60, 63 • HS: Analyzing and Interpreting Data, pp. 72, 85, 87, 96, 98 **IIC:** *NSTA Quick Reference Guide to the NGSS K–12:* Scientific and Engineering Practices • K–12: Goals by the End of Grade 12, pp. 11–12 • K–2: Analyzing and Interpreting Data, p. 88 • 3–5: Analyzing and Interpreting Data, p. 100 • 6–8: Analyzing and Interpreting Data, p. 117 • 9–12: Analyzing and Interpreting Data, p. 141
III. Curriculum, Instruction, and Formative Assessment	**IIIA:** *Framework for K–12 Science Education:* Progression Narrative • Ch. 3: Analyzing and Interpreting Data, p. 63 **IIIB:** *NSTA Quick Reference Guide to the NGSS K–12:* Progression Narrative • Practice 4: Analyzing and Interpreting Data, p. 12 **IIIC:** *Helping Students Make Sense of the World Using NGSS Practices* • Ch. 8: What Is Not Meant by the Practice of Analyzing and Interpreting Data, p. 162 • Ch. 8: Elementary Vignette, pp. 164–167; Middle School Vignette, pp. 167–171; High School Vignette, pp. 171–175 • Ch. 8: Common Challenges in Analyzing and Interpreting Data and Assessment, pp. 175–177 • Ch. 8: Equity With Regard to Analyzing and Interpreting Data, pp. 178–180
IV. Research on Commonly Held Ideas	**IVA:** *Benchmarks for Science Literacy:* Chapter 15 Research • 9B: Graphs, p. 351 • 9D: Summarizing Data, pp. 353–354 • 12E: Interpretation of Data, p. 361

Section and Outcome	Selected Sources and Readings for Study and Reflection Read and examine **related** parts of
V. **K–12 Articulation and Connections**	**VA:** *Next Generation Science Standards:* Appendices: Progression Chart • Appendix F: Practice 4: Analyzing and Interpreting Data, p. 9 **VB:** *NSTA Quick Reference Guide to the NGSS K–12:* Progression Chart • Practice 4: Analyzing and Interpreting Data, p. 53 **VC:** *The NSTA Atlas of the Three Dimensions:* Map Page • 1.1: Analyzing and Interpreting Data (DATA)
VI. **Assessment Expectation**	**VIA:** *State Standards* • Examine your state's standards **VIB:** *Next Generation Science Standards:* Performance Expectations • K: K-PS2-2, p. 5; K-LS1-1, p. 6; K-ESS2-1, p. 7; K-2-ETS1-3, p. 16 • Grade 1: 1-ESS-1, p. 11; K-2-ETS1-3, p. 16 • Grade 2: 2-PS1-2, p. 13; K-2-ETS1-3, p. 16 • Grade 3: 3-LS3-1, p. 19; 3-LS4-1, p. 20; 3-ESS2-1, p. 21 • Grade 4: 4-ESS2-2, p. 26 • Grade 5: 5-ESS1-2, p. 31 • MS: MS-PS1-2, p. 37; MS-PS3-1, p. 40; MS-LS2-1, p. 47; MS-LS4-3, MS-LS4-1, p. 52; MS-ESS1-3, p. 56; MS-ESS2-3, p. 57; MS-ESS3-2, p. 60; MS-ETS1-3, p. 63 • HS: HS-PS2-1, p. 72; HS-LS3-3, p. 85; HS-LS4-3, p. 87; HS-ESS2-2, p. 96; HS-ESS3-5, p. 98 **VIC:** *NSTA Quick Reference Guide to the NGSS K–12:* Performance Expectations • K: K-PS2-2, p. 93; K-LS1-1, K-ESS2-1, p. 92; K-2-ETS1-3, p. 98 • Grade 1: 1-ESS-1, p. 94; K-2-ETS1-3, p. 98 • Grade 2: 2-PS1-2, p. 97; K-2-ETS1-3, p. 98 • Grade 3: 3-LS3-1, 3-LS4-1, p. 104; 3-ESS2-1, p. 105 • Grade 4: 4-4-ESS2-2, p. 107 • Grade 5: 5-ESS1-2, p. 111 • MS: MS-PS1-2, p. 122; MS-PS3-1, p. 125; MS-LS2-1, p. 130, MS-LS4-3, p. 132; MS-LS4-1, p. 131; MS-ESS1-3, p. 133; MS-ESS2-3, p. 134; MS-ESS3-2, p. 136; MS-ETS1-3, p. 138 • HS: HS-PS2-1, p. 149; HS-LS3-3, p. 158; HS-LS4-3, p. 159; HS-ESS2-2, p. 163; HS-ESS3-5, p. 166

Review Chapter 2 instructions on how to use this guide.

Visit curriculumtopicstudy2.org for more information about CTS and additional resources.

NOTE: *Atlas* page numbers have not been provided because *The NSTA Atlas of the Three Dimensions* was produced concurrently with this edition. Titles and map codes are accurate.

Additional Readings:

 Available for download at www.curriculumtopicstudy2.org

Copyright © 2020 by Corwin Press, Inc. All rights reserved. Reprinted from *Science Curriculum Topic Study: Bridging the Gap Between Three-Dimensional Standards, Research, and Practice* (2nd ed.) by Page Keeley and Joyce Tugel. Thousand Oaks, CA: Corwin, www.corwin.com. Reproduction authorized for educational use by educators, local school sites, and/or noncommercial or nonprofit entities that have purchased the book.

Using Mathematics and Computational Thinking
Grades K–12 Standards- and Research-Based Study of a Curricular Topic

Section and Outcome	Selected Sources and Readings for Study and Reflection Read and examine **related** parts of
I. Content Knowledge	**IA:** *Science for All Americans* • Ch. 2: Mathematics, Science, and Technology, pp. 17–18 • Ch. 9: The Mathematical World, pp. 129–143 • Ch. 11: Models, pp. 171–172 (scroll down to Mathematical Models) **IB:** *Framework for K–12 Science Education:* Narrative Section • Ch. 3: Practice 5: Using Mathematics and Computational Thinking, pp. 64–65 **IC:** *NSTA Quick Reference Guide to the NGSS K–12:* Scientific and Engineering Practices • Practice 5: Using Mathematics and Computational Thinking, pp. 13–14 **ID:** *Helping Students Make Sense of the World Using NGSS Practices* • Ch. 8: Why Is the Practice of Using Mathematics and Computational Thinking Important? pp. 184–187 • Ch. 8: What Is the Practice of Using Mathematics and Computational Thinking and How Does It Relate to Other Practices? pp. 188–194 **IE:** *Next Generation Science Standards:* Appendices • Appendix F: Practice 5: Using Mathematics and Computational Thinking, p. 10 **IF:** *The NSTA Atlas of the Three Dimensions:* Narrative Page • 1.5: Using Mathematics and Computational Thinking (MATH)
II. Concepts, Core Ideas, or Practices	**IIA:** *Framework for K–12 Science Education:* Goals • Ch. 3: Practice 5: Using Mathematics and Computational Thinking, pp. 65–66 **IIB:** *Next Generation Science Standards:* Scientific and Engineering Practices Column • Grade 5: Using Mathematics and Computational Thinking, pp. 28, 30 • MS: Using Mathematics and Computational Thinking, pp. 42, 52 • HS: Using Mathematics and Computational Thinking, pp. 70, 72, 74, 76, 81, 83, 92, 99, 102 **IIC:** *NSTA Quick Reference Guide to the NGSS K–12:* Scientific and Engineering Practices • K–12: Goals by the End of Grade 12, p. 14 • K–2: Using Mathematics and Computational Thinking, p. 88 • 3–5: Using Mathematics and Computational Thinking, p. 101 • 6–8: Using Mathematics and Computational Thinking, p. 117 • 9–12: Using Mathematics and Computational Thinking, p. 141
III. Curriculum, Instruction, and Formative Assessment	**IIIA:** *Framework for K–12 Science Education:* Progression Narrative • Ch. 3: Using Mathematics and Computational Thinking, pp. 66–67 **IIIB:** *NSTA Quick Reference Guide to the NGSS K–12:* Progression Narrative • Using Mathematics and Computational Thinking, pp. 14–15 **IIIC:** *Uncovering Student Ideas:* Probe and Suggestions for Instruction • USI.PS1: How Far Did It Go? pp. 15, 17; Go Cart Test Run, pp. 31, 33–34; Checking the Speedometer, pp. 35, 37–38; Speed Units, pp. 39, 41; String Around the Earth, pp. 63, 65–66 **IIID:** *Helping Students Make Sense of the World Using NGSS Practices* • Ch. 9: Elementary Vignette, pp. 194–197; Middle School Vignette, pp. 200–202 • Ch. 9: What Is Not Included in This Practice, p. 194 • Ch. 9: Equity and Supporting and Assessing the Practice, pp. 198-200 • Ch. 9: How Do I Get Started? and Introducing the Practice in Your Classroom, pp. 200–204
IV. Research on Commonly Held Ideas	**IVA:** *Benchmarks for Science Literacy:* Chapter 15 Research • 9A: Numbers, p. 350 • 9B: Symbolic Relationships, pp. 350–352 • 9D: Uncertainty, pp. 353–354

Section and Outcome	Selected Sources and Readings for Study and Reflection Read and examine **related** parts of
	• 12B: Computation and Estimation, pp. 358–360 • 12C: Manipulation and Observation, p. 360 **IVB: *Uncovering Student Ideas:*** Related Research • USI.PS1: How Far Did It Go? p. 17; Go Cart Test Run, p. 33; Checking the Speedometer, p. 37; Speed Units, p. 41; String Around the Earth, p. 65
V. **K–12 Articulation and Connections**	**VA: *Next Generation Science Standards:*** Appendices: Progression Chart • Appendix F: Practice 5: Using Mathematics and Computational Thinking, p. 10 **VB: *NSTA Quick Reference Guide to the NGSS K–12:*** Progression Chart • Using Mathematics and Computational Thinking, Condensed Practices, p. 54 **VC: *The NSTA Atlas of the Three Dimensions:*** Map Page • 1.5: Using Mathematics and Computational Thinking (MATH)
VI. **Assessment Expectation**	**VIA: *State Standards*** • Examine your state's standards **VIB: *Next Generation Science Standards:*** Performance Expectations • Grade 5: 5-PS1-2, p. 28; 5-ESS2-2, p. 30 • MS: MS-PS4-1, p. 42; MS-LS4-6, p. 52 • HS: HS-PS1-7, p. 70; HS-PS2-2, HS-PS2-4, p. 72; HS-PS3-1, p. 74; HS-PS4-1, p. 76; HS-LS2-4, p. 81; HS-LS2-1, p. 81; HS-LS2-2, HS-LS4-6, p. 83; HS-ESS1-4, p. 92; HS-ESS3-3, HS-ESS3-6, p. 99; HS-ETS1-4, p. 102 **VIC: *NSTA Quick Reference Guide to the NGSS K–12:*** Performance Expectations • Grade 5: 5-PS1-2, 5-ESS2-2, p. 112 • MS: MS-PS4-1, p. 127; MS-LS4-6, p. 132 • HS: HS-PS1-7, p. 148; HS-PS2-2, p. 149; HS-PS2-4, HS-PS3-1, p. 150; HS-PS4-1, p. 152; HS-LS2-1, p. 155; HS-LS2-2, HS-LS2-4, p. 156; HS-LS2-6, p. 157; HS-ESS1-4, p. 162; HS-ESS3-3, HS-ESS3-6, p. 166; HS-ETS1-4, p. 167

Review Chapter 2 instructions on how to use this guide.

Visit curriculumtopicstudy2.org for more information about CTS and additional resources.

NOTE: *Atlas* page numbers have not been provided because *The NSTA Atlas of the Three Dimensions* was produced concurrently with this edition. Titles and map codes are accurate.

Additional Readings:

Available for download at www.curriculumtopicstudy2.org

Copyright © 2020 by Corwin Press, Inc. All rights reserved. Reprinted from *Science Curriculum Topic Study: Bridging the Gap Between Three-Dimensional Standards, Research, and Practice* (2nd ed.) by Page Keeley and Joyce Tugel. Thousand Oaks, CA: Corwin, www.corwin.com. Reproduction authorized for educational use by educators, local school sites, and/or noncommercial or nonprofit entities that have purchased the book.

Constructing Scientific Explanations
Grades K–12 Standards- and Research-Based Study of a Curricular Topic

Section and Outcome	Selected Sources and Readings for Study and Reflection Read and examine **related** parts of
I. Content Knowledge	**IA:** *Science for All Americans* • Ch. 1: Scientific Inquiry, Science Explains and Predicts, p. 6 **IB:** *Framework for K–12 Science Education:* Narrative Section • Ch. 3: Practice 6: Constructing Explanations and Designing Solutions, pp. 67–68 (stop at last paragraph) **IC:** *NSTA Quick Reference Guide to the NGSS K–12:* Scientific and Engineering Practices • Practice 6: Constructing Explanations and Designing Solutions, pp. 16–17 **ID:** *Helping Students Make Sense of the World Using NGSS Practices* • Ch. 10: Why Is Constructing Explanations Important? pp. 208–209 • Ch. 10: What Is a Scientific Explanation? pp. 209–216 • Ch. 10: How Does Constructing Explanations Relate to Other Practices? pp. 216–218 **IE:** *Next Generation Science Standards:* Appendices • Appendix F: Practice 6: Constructing Explanations and Designing Solutions, p. 11 **IF:** *The NSTA Atlas of the Three Dimensions:* Narrative Page • 1.6: Constructing Explanations and Designing Solutions (CEDS)
II. Concepts, Core Ideas, or Practices	**IIA:** *Framework for K–12 Science Education:* Goals • Ch. 3: Practice 6: Constructing Explanations and Designing Solutions, p. 69 (first 4 bullets) **IIB:** *Next Generation Science Standards:* Scientific and Engineering Practices Column • Grade 1: Constructing Explanations and Designing Solutions, pp. 9, 10 • Grade 2: Constructing Explanations and Designing Solutions, pp. 13, 15 • Grade 3: Constructing Explanations and Designing Solutions, p. 20 • Grade 4: Constructing Explanations and Designing Solutions, pp. 23, 26 • MS: Constructing Explanations and Designing Solutions, pp. 47, 49, 50, 52, 57, 58 • HS: Constructing Explanations and Designing Solutions, pp. 70, 80, 81, 87, 92, 94, 99 **IIC:** *NSTA Quick Reference Guide to the NGSS K–12:* Scientific and Engineering Practices • K–12: Goals by the End of Grade 12, p. 17 • K–2: Constructing Explanations and Designing Solutions, p. 89 • 3–5: Constructing Explanations and Designing Solutions, p. 101 • 6–8: Constructing Explanations and Designing Solutions, p. 117 • 9–12: Constructing Explanations and Designing Solutions, p. 141
III. Curriculum, Instruction, and Formative Assessment	**IIIA:** *Framework for K–12 Science Education:* Progression Narrative • Ch. 3: Constructing Explanations and Designing Solutions, pp. 69–70 (stop at Progression for Design) **IIIB:** *NSTA Quick Reference Guide to the NGSS K–12:* Progression Narrative • Constructing Explanations and Designing Solutions, pp. 17–18 (stop at Progression for Design) **IIIC:** *Helping Students Make Sense of the World Using NGSS Practices* • Ch. 10: Elementary Vignette, pp. 218–219; Middle School Vignette, pp. 219–220; High School Vignette, pp. 220–222 • Ch. 10: What Does Not Count as an Explanation in the *NGSS*, p. 212 • Ch. 10: Equity in Regard to This Practice, p. 222 • Ch. 10: How Can I Support and Assess This Practice? pp. 223–227
IV. Research on Commonly Held Ideas	**IVA:** *Benchmarks for Science Literacy:* Chapter 15 Research • 12E: Theory and Evidence, Interpretation of Data, p. 361

Section and Outcome	Selected Sources and Readings for Study and Reflection Read and examine **related** parts of
V. **K–12 Articulation and Connections**	**VA:** *Next Generation Science Standards:* Appendices: Progression Chart • Appendix F: Practice 6: Constructing Explanations and Designing Solutions (focus on explanations in science), pp. 11–12 **VB:** *NSTA Quick Reference Guide to the NGSS K–12:* Progression Chart • Constructing Explanations and Designing Solutions (focus on explanations in science), p. 55 **VC:** *The NSTA Atlas of the Three Dimensions:* Map Page • 1.6: Constructing Explanations and Designing Solutions (CEDS)
VI. **Assessment Expectation**	**VIA:** *State Standards* • Examine your state's standards **VIB:** *Next Generation Science Standards:* Performance Expectations • Grade 1: 1-PS4-2, p. 9; 1-LS3-1, p. 10 • Grade 2: 2-PS1-3, p. 13; 2-ESS1-1, p. 15 • Grade 3: 3-LS3-2, 3-LS4-2, p. 20 • Grade 4: 4-PS3-1, p. 23; 4-ESS1-1, p. 26 • MS: MS-LS1-6, p. 47; MS-LS2-2, p. 49; MS-LS1-5, p. 50; MS-LS4-2, MS-LS4-4, p. 52; MS-ESS1-4, MS-ESS2-2, p. 57; MS-ESS3-1, p. 58 • HS: HS-PS1-2, HS-PS1-5, p. 70; HS-LS1-1, p. 80; HS-LS1-6, HS-LS2-3, p. 81; HS-LS4-2, HS-LS4-4, p. 87; HS-ESS1-2, p. 92; HS-ESS1-6, p. 94; HS-ESS3-1, p. 99 **VIC:** *NSTA Quick Reference Guide to the NGSS K–12:* Performance Expectations • Grade 1: 1-PS4-2, p. 95; 1-LS3-1, p. 94 • Grade 2: 2-PS1-3, p. 97; 2-ESS1-1, p. 96 • Grade 3: 3-LS3-2, 3-LS4-2, p. 104 • Grade 4: 4-PS3-1, p. 108; 4-ESS1-1, p. 107 • MS: MS-LS1-5, MS-LS1-6, p. 129; MS-LS2-2, p. 130; MS-LS4-2, p. 131; MS-LS4-4, p. 132; MS-ESS1-4, MS-ESS2-2, p. 134; MS-ESS3-1, p. 136 • HS: HS-PS1-2, p. 147; HS-PS1-5, p. 148; HS-LS1-1, p. 154; HS-LS1-6, p. 155; HS-LS2-3, p. 156; HS-LS4-2, p. 159; HS-LS4-4, p. 160; HS-ESS1-2, p. 161; HS-ESS1-6, p. 162; HS-ESS3-1, p. 165

Review Chapter 2 instructions on how to use this guide.

Visit curriculumtopicstudy2.org for more information about CTS and additional resources.

NOTE: *Atlas* page numbers have not been provided because *The NSTA Atlas of the Three Dimensions* was produced concurrently with this edition. Titles and map codes are accurate.

Additional Readings:

Available for download at www.curriculumtopicstudy2.org

Copyright © 2020 by Corwin Press, Inc. All rights reserved. Reprinted from *Science Curriculum Topic Study: Bridging the Gap Between Three-Dimensional Standards, Research, and Practice* (2nd ed.) by Page Keeley and Joyce Tugel. Thousand Oaks, CA: Corwin, www.corwin.com. Reproduction authorized for educational use by educators, local school sites, and/or noncommercial or nonprofit entities that have purchased the book.

Designing Solutions (Engineering)
Grades K–12 Standards- and Research-Based Study of a Curricular Topic

Section and Outcome	Selected Sources and Readings for Study and Reflection Read and examine **related** parts of
I. Content Knowledge	**IA:** *Framework for K–12 Science Education:* Narrative Section • Ch. 3: Practice 6: Constructing Explanations and Designing Solutions, pp. 67–69 **IB:** *NSTA Quick Reference Guide to the NGSS K–12:* Scientific and Engineering Practices • Constructing Explanations and Designing Solutions, p. 17 (last paragraph before Goals) **IC:** *The NSTA Atlas of the Three Dimensions:* Narrative Page • 1.6: Constructing Explanations and Designing Solutions (CEDS) **ID:** *Next Generation Science Standards:* Appendices • Appendix F: Practice 6: Constructing Explanations and Designing Solutions, p. 11
II. Concepts, Core Ideas, or Practices	**IIA:** *Framework for K–12 Science Education:* Goals • Ch. 3: Practice 6: Constructing Explanations and Designing Solutions, p. 69 (last 4 bullets) **IIB:** *Next Generation Science Standards:* Scientific and Engineering Practices Column • K: Constructing Explanations and Designing Solutions, p. 7 • Grade 1: Constructing Explanations and Designing Solutions, pp. 9, 10 • Grade 2: Constructing Explanations and Designing Solutions, p. 15 • Grade 3: Constructing Explanations and Designing Solutions, p. 32 • Grade 4: Constructing Explanations and Designing Solutions, pp. 23, 24, 26, 32 • Grade 5: Constructing Explanations and Designing Solutions, p. 32 • MS: Constructing Explanations and Designing Solutions, pp. 37, 40, 60 • HS: Constructing Explanations and Designing Solutions, pp. 70, 72, 74–75, 83, 99, 102 **IIC:** *NSTA Quick Reference Guide to the NGSS K–12:* Scientific and Engineering Practices • K–12: Goals by the End of Grade 12, p. 17 • K–2: Constructing Explanations and Designing Solutions, p. 89 • 3–5: Constructing Explanations and Designing Solutions, p. 101 • 6–8: Constructing Explanations and Designing Solutions, p. 117 • 9–12: Constructing Explanations and Designing Solutions, p. 141
III. Curriculum, Instruction, and Formative Assessment	**IIIA:** *Framework for K–12 Science Education:* Progression Narrative • Ch. 3: Practice 6: Constructing Explanations and Designing Solutions, p. 70–71 (Start at Progression for Design) **IIIB:** *NSTA Quick Reference Guide to the NGSS K–12:* Progression Narrative • Constructing Explanations and Designing Solutions, pp. 18–19 (start at Progression for Design) **IIIC:** *Helping Students Make Sense of the World Using NGSS Practices* • Ch. 13: Constructing Explanations and Designing Solutions, pp. 288–290 • Ch. 13: Engineering Design and Equity, pp. 290–291 • Ch. 13: Elementary Vignette, pp. 291–303 • Ch. 13: High School Vignette, pp. 303–304 • Ch. 13: What an Engineering Activity Should Include, pp. 305–306
IV. Research on Commonly Held Ideas	Research for this topic is not available in the current CTS resources.
V. K–12 Articulation and Connections	**VA:** *Next Generation Science Standards:* Appendices: Progression Chart • Appendix F: Practice 6: Constructing Explanations and Designing Solutions (focus on designing solutions), pp. 11–12 **VB:** *NSTA Quick Reference Guide to the NGSS K–12:* Progression Chart • Constructing Explanations and Designing Solutions, Condensed Practices, p. 55 (last block for each grade span)

Section and Outcome	Selected Sources and Readings for Study and Reflection
	Read and examine **related** parts of
	VC: *The NSTA Atlas of the Three Dimensions:* Map Page • 1.6: Constructing Explanations and Designing Solutions (CEDS)
VI. Assessment Expectation	VIA: *State Standards* • Examine your state's standards VIB: *Next Generation Science Standards:* Performance Expectations • K: K-PS3-2, p. 7 • Grade 1: 1-PS4-4, p. 9; 1-LS1-1, p. 10 • Grade 2: 2-ESS2-1, p. 15 • Grade 3: 3-5-ETS1-2, p. 32 • Grade 4: 4-PS3-4, p. 23; 4-PS4-3, p. 24; 4-ESS3-2, p. 26; 3-5-ETS1-2, p. 32 • Grade 5: 3-5-ETS1-2, p. 32 • MS: MS-PS1-6, p. 37; MS-PS3-3, p. 40; ESS3-3, p. 60 • HS: HS-PS1-6, p. 70; HS-PS2-3, p. 72; PS3-3, p. 74; HS-LS2-7, p. 83; HS-ESS3-4, p. 99; HS-ETS1-2, ETS1-3, p. 102 VIC: *NSTA Quick Reference Guide to the NGSS K–12:* Performance Expectations (left column) • K: K-PS3-2, p. 93 • Grade 1: 1-LS1-1, p. 94; 1-PS4-4, p. 95, • Grade 2: 2-ESS2-1, p. 96 • Grade 3: 3-5-ETS1-2, p. 114. • Grade 4: 4-PS3-4, p. 109; 4-PS4-3, p. 110; 4-ESS3-2, p. 108; 3-5-ETS1-2, p. 114 • Grade 5: 3-5-ETS1-2, p. 114 • MS: MS-PS1-6, p. 124; MS-PS3-3, p. 126; ESS3-3, p. 136 • HS: HS-PS1-6, p. 148; HS-PS2-3, p. 149; HS-PS3-3, p. 151; HS-LS2-7, p. 157; HS-ESS3-4, p. 166; HS-ETS1-2, ETS1-3, p. 167

Review Chapter 2 instructions on how to use this guide.

Visit curriculumtopicstudy2.org for more information about CTS and additional resources.

NOTE: *Atlas* page numbers have not been provided because *The NSTA Atlas of the Three Dimensions* was produced concurrently with this edition. Titles and map codes are accurate.

Additional Readings:

 Available for download at www.curriculumtopicstudy2.org

Copyright © 2020 by Corwin Press, Inc. All rights reserved. Reprinted from *Science Curriculum Topic Study: Bridging the Gap Between Three-Dimensional Standards, Research, and Practice* (2nd ed.) by Page Keeley and Joyce Tugel. Thousand Oaks, CA: Corwin, www.corwin.com. Reproduction authorized for educational use by educators, local school sites, and/or noncommercial or nonprofit entities that have purchased the book.

Argumentation
Grades K–12 Standards- and Research-Based Study of a Curricular Topic

Section and Outcome	Selected Sources and Readings for Study and Reflection Read and examine **related** parts of
I. **Content Knowledge**	**IA:** *Science for All Americans* • Ch. 12E: Critical Response Skills, pp. 193–194 **IB:** *Framework for K–12 Science Education:* Narrative Section • Ch. 3: Practice 7: Engaging in Argument From Evidence, pp. 71–72 **IC:** *NSTA Quick Reference Guide to the NGSS K–12:* Scientific and Engineering Practices • Engaging in Argument From Evidence, p. 20 **ID:** *Helping Students Make Sense of the World Using NGSS Practices* • Ch. 11: Why Is the Practice of Engaging in Argument From Evidence Important? pp. 231–232 • Ch. 11: What Is the Practice of Engaging in Argument From Evidence All About? pp. 232–240 • Ch. 11: How Does the Practice of Engaging in Argument From Evidence Relate to Other Practices? pp. 241–242 **IE:** *Next Generation Science Standards:* Appendices • Appendix F: Practice 7: Engaging in Argument From Evidence, p. 13 **IF:** *The NSTA Atlas of the Three Dimensions:* Narrative Page • 1.7: Engaging in Argument From Evidence (ARG)
II. **Concepts, Core Ideas, or Practices**	**IIA:** *Framework for K–12 Science Education:* Goals • Ch. 3: Practice 7: Engaging in Argument From Evidence, pp. 72–73 (stop at Progression) **IIB:** *Next Generation Science Standards:* Scientific and Engineering Practices Column • K: Engaging in Argument From Evidence, p. 6 • Grade 2: Engaging in Argument From Evidence, p. 13 • Grade 3: Engaging in Argument From Evidence, pp. 19, 21 • Grade 4: Engaging in Argument From Evidence, p. 25 • Grade 5: Engaging in Argument From Evidence, pp. 29, 31 • MS: Engaging in Argument From Evidence, pp. 38, 40, 45, 47, 49, 50, 60, 63 • HS: Engaging in Argument From Evidence, pp. 76, 83, 85, 87, 94, 96, 99–100 **IIC:** *NSTA Quick Reference Guide to the NGSS K–12:* Scientific and Engineering Practices • K–12: Goals by the End of Grade 12, pp. 20–21 • K–2: Engaging in Argument From Evidence, p. 89 • 3–5: Constructing Explanations and Designing Solutions, p. 101 • 6–8: Constructing Explanations and Designing Solutions, p. 118 • 9–12: Constructing Explanations and Designing Solutions, p. 142
III. **Curriculum, Instruction, and Formative Assessment**	**IIIA:** *Framework for K–12 Science Education:* Progression Narrative • Ch. 3: Engaging in Argument From Evidence, pp. 73–74 **IIIB:** *NSTA Quick Reference Guide to the NGSS K–12:* Progression Narrative • Engaging in Argument From Evidence, pp. 21–22 **IIIC:** *Helping Students Make Sense of the World Using NGSS Practices* • Ch. 11: Elementary Vignette, pp. 243–245; Middle School Vignette, pp. 245–247; High School Vignette, pp. 248–249 • Ch. 11: What Does Not Count as Scientific Argumentation? pp. 250–251 • Ch. 11: Equity and Supporting and Assessing the Practice, pp. 251–256 • Ch. 11: Summary of Strategies for Supporting Scientific Argumentation, p. 255
IV. **Research on Commonly Held Ideas**	**IVA:** *Benchmarks for Science Literacy:* Chapter 15 Research • 12E: Inadequacies in Arguments, p. 361

Section and Outcome	Selected Sources and Readings for Study and Reflection Read and examine **related** parts of
V. **K–12 Articulation and Connections**	**VA:** *Next Generation Science Standards:* Appendices: Progression • Appendix F: Practice 7: Engaging in Argument From Evidence, pp. 13–14 **VB:** *NSTA Quick Reference Guide to the NGSS K–12:* Progression Chart • Engaging in Argument From Evidence, Condensed Practices, p. 56 **VC:** *The NSTA Atlas of the Three Dimensions:* Map Page • 1.7: Engaging in Argument From Evidence (ARG)
VI. **Assessment Expectation**	**VIA:** *State Standards* • Examine your state's standards **VIB:** *Next Generation Science Standards:* Performance Expectations • K: K-ESS2-2, p. 6 • Grade 2: 2-PS1-4, p. 13 • Grade 3: 3-LS2-1, LS4-3, LS4-4, p. 19; ESS3-1, p. 21 • Grade 4: 4-LS1-1, p. 25 • Grade 5: 5-PS2-1, p. 31; 5-LS1-1, p. 29; 5-ESS1-1, p. 31 • MS: MS-PS2-4, p. 38; MS-PS3-5, p. 40; MS-LS1-3, p. 45; MS-LS1-4, p. 50; MS-LS2-4, p. 47; MS-LS2-5, p. 49; MS-ESS3-4, p. 60, MS-ETS1-2, p. 63 • HS: HS-PS4-3, p. 76; HS-LS2-6, HS-LS2-8, p. 83; HS-LS3-2, p. 85; HS-LS4-5, p. 87; HS-ESS1-5, p. 94; HS-ESS2-7, p. 96; HS-ESS3-2, p. 99 **VIC:** *NSTA Quick Reference Guide to the NGSS K–12:* Performance Expectations • K: K-ESS2-2, p. 92 • Grade 2: 2-PS1-4, p. 97 • Grade 3: 3-LS2-1, p. 104; LS4-3, LS4-4, ESS3-1, p. 105 • Grade 4: 4-LS1-1, 4-LS1-2, p. 107 • Grade 5: 5-PS2-1, p. 113; 5-LS1-1, 5-ESS1-1, p. 111 • MS: MS-PS2-4, p. 125; MS-PS3-5, p. 126; MS-LS1-3, MS-LS1-4, p. 128; MS-LS2-4, p. 130; MS-ESS3-4, p. 137, MS-ETS1-2, p. 138 • HS: HS-PS4-3, p. 152; HS-LS2-6, p. 157; HS-LS2-8, HS-LS3-2, p. 158; HS-LS4-5, p. 160; HS-ESS1-5, p. 162; HS-ESS2-7, HS-ESS3-2, p. 165

Review Chapter 2 instructions on how to use this guide.

Visit curriculumtopicstudy2.org for more information about CTS and additional resources.

NOTE: *Atlas* page numbers have not been provided because *The NSTA Atlas of the Three Dimensions* was produced concurrently with this edition. Titles and map codes are accurate.

Additional Readings:

 Available for download at www.curriculumtopicstudy2.org

Copyright © 2020 by Corwin Press, Inc. All rights reserved. Reprinted from *Science Curriculum Topic Study: Bridging the Gap Between Three-Dimensional Standards, Research, and Practice* (2nd ed.) by Page Keeley and Joyce Tugel. Thousand Oaks, CA: Corwin, www.corwin.com. Reproduction authorized for educational use by educators, local school sites, and/or noncommercial or nonprofit entities that have purchased the book.

Obtaining, Evaluating, and Communicating Information
Grades K–12 Standards- and Research-Based Study of a Curricular Topic

Section and Outcome	Selected Sources and Readings for Study and Reflection Read and examine **related** parts of
I. Content Knowledge	**IA:** *Science for All Americans* • Ch. 12D: Communication Skills, pp. 192–193 **IB:** *Framework for K–12 Science Education:* Narrative Section • Ch. 3: Practice 8: Obtaining, Evaluating, and Communicating Information, pp. 74–75 **IC:** *NSTA Quick Reference Guide to the NGSS K–12:* Scientific and Engineering Practices • Obtaining, Evaluating, and Communicating Information, pp. 23–24 **ID:** *Helping Students Make Sense of the World Using NGSS Practices* • Ch. 12: What Is the Practice of Obtaining, Evaluating, and Communicating Information All About and Why Is It Important? pp. 260–263 **IE:** *Next Generation Science Standards:* Appendices • Appendix F: Practice 8: Obtaining, Evaluating, and Communicating Information, p. 15 **IF:** *The NSTA Atlas of the Three Dimensions:* Narrative Page • 1.8 Obtaining, Evaluating, and Communicating Information (INFO)
II. Concepts, Core Ideas, or Practices	**IIA:** *Framework for K–12 Science Education:* Goals • Ch. 3: Practice 8: Obtaining, Evaluating, and Communicating Information, pp. 75–76 **IIB:** *Next Generation Science Standards:* Scientific and Engineering Practices Column • K: Obtaining, Evaluating, and Communicating Information, pp. 6, 7 • Grade 1: Obtaining, Evaluating, and Communicating Information, p. 10 • Grade 2: Obtaining, Evaluating, and Communicating Information, p. 15 • Grade 3: Obtaining, Evaluating, and Communicating Information, p. 21 • Grade 4: Obtaining, Evaluating, and Communicating Information, p. 23 • Grade 5: Obtaining, Evaluating, and Communicating Information, p. 30 • MS: Obtaining, Evaluating, and Communicating Information, pp. 35, 42, 45, 50 • HS: Obtaining, Evaluating, and Communicating Information, pp. 68, 76, 87, 92 **IIC:** *NSTA Quick Reference Guide to the NGSS K–12:* Scientific and Engineering Practices • K-12: Goals by the End of Grade 12, p. 24 • K–2: Obtaining, Evaluating, and Communicating Information, p. 89 • 3–5: Obtaining, Evaluating, and Communicating Information, p. 101 • 6–8: Obtaining, Evaluating, and Communicating Information, p. 118 • 9–12: Obtaining, Evaluating, and Communicating Information, p. 142
III. Curriculum, Instruction, and Formative Assessment	**IIIA:** *Framework for K–12 Science Education:* Progression Narrative • Ch. 3: Practice 8: Obtaining, Evaluating, and Communicating Information, pp. 76–77 **IIIB:** *NSTA Quick Reference Guide to the NGSS K–12:* Progression Narrative • Obtaining, Evaluating, and Communicating Information, pp. 24–25 **IIIC:** *Helping Students Make Sense of the World Using NGSS Practices* • Ch. 12: How Can I Support the Practice of Obtaining, Evaluating, and Communicating Information? pp. 263–270 • Ch. 12: What Is Not Intended by This Practice, p. 269 • Ch. 12: Working Toward Equity in Regard to This Practice, p. 270 • Ch. 12: PreK Vignette, pp. 271–273; Elementary Vignette, pp. 273–275; Middle School Vignette, pp. 275–277 • Ch. 12: How Can I Get Started on Supporting This Practice? pp. 277–279
IV. Research on Commonly Held Ideas	Research not available in current CTS resources.

Section and Outcome	Selected Sources and Readings for Study and Reflection Read and examine **related** parts of
V. **K–12 Articulation and Connections**	**VA:** *Next Generation Science Standards:* Appendices: Progression Chart • Appendix F: Practice 8: Obtaining, Evaluating, and Communicating Information, p. 15 **VB:** *NSTA Quick Reference Guide to the NGSS K–12:* Progression Chart • Obtaining, Evaluating, and Communicating Information, Condensed Practices, p. 57 **VC:** *The NSTA Atlas of the Three Dimensions:* Map Page • 1.8 Obtaining, Evaluating, and Communicating Information (INFO)
VI. **Assessment Expectation**	**VIA:** *State Standards* • Examine your state's standards **VIB:** *Next Generation Science Standards:* Performance Expectations • K: K-ESS3-2, p. 7; K-ESS3-3, p. 6 • Grade 1: LS1-2, p. 10 • Grade 2: 2-ESS2-3, p. 15 • Grade 3: 3-ESS2-2, p. 21 • Grade 4: 4-ESS3-1, p. 23 • Grade 5: 5-ESS3-1, p. 30 • MS: MS-PS1-3, p. 35; MS-PS4-3, p. 42; MS-LS1-8, p. 45; MS-LS4-5, p. 50 • HS: HS-PS2-6, p. 68; HS-PS4-4, HS-PS4-5, p. 76; HS-LS4-1, p. 87; HS-ESS1-3, p. 92 **VIC:** *NSTA Quick Reference Guide to the NGSS K–12:* Performance Expectations (left column) • K: K-ESS3-2, K-ESS3-3, pp. 92, 93 • Grade 1: LS1-2, p. 94 • Grade 2: 2-ESS2-3, p. 96 • Grade 3: 3-ESS2-2, p. 105 • Grade 4: 4-ESS3-1, p. 108 • Grade 5: 5-ESS3-1, p. 112 • MS: MS-PS1-3, p. 122; MS-PS4-3, p. 127; MS-LS1-8, p. 129; MS-LS4-5, p. 132 • HS: HS-PS2-6, p. 150; HS-PS4-4, HS-PS4-5, p. 153; HS-LS4-1, p. 159; HS-ESS1-3, p. 161

Review Chapter 2 instructions on how to use this guide.

Visit curriculumtopicstudy2.org for more information about CTS and additional resources.

NOTE: *Atlas* page numbers have not been provided because *The NSTA Atlas of the Three Dimensions* was produced concurrently with this edition. Titles and map codes are accurate.

Additional Readings:

Available for download at www.curriculumtopicstudy2.org

Copyright © 2020 by Corwin Press, Inc. All rights reserved. Reprinted from *Science Curriculum Topic Study: Bridging the Gap Between Three-Dimensional Standards, Research, and Practice* (2nd ed.) by Page Keeley and Joyce Tugel. Thousand Oaks, CA: Corwin, www.corwin.com. Reproduction authorized for educational use by educators, local school sites, and/or noncommercial or nonprofit entities that have purchased the book.

Notes

Category E: Crosscutting Concepts Guides

The seven CTS guides in this section focus on the crosscutting concepts that bridge disciplinary boundaries. They provide an organizational framework for coherently connecting ideas from science and engineering. The guides in this section are arranged in the order they appear in the *Framework for K–12 Science Education* (NRC, 2012) and include

- Patterns
- Cause and Effect
- Scale, Proportion, and Quantity
- Systems and System Models
- Energy and Matter: Flows, Cycles, and Conservation
- Structure and Function
- Stability and Change

NOTES FOR USING CATEGORY E GUIDES

Overall

- *Atlas* page numbers have not been provided because *The NSTA Atlas of the Three Dimensions* was produced concurrently with this edition. Titles and map codes are accurate.
- Eliminate redundancy. Some readings include the exact same information. However, even when the information is the same, there may be an advantage in how the information is presented. Select the reading based on the resources you have available to use for CTS and/or the advantage of using one over the other.

Section I

- Readings from the *Framework*, *NSTA Quick Reference Guide*, and *Atlas* narrative are exactly the same. Choose one of these resources for this section.
- Read only the narrative sections beneath the title of the crosscutting concept in the *Framework* or the *NSTA Quick Reference Guide*. Stop at Progression.
- In Appendix G, read only the narrative section. Stop at the chart.

Section II

- Readings from the *NGSS* and the *NSTA Quick Reference Guide* goals are exactly the same. The difference is that the *NGSS* breaks them down by specific grade and the *NSTA Quick Reference Guide* breaks them down by grade span, K–2, 3–5, 6–8, 9–12. Choose the level of specificity (grade or grade span) when deciding which of these resources to use.

Section III

- Readings from the *Framework* and the *NSTA Quick Reference Guide* are exactly the same. Choose one of these resources for this section. Start at Progression.

- Readings from the *Framework* and the *NSTA Quick Reference Guide* are narrative descriptions of what the crosscutting concept looks like at different grade spans, including the kinds of instructional experiences students should have. This is different from section V, which shows the progression of specific elements of what students are expected to know about the crosscutting concept at different grade spans.

Section IV

- Except for *Benchmarks for Science Literacy:* Chapter 15, research summaries for the crosscutting concepts are not included in the common resources used for this section. If you have access to professional journals, you may search for research articles on difficulties students encounter when they use the crosscutting concepts and add these to the additional readings at the end of the guides.

Section V

- Readings from the *NGSS* Appendix progression chart summarize the crosscutting concepts at each grade span with an example from the performance expectations. It does not list the specific crosscutting concepts.

- The *NSTA Quick Reference Guide* progression chart lists the specific crosscutting concepts for each grade span.

- The *Atlas* includes the same information as the *NSTA Quick Reference Guide* but the visual mapping of the concepts allows you to see precursor ideas and connections, as well as connections to other maps. Be sure to read the front matter of the *Atlas* before using the maps. The information will help you use the maps effectively.

Section VI

- The *NGSS* and the *NSTA Quick Reference Guide* provide the exact same information. Clarifications and assessment boundaries are also included in both. Choose one of these resources based on the advantages listed below or the resources you have available.

- An advantage to using the *NSTA Quick Reference Guide* is that the performance expectation is listed next to the disciplinary core idea that uses that crosscutting concept. A disadvantage is that you have to identify which crosscutting concept is in the performance expectation. It is not listed separately.

- An advantage to using the *NGSS* that is not evident in the *NSTA Quick Reference Guide* is that the crosscutting concept chart, included below the performance expectations, shows which crosscutting concept is part of the performance expectation. It also shows which scientific or engineering practice is combined with the crosscutting concept.

Patterns
Grades K–12 Standards- and Research-Based Study of a Curricular Topic

Section and Outcome	Selected Sources and Readings for Study and Reflection Read and examine **related** parts of
I. **Content Knowledge**	**IA:** *Framework for K–12 Science Education:* Narrative • Ch. 4: Patterns, pp. 85–86 **IB:** *NSTA Quick Reference Guide to the NGSS K–12:* Narrative • Crosscutting Concept 1: Patterns, p. 26 **IC:** *Next Generation Science Standards:* Appendices Narrative • Appendix G: Patterns, pp. 3–4 **ID:** *The NSTA Atlas of the Three Dimensions:* Narrative Page • 2.1: Patterns (PAT)
II. **Concepts, Core Ideas, or Practices**	**IIA:** *Next Generation Science Standards:* Crosscutting Concepts Column • K: Patterns, pp. 6, 7 • Grade 1: Patterns, pp. 10, 11 • Grade 2: Patterns, pp. 13, 15 • Grade 3: Patterns, pp. 18, 20, 21 • Grade 4: Patterns, pp. 24, 26 • Grade 5: Patterns, p. 31 • MS: Patterns, pp. 37, 42, 49, 52, 56, 57, 60 • HS: Patterns, pp. 68, 70, 72, 87, 94 **IIB:** *NSTA Quick Reference Guide to the NGSS K–12:* Crosscutting Concepts • K–2: Patterns, p. 90 • 3–5: Patterns, p. 102 • 6–8: Patterns, p. 119 • 9–12: Patterns, p. 143
III. **Curriculum, Instruction, and Formative Assessment**	**IIIA:** *Framework for K–12 Science Education:* Progression • Ch. 4: Patterns, pp. 86–87 **IIIB:** *NSTA Quick Reference Guide to the NGSS K–12:* Progression • Patterns, pp. 26–27
IV. **Research on Commonly Held Ideas**	Research on *Patterns* is not available in current CTS resources.
V. **K–12 Articulation and Connections**	**VA:** *Next Generation Science Standards:* Appendices: Progression Chart • Appendix G: Patterns, pp. 4–5 • Appendix G: Patterns Connections, p. 11 **VB:** *NSTA Quick Reference Guide to the NGSS K–12:* Progression Chart • Patterns, p. 58 **VC:** *The NSTA Atlas of the Three Dimensions:* Map Page • 2.1: Patterns (PAT)

Section and Outcome	Selected Sources and Readings for Study and Reflection Read and examine **related** parts of
VI. **Assessment Expectation**	**VIA: State Standards** • Examine your state's standards **VIB: *Next Generation Science Standards:* Performance Expectations** • K: K-LS1-1, p. 6; K-ESS2-1, p. 7 • Grade 1: 1-LS1-2, 1-LS3-1, p. 10; ESS1-1, ESS1-2, p. 11 • Grade 2: 2-PS1-1, p. 13; 2-ESS2-2, ESS2-3, p. 15 • Grade 3: 3-PS2-2, p. 18; 3-LS1-1, 3-LS3-1, p. 20; 3-ESS2-1, ESS2-2, p. 21 • Grade 4: 4-PS4-1, PS4-3, p. 24; 4-ESS1-1, 4-ESS2-2, p. 26 • Grade 5: 5-ESS1-2, p. 31 • MS: MS-PS1-2, p. 37; MS-PS4-1, p. 42; MS-LS2-2, p. 49; MS-LS4-2, MS-LS4-1, MS-LS4-3, p. 52; MS-ESS1-1, p. 56; MS-ESS2-3, p. 57; MS-ESS3-2, p. 60 • HS: HS-PS1-1, HS-PS1-3, p. 68; HS-PS1-2, HS-PS1-5, p, 70; HS-PS2-4, p. 72; HS-LS4-1, HS-LS4-3, p. 87; HS-ESS1-5, p. 94 **VIC: *NSTA Quick Reference Guide to the NGSS K–12:* Performance Expectations** • K: K-LS1-1, K-ESS2-1, p. 92 • Grade 1: 1-LS1-2, 1-LS3-1, 1-ESS1-1, p. 94; 1-ESS1-2, p. 95 • Grade 2: 2-PS1-1, 2-ESS2-2, ESS2-3, p. 96 • Grade 3: 3-PS2-2, p. 106; 3-LS1-1, 3-LS3-1, p. 104; 3-ESS2-1, ESS2-2, p. 105 • Grade 4: 4-PS4-1, 4-PS4-3, p. 109; 4-ESS1-1, 4-ESS2-2, p. 107 • Grade 5: 5-ESS1-2, p. 111 • MS: MS-PS1-2, p. 122; MS-PS4-1, p. 127; MS-LS2-2, p. 130; MS-LS4-2, MS-LS4-1, p. 131; MS-LS4-3, p. 132; MS-ESS1-1, p. 133; MS-ESS2-3, p. 134; ESS3-2, p. 136 • HS: HS-PS1-1, HS-PS1-2, HS-PS1-3, p. 147; HS-PS1-5, p. 148; HS-PS2-4, p. 149; HS-LS4-1, HS-LS4-3, p. 159; HS-ESS1-5, p. 162

Review Chapter 2 instructions on how to use this guide.

Visit curriculumtopicstudy2.org for more information about CTS and additional resources.

NOTE: *Atlas* page numbers have not been provided because *The NSTA Atlas of the Three Dimensions* was produced concurrently with this edition. Titles and map codes are accurate.

Additional Readings:

 Available for download at www.curriculumtopicstudy2.org

Copyright © 2020 by Corwin Press, Inc. All rights reserved. Reprinted from *Science Curriculum Topic Study: Bridging the Gap Between Three-Dimensional Standards, Research, and Practice* (2nd ed.) by Page Keeley and Joyce Tugel. Thousand Oaks, CA: Corwin, www.corwin.com. Reproduction authorized for educational use by educators, local school sites, and/or noncommercial or nonprofit entities that have purchased the book.

Cause and Effect
K–12 Standards- and Research-Based Study of a Curricular Topic

Section and Outcome	Selected Sources and Readings for Study and Reflection Read and examine **related** parts of
I. Content Knowledge	**IA:** *Framework for K–12 Science Education:* Narrative • Ch. 4: Cause and Effect, pp. 87–88 **IB:** *NSTA Quick Reference Guide to the NGSS K–12:* Narrative • Crosscutting Concept 2: Cause and Effect, pp. 28–29 **IC:** *Next Generation Science Standards:* Appendices Narrative • Appendix G: Cause and Effect, p. 5 **ID:** *The NSTA Atlas of the Three Dimensions:* Narrative Page • 2.2: Cause and Effect: Mechanism and Explanation (CE)
II. Concepts, Core Ideas, or Practices	**IIA:** *Next Generation Science Standards:* Crosscutting Concepts Column • K: Cause and Effect, pp. 5, 6, 7 • Grade 1: Cause and Effect, p. 9 • Grade 2: Cause and Effect, pp. 13, 14 • Grade 3: Cause and Effect, pp. 18, 19, 20, 21 • Grade 4: Cause and Effect, pp. 23, 25, 26 • Grade 5: Cause and Effect, pp. 28, 31 • MS: Cause and Effect, pp. 35, 38, 45, 47, 50, 52, 58, 59, 60 • HS: Cause and Effect, pp. 72, 74, 76, 83, 85, 87, 98, 99 **IIB:** *NSTA Quick Reference Guide to the NGSS K–12:* Crosscutting Concepts • K–2: Cause and Effect, p. 90 • 3–5: Cause and Effect, p. 102 • 6–8: Cause and Effect, p. 119 • 9–12: Cause and Effect, p. 143
III. Curriculum, Instruction, and Formative Assessment	**IIIA:** *Framework for K–12 Science Education:* Progression • Ch. 4: Cause and Effect, pp. 88–89 **IIIB:** *NSTA Quick Reference Guide to the NGSS K–12:* Progression • Cause and Effect, p. 29
IV. Research on Commonly Held Ideas	Research on *Cause and Effect* is not available in current CTS resources.
V. K–12 Articulation and Connections	**VA:** *Next Generation Science Standards:* Appendices: Progression Chart • Appendix G: Cause and Effect, pp. 5–6 • Appendix G: Cause and Effect Connections, pp. 11–12 **VB:** *NSTA Quick Reference Guide to the NGSS K–12:* Progression Chart • Cause and Effect, p. 58 **VC:** *The NSTA Atlas of the Three Dimensions:* Map Page • 2.2: Cause and Effect: Mechanism and Explanation (CE)
VI. Assessment Expectation	**VIA:** *State Standards* • Examine your state's standards

Section and Outcome	Selected Sources and Readings for Study and Reflection Read and examine **related** parts of
	VIB: *Next Generation Science Standards:* Performance Expectations • K: K-ESS3-3, p. 6; K-PS2-1, K-PS2-2, p. 5; K-PS3-1, K-PS3-2, K-ESS3-2, p. 7 • Grade 1: 1-PS4-1, 1-PS4-2, 1-PS4-3, p. 9 • Grade 2: 2-PS1-4, 2-PS1-2, p. 13; 2-LS2-1, p. 14 • Grade 3: 3-PS2-1, 3-PS2-3, p. 18; 3-LS2-1, 3-LS4-3, p. 19; 3-LS3-2, 3-LS42-2, p. 20; 3-ESS3-1, p. 21 • Grade 4: 4-ESS3-1, p. 23; 4-PS4-2, p. 25; 4-ESS2-1, 4-ESS3-2, p. 26 • Grade 5: 5-PS1-4, p. 28; 5-PS2-1, p. 31 • MS: MS-PS1-4, p. 35; MS-PS2-3, MS-PS2-5, p. 38; MS-LS1-8, p. 45; MS-LS2-1, p. 47; MS-LS3-2, p. 50; MS-LS4-4, MS-LS4-6, p. 52; MS-ESS3-1, p. 58; MS-ESS2-5, p. 59; MS-ESS3-3, MS-ESS3-4, p. 60 • HS: HS-PS2-2, HS-PS2-5, HS-PS2-3, p. 72; HS-PS3-5, p. 74; HS-PS4-1, HS-PS2-3, p. 149; HS-PS2-5, p. 150, p. 76; HS-LS2-8, HS-LS4-6, p. 83; HS-LS3-1, HS-LS3-2, p. 85; HS-LS4-2, HS-LS4-4, HS-LS4-5, p. 87; HS-ESS2-4, p. 98; HS-ESS3-1, p. 99 **VIC: *NSTA Quick Reference Guide to the NGSS K–12:*** Performance Expectations • K: K-ESS3-3, K-PS2-1, K-PS2-2, K-PS3-1, K-PS3-2, p. 93 • Grade 1: 1-PS4-1, 1-PS4-2, 1-PS4-3, p. 95 • Grade 2: 2-PS1-4, 2-PS1-2, p. 97; 2-LS2-1, p. 96 • Grade 3: 3-PS2-1, p. 105; 3-PS2-3, p. 106; 3-LS2-1; 3-LS3-2; 3-LS4-2, p. 104; 3-LS4-3; 3-ESS3-1, p. 105 • Grade 4: 4-ESS2-1, p. 107; 4-ESS3-1, 4-ESS3-2, p. 108; 4-PS4-2, p. 110 • Grade 5: 5-PS1-4, 5-PS2-1, p. 113 • MS: MS-PS1-2, p. 122; MS-PS2-3, p. 124; MS-PS2-5, p. 125; MS-LS1-4, p. 128; MS-LS1-5, p. 129; MS-LS2-1, p. 130; MS-LS3-2, p. 131; MS-LS4-4, MS-LS4-5, MS-LS4-6, p. 132; MS-ESS2-5, p. 135; MS-ESS3-2, p. 136 • HS: HS-PS2-1, HS-PS2-3, p. 149; HS-PS2-5, p. 150; HS-PS3-5, HS-PS4-1, p. 152; HS-PS4-4, HS-PS4-5, p. 153; HS-LS2-8, HS-LS3-1, HS-LS3-2, p. 158; HS-LS4-2, p. 159; HS-LS4-4, HS-LS4-5, HS-LS4-6, p. 160; HS-ESS2-4, p. 164; HS-ESS3-1, p. 165

Review Chapter 2 instructions on how to use this guide.

Visit curriculumtopicstudy2.org for more information about CTS and additional resources.

NOTE: *Atlas* page numbers have not been provided because *The NSTA Atlas of the Three Dimensions* was produced concurrently with this edition. Titles and map codes are accurate.

Additional Readings:

 Available for download at www.curriculumtopicstudy2.org

Copyright © 2020 by Corwin Press, Inc. All rights reserved. Reprinted from *Science Curriculum Topic Study: Bridging the Gap Between Three-Dimensional Standards, Research, and Practice* (2nd ed.) by Page Keeley and Joyce Tugel. Thousand Oaks, CA: Corwin, www.corwin.com. Reproduction authorized for educational use by educators, local school sites, and/or noncommercial or nonprofit entities that have purchased the book.

Scale, Proportion, and Quantity
Grades K–12 Standards- and Research-Based Study of a Curricular Topic

Section and Outcome	Selected Sources and Readings for Study and Reflection Read and examine **related** parts of
I. Content Knowledge	**IA: *Science for All Americans*** • Ch. 11: Scale, pp. 179–181 **IB: *Framework for K–12 Science Education:*** Narrative • Ch. 4: Scale, Proportion, and Quantity, pp. 89–90 **IC: *NSTA Quick Reference Guide to the NGSS K–12:*** Narrative • Crosscutting Concept 3: Scale, Proportion, and Quantity, p. 30 **ID: *Next Generation Science Standards:*** Appendices Narrative • Appendix G: Scale, Proportion, and Quantity, p. 6 **IE: *The NSTA Atlas of the Three Dimensions:*** Narrative Page • 2.3: Scale, Proportion, and Quantity (SPQ)
II. Concepts, Core Ideas, or Practices	**IIA: *Next Generation Science Standards:*** Crosscutting Concepts Column • Grade 3: Scale, Proportion, and Quantity, p. 19 • Grade 5: Scale, Proportion, and Quantity, pp. 28, 30, 31 • MS: Scale, Proportion, and Quantity, pp. 35, 40, 45, 56, 57 • HS: Scale, Proportion, and Quantity, pp. 83, 85, 92 **IIB: *NSTA Quick Reference Guide to the NGSS K–12:*** Crosscutting Concepts • K–2: Scale, Proportion, and Quantity, p. 90 • 3–5: Scale, Proportion, and Quantity, p. 102 • 6–8: Scale, Proportion, and Quantity, p. 119 • 9–12: Scale, Proportion, and Quantity, p. 143
III. Curriculum, Instruction, and Formative Assessment	**IIIA: *Framework for K–12 Science Education:*** Progression • Ch. 4: Scale, Proportion, and Quantity, pp. 90–91 **IIIB: *NSTA Quick Reference Guide to the NGSS K–12:*** Progression • Scale, Proportion, and Quantity, p. 31
IV. Research on Commonly Held Ideas	**IVA: *Benchmarks for Science Literacy:*** Chapter 15 Research • 12B: Proportional Reasoning, p. 360
V. K–12 Articulation and Connections	**VA: *Next Generation Science Standards:*** Appendices: Progression Chart • Appendix G: Scale, Proportion, and Quantity, p. 7 • Appendix G: Scale, Proportion, and Quantity Connections, p. 12 **VB: *NSTA Quick Reference Guide to the NGSS K–12:*** Progression Chart • Scale, Proportion, and Quantity, p. 59 **VC: *The NSTA Atlas of the Three Dimensions:*** Map Page • 2.3: Scale, Proportion, and Quantity (SPQ)

Section and Outcome	Selected Sources and Readings for Study and Reflection Read and examine **related** parts of
VI. **Assessment Expectation**	**VIA:** *State Standards* • Examine your state's standards **VIB:** *Next Generation Science Standards:* Performance Expectations • Grade 3: 3-LS4-1, p. 19 • Grade 5: 5-PS1-1, 5-PS1-2, 5-PS1-3, p. 28; 5-ESS2-2, p. 30; 5-ESS1-1, p. 31 • MS: MS-PS1-1, p. 35; MS-PS3-1, MS-PS3-4, p. 40; MS-LS1-1, p. 45; MS-ESS1-3, p. 56; MS-ESS1-4, MS-ESS2-2, p. 57 • HS: HS-LS2-1, HS-LS2-2, p. 83; HS-LS3-3, p. 85; HS-ESS1-1, HS-ESS1-4, p. 92 **VIC:** *NSTA Quick Reference Guide to the NGSS K–12:* Performance Expectations • Grade 3: 3-LS4-1, p. 104 • Grade 5: 5-PS1-1, 5-PS1-2, p. 112; 5-PS1-3, p. 113; 5-ESS1-1, p. 111; 5-ESS2-2, p. 112 • MS: MS-PS1-1, p. 122; MS-PS3-1, p. 125; MS-PS3-4, p. 126; MS-LS1-1, p. 128; MS-ESS1-3, p. 133; MS-ESS1-4, MS-ESS2-2, p. 134 • HS: HS-LS2-1, p. 155; HS-LS2-2, p. 156; HS-LS3-3, p. 158; HS-ESS1-1, p. 161; HS-ESS1-4, p. 162

Review Chapter 2 instructions on how to use this guide.

Visit curriculumtopicstudy2.org for more information about CTS and additional resources.

NOTE: *Atlas* page numbers have not been provided because *The NSTA Atlas of the Three Dimensions* was produced concurrently with this edition. Titles and map codes are accurate.

Additional Readings:

Available for download at www.curriculumtopicstudy2.org

Copyright © 2020 by Corwin Press, Inc. All rights reserved. Reprinted from *Science Curriculum Topic Study: Bridging the Gap Between Three-Dimensional Standards, Research, and Practice* (2nd ed.) by Page Keeley and Joyce Tugel. Thousand Oaks, CA: Corwin, www.corwin.com. Reproduction authorized for educational use by educators, local school sites, and/or noncommercial or nonprofit entities that have purchased the book.

Systems and Systems Models
Grades K–12 Standards- and Research-Based Study of a Curricular Topic

Section and Outcome	Selected Sources and Readings for Study and Reflection Read and examine **related** parts of
I. Content Knowledge	**IA:** *Science for All Americans* • Ch. 11: Systems, pp. 166–168 **IB:** *Framework for K–12 Science Education:* Narrative • Ch. 4: Systems and System Models, pp. 91–93 **IC:** *NSTA Quick Reference Guide to the NGSS K–12:* Narrative • Crosscutting Concept 4: Systems and System Models, pp. 32–33 **ID:** *Next Generation Science Standards:* Appendices Narrative • Appendix G: Systems and System Models, pp. 7–8 **IE:** *The NSTA Atlas of the Three Dimensions:* Narrative Page • 2.4: Systems and System Models (SYS)
II. Concepts, Core Ideas, or Practices	**IIA:** *Next Generation Science Standards:* Crosscutting Concepts Column • K: Systems and System Models, p. 6 • Grade 3: Systems and System Models, p. 19 • Grade 4: Systems and System Models, p. 25 • Grade 5: Systems and System Models, pp. 29, 30 • MS: Systems and System Models, pp. 38, 40, 45, 56, 59 • HS: Systems and System Models, pp. 72, 74, 76, 80, 81, 85, 99, 102 **IIB:** *NSTA Quick Reference Guide to the NGSS K–12:* Crosscutting Concepts • K–2: Systems and System Models, p. 90 • 3–5: Systems and System Models, p. 102 • 6–8: Systems and System Models, p. 119 • 9–12: Systems and System Models, p. 143
III. Curriculum, Instruction, and Formative Assessment	**IIIA:** *Framework for K–12 Science Education:* Progression • Ch. 4: Systems and System Models, pp. 93–94 **IIIB:** *NSTA Quick Reference Guide to the NGSS K–12:* Progression • Systems and System Models, pp. 33–34 **IIIC:** *Uncovering Student Ideas:* Assessment Probe and Suggestions for Instruction • USI.4: Is It a System? pp. 81, 85–86
IV. Research on Commonly Held Ideas	**IVA:** *Benchmarks for Science Literacy:* Chapter 15 Research • 11A: Systems, pp. 355–356 **IVB:** *Uncovering Student Ideas:* Related Research • USI.4: Is It a System? p. 85
V. K–12 Articulation and Connections	**VA:** *Next Generation Science Standards:* Appendices: Progression Chart • Appendix G: Systems and System Models, p. 8 • Appendix G: Systems and Systems Models Connections, p. 12 **VB:** *NSTA Quick Reference Guide to the NGSS K–12:* Progression Chart • Systems and System Models, p. 59 **VC:** *The NSTA Atlas of the Three Dimensions:* Map Page • 2.4: Systems and System Models (SYS)

Section and Outcome	Selected Sources and Readings for Study and Reflection Read and examine **related** parts of
VI. **Assessment** **Expectation**	**VIA:** *State Standards* • Examine your state's standards **VIB:** *Next Generation Science Standards:* Performance Expectations • K: K-ESS2-2, K-ESS3-1, p. 6 • Grade 3: 3-LS4-4, p. 19 • Grade 4: 4-LS1-1, LS1-2, p. 25 • Grade 5: 5-LS2-1, p. 29; 5-ESS2-1, 5-ESS3-1, p. 30 • MS: MS-PS2-1, MS-PS2-4, p. 38; MS-PS3-2, p. 40; MS-LS1-3, p. 45; MS-ESS1-2, p. 56; MS-ESS2-6, p. 59 • HS: HS-PS2-2, p. 72; HS-PS3-4, HS-PS3-1; p. 74; HS-PS4-3, p. 76; HS-LS1-2, p. 80; HS-LS2-5, p. 81; HS-LS1-4, p. 85; HS-ESS3-6, p. 99; HS-ETS1-4, p. 102 **VIC:** *NSTA Quick Reference Guide to the NGSS K–12:* Performance Expectations • K: K-ESS2-2, K-ESS3-1, p. 92 • Grade 3: 3-LS4-4, p. 105 • Grade 4: 4-LS1-1, 4-LS1-2, p. 107 • Grade 5: 5-LS2-1, p. 111; 5-ESS2-1, 5-ESS3-1, p. 112 • MS: MS-PS2-1, p. 124; MS-PS2-4, MS-PS3-2, p. 125; MS-LS1-3, p. 128; MS-ESS1-2, p. 133; MS-ESS2-6, p. 135 • HS: HS-PS2-2, p. 149; HS-PS3-4, p. 151; HS-PS3-1; p. 150; HS-PS4-3, p. 152; HS-LS1-2, HS-LS1-4, HS-LS1-5, p. 154; HS-ESS3-6, p. 166; HS-ETS1-4, p. 167

Review Chapter 2 instructions on how to use this guide.

Visit curriculumtopicstudy2.org for more information about CTS and additional resources.

NOTE: *Atlas* page numbers have not been provided because *The NSTA Atlas of the Three Dimensions* was produced concurrently with this edition. Titles and map codes are accurate.

Additional Readings:

Available for download at www.curriculumtopicstudy2.org

Copyright © 2020 by Corwin Press, Inc. All rights reserved. Reprinted from *Science Curriculum Topic Study: Bridging the Gap Between Three-Dimensional Standards, Research, and Practice* (2nd ed.) by Page Keeley and Joyce Tugel. Thousand Oaks, CA: Corwin, www.corwin.com. Reproduction authorized for educational use by educators, local school sites, and/or noncommercial or nonprofit entities that have purchased the book.

Energy and Matter: Flows, Cycles, and Conservation
K–12 Standards- and Research-Based Study of a Curricular Topic

Section and Outcome	Selected Sources and Readings for Study and Reflection Read and examine **related** parts of
I. **Content Knowledge**	**IA:** *Framework for K–12 Science Education:* Narrative • Ch. 4: Energy and Matter: Flows, Cycles, and Conservation, pp. 94–95 **IB:** *NSTA Quick Reference Guide to the NGSS K–12:* Narrative • Crosscutting Concept 5: Energy and Matter: Flows, Cycles, and Conservation, p. 35 **IC:** *Next Generation Science Standards:* Appendices Narrative • Appendix G: Energy and Matter: Flows, Cycles, and Conservation, pp. 8–9 **ID:** *The NSTA Atlas of the Three Dimensions:* Narrative Page • 2.5: Energy and Matter: Flows, Cycles, and Conservation (EM)
II. **Concepts, Core Ideas, or Practices**	**IIA:** *Next Generation Science Standards:* Crosscutting Concepts Column • Grade 2: Energy and Matter, p. 13 • Grade 4: Energy and Matter, p. 23 • Grade 5: Energy and Matter, p. 29 • MS: Energy and Matter, pp. 37, 40, 47, 58 • HS: Energy and Matter, pp. 68, 70, 74, 81, 92, 96 **IIB:** *NSTA Quick Reference Guide to the NGSS K–12:* Crosscutting Concepts • K–2: Energy and Matter: Flows, Cycles, and Conservation, p. 90 • 3–5: Energy and Matter: Flows, Cycles, and Conservation, p. 102 • 6–8: Energy and Matter: Flows, Cycles, and Conservation, p. 119 • 9–12: Energy and Matter: Flows, Cycles, and Conservation, p. 143
III. **Curriculum, Instruction, and Formative Assessment**	**IIIA:** *Framework for K–12 Science Education:* Progression • Ch. 4: Energy and Matter: Flows, Cycles, and Conservation, pp. 95–96 **IIIB:** *NSTA Quick Reference Guide to the NGSS K–12*: Progression • Energy and Matter: Flows, Cycles, and Conservation Progression, pp. 35–36 **IIIC:** *Uncovering Student Ideas:* Assessment Probe and Suggestions for Instruction • USI.1: Ice Cubes in a Bag, pp. 45, 51–51; Lemonade, pp. 55, 58–59; Cookie Crumbles, pp. 61, 65; Seedlings in a Jar, pp. 67, 71–72 • USI.2: Sequoia Tree, pp. 121, 126–127 • USI.3: Earth's Mass, pp. 147, 153 • USI.4: Burning Paper, pp. 23, 28 • USI.LS: Ecosystem Cycles, pp. 97, 101
IV. **Research on Commonly Held Ideas**	**IVA:** *Benchmarks for Science Literacy:* Chapter 15 Research • 4D: Conservation of Matter, pp. 336–337 • 4E: Energy Conservation, p. 338 • 5E: Flow of Matter and Energy, pp. 119–121 **IVB:** *Uncovering Student Ideas:* Related Research • USI.1: Ice Cubes in a Bag, pp. 49–50; Lemonade, pp. 57–58; Cookie Crumbles, p. 64; Seedlings in a Jar, p. 71 • USI.2: Sequoia Tree, p. 126 • USI.3: Earth's Mass, pp. 152–153 • USI.4: Burning Paper, pp. 27–28 • USI.LS: Ecosystem Cycles, p. 100

Section and Outcome	Selected Sources and Readings for Study and Reflection Read and examine **related** parts of
V. **K–12 Progression and Connections**	**VA:** *Next Generation Science Standards:* Appendices: Progression Chart • Appendix G: Energy and Matter: Flows, Cycles, and Conservation, p. 9 • Appendix G: Energy and Matter Connections, p. 12 **VB:** *NSTA Quick Reference Guide to the NGSS K–12:* Progression Chart • Energy and Matter: Flows, Cycles, and Conservation, p. 60 **VC:** *The NSTA Atlas of the Three Dimensions:* Map Page • 2.5: Energy and Matter: Flows, Cycles, and Conservation (EM)
VI. **Assessment Expectation**	**VIA:** *State Standards* • Examine your state's standards **VIB:** *Next Generation Science Standards:* Performance Expectations • Grade 2: 2-PS1-3, p. 13 • Grade 4: 4-PS3-1, 4-PS3-2, 4-PS3-3, 4-PS3-4, p. 23 • Grade 5: 5-PS3-1, 5-LS1-1, p. 29 • MS: MS-PS1-5, MS-PS1-6, p. 37; MS-PS3-3, MS-PS3-5, p. 40; MS-LS1-6, MS-LS1-7, MS-LS2-3, p. 47; MS-ESS2-4, p. 58 • HS-PS1-8, p. 68; HS: HS-PS1-4, HS-PS1-7, p. 70; HS-PS3-2, HS-PS3-3, p. 74; HS-LS1-5, HS-LS1-6, HS-LS1-7, HS-LS2-3, HS-LS2-4, p. 81; HS-ESS1-2, HS-ESS1-3, p. 92; HS-ESS2-3, HS-ESS2-6, p. 96 **VIC:** *NSTA Quick Reference Guide to the NGSS K–12:* Performance Expectations • Grade 2: 2-PS1-3, p. 97 • Grade 4: 4-PS3-1, 4-PS3-2, p. 108; 4-PS3-3, 4-PS3-4, p. 109 • Grade 5: 5-PS3-1, p. 113; 5-LS1-1, p. 111 • MS: MS-PS1-5, p. 123; MS-PS1-6, p. 124; MS-PS3-3, MS-PS3-5, p. 126; MS-LS1-6, MS-LS1-7, p. 129; MS-LS2-3, p. 130; MS-ESS2-4, p. 135 • HS: HS-PS1-4, p. 147; HS-PS1-7, HS-PS1-8, p. 148; HS-PS3-2, HS-PS3-3, p. 151; HS-LS1-5, p. 154; HS-LS1-6, HS-LS1-7, p. 155; HS-LS2-3, HS-LS2-4, p. 156; HS-ESS1-2, HS-ESS1-3, p. 161; HS-ESS2-3, p. 163; HS-ESS2-6, p. 164

Review Chapter 2 instructions on how to use this guide.

Visit curriculumtopicstudy2.org for more information about CTS and additional resources.

SEE ALSO: CTS guide Cycling of Matter and Flow of Energy in Ecosystems.

NOTE: *Atlas* page numbers have not been provided because *The NSTA Atlas of the Three Dimensions* was produced concurrently with this edition. Titles and map codes are accurate.

Additional Readings:

 Available for download at www.curriculumtopicstudy2.org

Copyright © 2020 by Corwin Press, Inc. All rights reserved. Reprinted from *Science Curriculum Topic Study: Bridging the Gap Between Three-Dimensional Standards, Research, and Practice* (2nd ed.) by Page Keeley and Joyce Tugel. Thousand Oaks, CA: Corwin, www.corwin.com. Reproduction authorized for educational use by educators, local school sites, and/or noncommercial or nonprofit entities that have purchased the book.

Structure and Function
Grades K–12 Standards- and Research-Based Study of a Curricular Topic

Section and Outcome	Selected Sources and Readings for Study and Reflection Read and examine **related** parts of
I. **Content Knowledge**	**IA: *Framework for K–12 Science Education:*** Narrative • Ch. 4: Structure and Function, pp. 96–97 **IB: *NSTA Quick Reference Guide to the NGSS K–12:*** Narrative • Crosscutting Concept 6: Structure and Function, p. 37 **IC: *Next Generation Science Standards:*** Appendices Narrative • Appendix G: Structure and Function, p. 9 **ID: *The NSTA Atlas of the Three Dimensions:*** Narrative Page • 2.6: Structure and Function (SF)
II. **Concepts, Core Ideas, or Practices**	**IIA: *Next Generation Science Standards:*** Crosscutting Concepts Column • K: Structure and Function, p. 16 • Grade 1: Structure and Function, pp. 10, 16 • Grade 2: Structure and Function, pp. 14, 16 • MS: Structure and Function, pp. 35, 42, 45, 50 • HS: Structure and Function, pp. 68, 80, 96 **IIB: *NSTA Quick Reference Guide to the NGSS K–12:*** Crosscutting Concepts • K–2: Structure and Function, p. 90 • 3–5: Structure and Function, p. 102 • 6–8: Structure and Function, p. 119 • 9–12: Structure and Function, p. 143
III. **Curriculum, Instruction, and Formative Assessment**	**IIIA: *Framework for K–12 Science Education:*** Progression • Ch. 4: Structure and Function, pp. 97–98 **IIIB: *NSTA Quick Reference Guide to the NGSS K–12:*** Progression • Structure and Function, pp. 37–38
IV. **Research on Commonly Held Ideas**	Research on *Structure and Function* is not available in current CTS resources.
V. **K–12 Articulation and Connections**	**VA: *Next Generation Science Standards:*** Appendices: Progression Chart • Appendix G: Structure and Function, p. 10 • Appendix G: Structure and Function Connections, p. 12 **VB: *NSTA Quick Reference Guide to the NGSS K–12:*** Progression Chart • Structure and Function, p. 60 **VC: *The NSTA Atlas of the Three Dimensions:*** Map Page • 2.6: Structure and Function (SF)

Section and Outcome	Selected Sources and Readings for Study and Reflection Read and examine **related** parts of
VI. **Assessment** **Expectation**	**VIA:** *State Standards* • Examine your state's standards **VIB:** *Next Generation Science Standards:* Performance Expectations • K: K–2-ETS1-2, p. 16 • Grade 1: 1: 1-LS1-1, p. 10; K–2-ETS1-2, p. 16 • Grade 2: 2-LS2-2, p. 14; K–2-ETS1-2, p. 16 • MS: MS-PS1-3, p. 35; MS-PS4-2, MS-PS4-3, p. 42; MS-LS1-2, p. 45; MS-LS3-1, p. 50 • HS: HS-PS2-6, p. 68; HS-LS1-1, p. 80; HS-ESS2-5, p. 96 **VIC:** *NSTA Quick Reference Guide to the NGSS K–12:* Performance Expectations • K: K–2-ETS1-2, p. 98 • Grade 1: 1-LS1-1, p. 94; K–2-ETS1-2, p. 98 • Grade 2: 2-LS2-2, p. 96; K–2-ETS1-2, p. 98 • MS: MS-PS1-3, p. 122; MS-PS4-2, MS-PS4-3, p. 127; MS-LS1-2, p. 128; MS-LS3-1, p. 131 • HS: HS-PS2-6, p. 150; HS-LS1-1, p. 154; HS-ESS2-5, p. 164

Review Chapter 2 instructions on how to use this guide.

Visit curriculumtopicstudy2.org for more information about CTS and additional resources.

NOTE: *Atlas* page numbers have not been provided because *The NSTA Atlas of the Three Dimensions* was produced concurrently with this edition. Titles and map codes are accurate.

Additional Readings:

Available for download at www.curriculumtopicstudy2.org

Copyright © 2020 by Corwin Press, Inc. All rights reserved. Reprinted from *Science Curriculum Topic Study: Bridging the Gap Between Three-Dimensional Standards, Research, and Practice* (2nd ed.) by Page Keeley and Joyce Tugel. Thousand Oaks, CA: Corwin, www.corwin.com. Reproduction authorized for educational use by educators, local school sites, and/or noncommercial or nonprofit entities that have purchased the book.

Stability and Change
Grades K–12 Standards- and Research-Based Study of a Curricular Topic

Section and Outcome	Selected Sources and Readings for Study and Reflection Read and examine **related** parts of
I. **Content Knowledge**	**IA:** *Framework for K–12 Science Education:* Narrative • Ch. 4: Stability and Change, pp. 98–100 **IB:** *NSTA Quick Reference Guide to the NGSS K–12:* Narrative • Crosscutting Concept 7: Stability and Change, pp. 39–41 **ID:** *Next Generation Science Standards:* Appendices Narrative • Appendix G: Stability and Change, pp. 10–11 **IC:** *The NSTA Atlas of the Three Dimensions:* Narrative Page • 2.7: Stability and Change (SC)
II. **Concepts, Core Ideas, or Practices**	**IIA:** *Next Generation Science Standards:* Crosscutting Concepts Column • Grade 2: Stability and Change, p. 15 • MS: Stability and Change, pp. 38, 47, 49, 58, 59 • HS: Stability and Change, pp. 70, 76, 80, 83, 94, 96, 98, 99 **IIB:** *NSTA Quick Reference Guide to the NGSS K–12:* Crosscutting Concepts • K–2: Stability and Change, p. 90 • 3–5: Stability and Change, p. 102 • 6–8: Stability and Change, p. 120 • 9–12: Stability and Change, p. 144
III. **Curriculum, Instruction, and Formative Assessment**	**IIIA:** *Framework for K–12 Science Education:* Progression • Ch. 4: Stability and Change, pp. 100–101 **IIIB:** *NSTA Quick Reference Guide to the NGSS K–12:* Progression • Stability and Change, p. 41
IV. **Research on Commonly Held Ideas**	**IVA:** *Benchmarks for Science Literacy:* Chapter 15 Research • 11C: Constancy and Change, pp. 357–358
V. **K–12 Articulation and Connections**	**VA:** *Next Generation Science Standards:* Appendices: Progression Chart • Appendix G: Stability and Change, p. 11 • Appendix G: Stability and Change Connections, p. 12 **VB:** *NSTA Quick Reference Guide to the NGSS K–12:* Progression Chart • Stability and Change, p. 60 **VC:** *The NSTA Atlas of the Three Dimensions:* Map Page • 2.7: Stability and Change (SC)
VI. **Assessment Expectation**	**VIA:** *State Standards* • Examine your state's standards **VIB:** *Next Generation Science Standards:* Performance Expectations • Grade 2: 2-ESS1-1, 2-ESS2-1, p. 15 • MS: MS-PS2-2, p. 38; MS-LS2-4, p. 47; MS-LS2-5, p. 49; MS-ESS2-1, p. 58; MS-ESS3-5, p. 59

Section and Outcome	Selected Sources and Readings for Study and Reflection Read and examine **related** parts of
	• HS: HS-PS1-6, p. 70; HS-PS4-2, p. 76; HS-LS1-3, p. 80; HS-LS2-6, HS-LS2-7, p. 83; HS-ESS1-6, HS-ESS2-1, p. 94; HS-ESS2-2, HS-ESS2-7, p. 96; HS-ESS3-5, p. 98; HS-ESS3-3, HS-ESS3-4, p. 99 **VIC: *NSTA Quick Reference Guide to the NGSS K–12:*** Performance Expectations • Grade 2: 2-ESS1-1, 2-ESS2-1, p. 96 • MS: MS-PS2-2, p. 124; MS-LS2-4, MS-LS2-5, p. 130; MS-ESS2-1, p. 134; MS-ESS3-5, p. 137 • HS: HS-PS1-6, p. 148; HS-PS4-2, p. 152; HS-LS1-3, p. 154; HS-LS2-6, HS-LS2-7, p. 157; HS-ESS1-6, p. 162; HS-ESS2-1, HS-ESS2-2, p. 163; HS-ESS2-7, p. 165; HS-ESS3-3, HS-ESS3-4, HS-ESS3-5, p. 166

Review Chapter 2 instructions on how to use this guide.

Visit curriculumtopicstudy2.org for more information about CTS and additional resources.

NOTE: *Atlas* page numbers have not been provided because *The NSTA Atlas of the Three Dimensions* was produced concurrently with this edition. Titles and map codes are accurate.

Additional Readings:

 Available for download at **www.curriculumtopicstudy2.org**

Copyright © 2020 by Corwin Press, Inc. All rights reserved. Reprinted from *Science Curriculum Topic Study: Bridging the Gap Between Three-Dimensional Standards, Research, and Practice* (2nd ed.) by Page Keeley and Joyce Tugel. Thousand Oaks, CA: Corwin, www.corwin.com. Reproduction authorized for educational use by educators, local school sites, and/or noncommercial or nonprofit entities that have purchased the book.

Notes

Category F: STEM Connections Guides

The twelve CTS guides in this section are divided into two subsections. The guides focus on connections between and within the three closely related STEM disciplines of science, technology, and engineering.

Engineering, Technology, and Applications of Science Guides focus on understanding the engineering design process, including how science is utilized in the process and how new technologies result. It includes both the distinctions of and the relationships between engineering, science, and technology and their impacts on society and the natural world we live in. The alphabetically arranged guides in this section include

- Defining and Delimiting an Engineering Problem
- Developing Possible Design Solutions
- Engineering Design
- Improving Proposed Designs
- Influence of Science, Technology, and Engineering on Society and the Natural World
- Interdependence of Science, Engineering, and Technology

Nature of Science Guides focus on understanding the nature of scientific knowledge. It is distinct from engaging in the scientific practices as the focus is more on understanding the enterprise of science. Some guides in this section connect to the scientific practices; others connect to the crosscutting concepts. The alphabetically arranged guides in this section include

- Hypotheses, Theories, and Laws
- Methods of Scientific Investigations
- Science Addresses Questions
- Science as a Human Endeavor
- Science as a Way of Knowing
- Scientific Knowledge Demands Evidence

NOTES FOR USING CATEGORY F GUIDES

Overall

- *Atlas* page numbers have not been provided because *The NSTA Atlas of the Three Dimensions* was produced concurrently with this edition. Titles and map codes are accurate.
- Nature of Science is not included in the *Framework*. It is listed in the *NGSS* but is considered a connection rather than one of the three dimensions. Engineering, Technology, and Applications of Science is included as a Core

Idea in the *Framework* and *NGSS* and has its own performance expectations in K–2 and 3–5.

- Eliminate redundancy. See below for readings that include the exact same information. Even when the information is the same, there may be an advantage in how the information is presented. Select the reading based on the resources you have available to use for CTS and/or the advantage of using one over the other.

Section I

- Readings from the *Framework* and *Atlas* narrative are exactly the same for Engineering, Technology, and Applications of Science. Choose one of these resources for this section. There are no Nature of Science readings in the *Framework*.

- When reading the *Framework*, stop at Grade Band Endpoints.

- Some *Atlas* maps combine topics. When using the *Atlas* narrative for this section, if there is more than one core idea included in the narrative, focus on the one that is listed on your CTS guide.

- When reading *Disciplinary Core Ideas* for this section, focus on the content that helps you understand this topic. There are also suggestions for instruction embedded in this reading that can be added to CTS section III.

Section II

- Readings from the *Framework*, *NGSS*, and *NSTA Quick Reference Guide* are practically the same for Engineering, Technology, and Applications of Science. In a few cases, a *Framework* idea was not included in the *NGSS* or was moved to a different grade span. In the *Framework* goals are described in grade bands K–2, 3–5, 6–8, and 9–12. In the *NGSS* and *NSTA Quick Reference Guide*, they are phrased as disciplinary core ideas and designate a specific grade.

- The readings from the *NGSS* and the *NSTA Quick Reference Guide* are exactly the same for Engineering, Technology, and Applications of Science, except the *NGSS* also includes disciplinary core ideas in a separate K–2 and 3–5 section. Choose one of these resources. An advantage to using the *NSTA Quick Reference Guide* is that the disciplinary core idea is matched to the performance expectation (section VI) on the chart.

Section III

- This category of guides does not have as many resources to choose from as the other categories. You may find additional resources to supplement this that can be added at the end of the guide.

- There are always new books coming out in the *Uncovering Student Ideas in Science* series that may include engineering, technology, and nature of science probes that are not listed on the topic guides. Check the CTS website for a list of new probes published after 2019.

Section IV

- There are fewer research summaries from the collection of CTS resources for this category. However, there is extensive research available, especially for the nature of science. If you have access to professional journals, you can include articles as additional readings listed at the end of each guide.

Section V

- The *NGSS* Appendix I is a summary of the components of the engineering design process for K–2, 3–5, 6–8, and 9–12 with an accompanying graphic.

- The *Atlas* includes the same information as the *NSTA Quick Reference Guide* but the visual mapping of the ideas allows you to see precursor ideas and connections, as well as connections to other ideas. Be sure to read the front matter of the *Atlas* before using the maps. The information will help you use the maps effectively.

Section VI

- The *NGSS* and the *NSTA Quick Reference Guide* provide the exact same information. Clarifications and assessment boundaries are also included in both. Choose one of these resources based on the advantages listed below or the resources you have available.

- An advantage to using the *NSTA Quick Reference Guide* is that the performance expectation for Engineering, Technology, and Applications of Science is listed next to the disciplinary core idea included in that performance expectation.

- An advantage to using the *NGSS* that is not evident in the *NSTA Quick Reference Guide* is that the crosscutting concept chart, included below the performance expectations, shows the third dimension that is part of the performance expectation.

- Nature of science is not included as one of the three dimensions in the performance expectations.

Defining and Delimiting an Engineering Problem
Grades K–12 Standards- and Research-Based Study of a Curricular Topic

Section and Outcome	Selected Sources and Readings for Study and Reflection Read and examine **related** parts of
I. Content Knowledge	**IA:** *Science for All Americans* • Ch. 3: The Essence of Engineering Is Design Under Constraint, pp. 28–29 **IB:** *Framework for K–12 Science Education:* Narrative • Ch. 8: ETS1.A: Defining and Delimiting an Engineering Problem, pp. 204–205 **IC:** *Disciplinary Core Ideas: Reshaping Teaching and Learning* • Ch. 13: ETS1.A: Defining and Delimiting an Engineering Problem, pp. 248–249 **ID:** *The NSTA Atlas of the Three Dimensions:* Narrative Page • 6.1: Defining and Delimiting an Engineering Problem (ETS1.A)
II. Concepts, Core Ideas, or Practices	**IIA:** *Framework for K–12 Science Education:* Grade Band Endpoints • Ch. 8: ETS1.A: Defining and Delimiting an Engineering Problem, pp. 205–206 **IIB:** *Next Generation Science Standards:* Disciplinary Core Ideas Column • K: ETS1.A, pp. 5, 7 • K–2: ETS1.A, p. 16 • Grade 4: ETS1.A, p 23 • 3–5: ETS1.A, p. 32 • MS: ETS1.A, pp. 40, 63 • HS: ETS1.A, pp. 72, 75, 102 **IIC:** *NSTA Quick Reference Guide to the NGSS K–12:* Disciplinary Core Ideas • K: ETS1A, pp. 92, 93 • K–2: ETS1.A, p. 98 • Grade 4: ETS1.A, p. 109 • 3–5: ETS1.A, p. 114 • MS: ETS1.A, pp. 126, 138 • HS: ETS1.A, pp. 149, 151, 167
III. Curriculum, Instruction, and Formative Assessment	**IIIA:** *Disciplinary Core Ideas: Reshaping Teaching and Learning* • Ch. 13: Teaching Strategies for Defining and Delimiting an Engineering Problem, p. 258
IV. Research on Commonly Held Ideas	**IVA:** *Benchmarks for Science Literacy:* Chapter 15 Research • 3B: Design and Systems, pp. 334–335 **IVB:** *Disciplinary Core Ideas: Reshaping Teaching and Learning* • Ch. 13: What Do Students Find Challenging About Engineering Design? pp. 255–256
V. K–12 Articulation and Connections	**VA:** *Next Generation Science Standards:* Appendices: Progression Graphics • Appendix I: Engineering Design, pp. 3–6 **VB:** *NSTA Quick Reference Guide to the NGSS K–12:* Progression Chart • ETS1.A: Defining and Delimiting an Engineering, p. 80 **VC:** *The NSTA Atlas of the Three Dimensions:* Map Page • 6.1: Defining and Delimiting an Engineering Problem (ETS1.A)

Section and Outcome	Selected Sources and Readings for Study and Reflection Read and examine **related** parts of
VI. **Assessment** **Expectation**	**VIA:** *State Standards* • Examine your state's standards **VIB:** *Next Generation Science Standards:* Performance Expectations • K: K-PS2-2, p. 5; K-ESS3-2, p. 7 • K–2, ETS-1, p. 16 • Grade 4: 4-PS3-4, p. 23 • 3–5, ETS1-1, p. 32 • MS: MS-PS3-3, p. 40; MS-ETS1-1, p. 63 • HS: HS-PS2-3, p. 72; HS-PS3-3, p. 74; HS-ETS1-1, p. 102 **VIC:** *NSTA Quick Reference Guide to the NGSS K–12:* Performance Expectations • K: K-ESS3-2, p. 92; K-PS2-2, p. 93 • K–2: ETS-1, p. 98 • Grade 4: 4-PS3-4, p. 109 • 3–5: 3-5-ETS1-1, p. 114 • MS: MS-PS3-3, p. 126; MS-ETS1-1, p. 138 • HS-PS2-3, p. 149; HS-PS3-3, p. 151, HS-ETS1-1, p. 167

Review Chapter 2 instructions on how to use this guide.

Visit curriculumtopicstudy2.org for more information about CTS and additional resources.

NOTE: *Atlas* page numbers have not been provided because *The NSTA Atlas of the Three Dimensions* was produced concurrently with this edition. Titles and map codes are accurate.

Additional Readings:

online resources — Available for download at **www.curriculumtopicstudy2.org**

Copyright © 2020 by Corwin Press, Inc. All rights reserved. Reprinted from *Science Curriculum Topic Study: Bridging the Gap Between Three-Dimensional Standards, Research, and Practice* (2nd ed.) by Page Keeley and Joyce Tugel. Thousand Oaks, CA: Corwin, www.corwin.com. Reproduction authorized for educational use by educators, local school sites, and/or noncommercial or nonprofit entities that have purchased the book.

Developing Possible Design Solutions
Grades K–12 Standards- and Research-Based Study of a Curricular Topic

Section and Outcome	Selected Sources and Readings for Study and Reflection Read and examine **related** parts of
I. Content Knowledge	**IA:** *Science for All Americans* • Ch. 3: Engineering Combines Scientific Inquiry and Practical Values, p. 27 (first paragraph) **IA:** *Framework for K–12 Science Education:* Narrative • Ch. 8: ETS1.B: Developing Possible Solutions, pp. 206–207 **IB:** *Disciplinary Core Ideas: Reshaping Teaching and Learning* • Ch. 13: ETS1.B: Developing Possible Solutions, pp. 249–251 **IC:** *The NSTA Atlas of the Three Dimensions:* Narrative Page • 6.2: Developing and Optimizing Design Solutions (ETS1.B)
II. Concepts, Core Ideas, or Practices	**IIA:** *Framework for K–12 Science Education:* Grade Band Endpoints • Ch. 8: ETS1.B: Developing Possible Solutions, pp. 207–208 **IIB:** *Next Generation Science Standards:* Disciplinary Core Ideas Column • K: ETS1.B, p. 6 • Grade 2: ETS1.B, p. 14 • K–2: K-2-ETS1.B, p. 16 • Grade 4: ETS1.B, p. 26 • 3–5, 3-5-ETS1.B, p. 32 • MS: ETS1.B, pp. 37, 40, 49, 63 • HS: ETS1.B, pp. 84, 99, 102 **IIC:** *NSTA Quick Reference Guide to the NGSS K–12:* Disciplinary Core Ideas • K: ETS1.B, p. 93 • Grade 2: ETS1.B, p. 96 • K–2: K-2-ETS1.B, p. 98 • Grade 4: ETS1.B, p. 108 • 3–5: 3-5-ETS1.B, p. 114 • MS: ETS1.B, pp. 124, 130, 138 • HS: ETS1.B, pp. 157, 160, 165, 166, 167
III. Curriculum, Instruction, and Formative Assessment	**IIIA:** *Disciplinary Core Ideas: Reshaping Teaching and Learning* • Ch. 13: Teaching Strategies for Developing Possible Solutions, p. 258
IV. Research on Commonly Held Ideas	**IVA:** *Disciplinary Core Ideas: Reshaping Teaching and Learning* • Ch. 13: What Do Students Find Challenging About Engineering Design? pp. 255–256
V. K–12 Articulation, and Connections	**VA:** *Next Generation Science Standards:* Appendices: Progression Graphics • Appendix I: Engineering Design, pp. 3–6 **VB:** *NSTA Quick Reference Guide to the NGSS K–12:* Progression Chart • ETS1.B: Developing Possible Solutions, p. 81

Section and Outcome	Selected Sources and Readings for Study and Reflection Read and examine **related** parts of
	VC: *The NSTA Atlas of the Three Dimensions:* Map Page • 6.2: Developing and Optimizing Design Solutions (ETS1.B & ETS1.C)
VI. **Assessment** **Expectation**	**VIA:** *State Standards* • Examine your state's standards **VIB:** *Next Generation Science Standards:* Performance Expectations • K: K-ESS3-3, p. 6 • Grade 2: 2-LS2-2, p. 14 • K–2-ETS1-2, p. 16 • Grade 4: 4-ESS3-2, p. 26 • 3–5: 3-5-ETS1-2, p. 32 • MS: MS-PS1-6, p. 37; MS-PS3-3, p. 40; MS-LS2-5, p. 49; MS-ETS1-2, MS-ETS1-3, p. 63 • HS: HS-LS2-7, HS-LS4-6, p. 83; HS-ESS3-2, p. 99; HS-ETS1-3, HS-ETS1-4, p. 102 **VIC:** *NSTA Quick Reference Guide to the NGSS K–12:* Performance Expectations • K: K-ESS3-3, p. 93 • Grade 2: 2-LS2-2, p. 96 • K–2: K-2-ETS1-2, p. 98 • Grade 4: 4-ESS3-2, p. 108 • 3–5: 3-5-ETS1-2, p. 114 • MS: MS-PS1-6, p. 124; MS-PS3-3, p. 126; MS-LS2-5, p. 130; MS-ETS1-2, MS-ETS1-3, p. 138 • HS: HS-LS2-7, p. 157, HS-LS4-6, p. 160, HS-ESS3-2, p. 165; HS-ETS1-3, HS-ETS1-4, p. 167

Review Chapter 2 instructions on how to use this guide.

Visit curriculumtopicstudy2.org for more information about CTS and additional resources.

NOTE: *Atlas* page numbers have not been provided because *The NSTA Atlas of the Three Dimensions* was produced concurrently with this edition. Titles and map codes are accurate.

Additional Readings:

 Available for download at www.curriculumtopicstudy2.org

Copyright © 2020 by Corwin Press, Inc. All rights reserved. Reprinted from *Science Curriculum Topic Study: Bridging the Gap Between Three-Dimensional Standards, Research, and Practice* (2nd ed.) by Page Keeley and Joyce Tugel. Thousand Oaks, CA: Corwin, www.corwin.com. Reproduction authorized for educational use by educators, local school sites, and/or noncommercial or nonprofit entities that have purchased the book.

Engineering Design
Grades K–12 Standards- and Research-Based Study of a Curricular Topic

Section and Outcome	Selected Sources and Readings for Study and Reflection Read and examine **related** parts of
I. **Content Knowledge**	**IA:** *Science for All Americans* • Ch. 3: Engineering Combines Scientific Inquiry and Practical Values, p. 27 (first paragraph) • Ch. 3: Design and Systems, pp. 28–32 **IB:** *Framework for K–12 Science Education:* Narrative • Ch. 8: Core Idea ETS1: Engineering Design, p. 204 • Ch. 8: ETS1.A Defining and Delimiting an Engineering Problem, pp. 204–205 • Ch. 8: ETS1.B Developing Possible Solutions, pp. 206–207 • Ch. 8: ETS1.C Optimizing the Design Solution, pp. 208–209 **IC:** *Disciplinary Core Ideas: Reshaping Teaching and Learning* • Ch. 13: What Is Engineering Design, and Why Is It a Disciplinary Core Idea? pp. 245–246 • Ch. 13: What Are the Component Ideas of ETS1? pp. 248–251 **ID:** *The NSTA Atlas of the Three Dimensions:* Narrative Page • 6.1: Defining and Delimiting an Engineering Problem (ETS1.A) • 6.2: Developing and Optimizing Design Solutions (ETS1.B & ETS1.C)
II. **Concepts, Core Ideas, or Practices**	**IIA:** *Framework for K–12 Science Education:* Grade Band Endpoints • Ch. 8: ETS1.A Defining and Delimiting an Engineering Problem, pp. 205–206 • Ch. 8: ETS1.B Developing Possible Solutions, pp. 207–208 • Ch. 8: ETS1.C Optimizing the Design Solution, pp. 209–210 **IIB:** *Next Generation Science Standards:* Disciplinary Core Ideas Column • K: ETS1.A, pp. 5, 7; ETS1.B, p. 6 • Grade 2: ETS1.B, p. 14; ETS1.C, p. 15 • K–2: ETS1.A, ETS1.B, ETS1.C, p. 16 • Grade 4: ETS1A, p. 23; ETS1.B, p. 25; ETS1.C, p. 24 • 3–5, ETS1.A, ETS1.B, ETS1.C, p. 32 • MS: ETS1.A, pp. 40, 63; ETS1.B, pp. 37, 40, 49, 63; ETS1.C, pp. 37, 63 • HS: ETS1.A, pp. 74, 99, 102; ETS1.B, pp. 83, 99, 102; ETS1.C, pp. 70, 72, 102 **IIC:** *NSTA Quick Reference Guide to the NGSS K–12:* Disciplinary Core Ideas Column • K: ETS1A, pp. 92, 93; ETS1.B, p. 93 • Grade 2: ETS1.B, p. 96; ETS1.C, p. 96 • K–2: K-2-ETS1: Engineering Design, p. 98 • Grade 4: ETS1A, p. 109; ETS1.B, p. 108; ETS1.C, p. 110 • 3–5, 3-5-ETS1: Engineering Design, p. 114 • MS: ETS1A, pp. 126, 138; ETS1.B, pp. 124, 130, 138; ETS1.C, pp. 124, 138 • HS: ETS1A, pp. 149, 151, 167; ETS1.B, pp. 157, 160, 165, 166, 167; ETS1.C, pp. 148, 149, 167
III. **Curriculum, Instruction, and Formative Assessment**	**IIIA:** *Disciplinary Core Ideas: Reshaping Teaching and Learning* • Ch. 13: Elementary, p. 252 • Ch. 13: Middle School, pp. 252–254 • Ch. 13: High School, pp. 254–255 • Ch. 13: What Approaches Can We Use to Teach About This Disciplinary Core Idea? pp. 256–259
IV. **Research on Commonly Held Ideas**	**IVA:** *Benchmarks for Science Literacy:* Chapter 15 Research • 3A: Technology and Science, p. 334 • 3B: Design and Systems, pp. 334–335 **IVB:** *Disciplinary Core Ideas: Reshaping Teaching and Learning* • Ch. 13: What Do Students Find Challenging About Engineering Design? pp. 255–256

Section and Outcome	Selected Sources and Readings for Study and Reflection Read and examine **related** parts of
V. **K–12 Articulation and Connections**	**VA:** *Next Generation Science Standards:* Appendices: Progression Graphics • Appendix I: Engineering Design, pp. 3–6 **VB:** *NSTA Quick Reference Guide to the NGSS K–12:* Progression Chart • ETS1: Engineering Design, pp. 80–81 **VC:** *The NSTA Atlas of the Three Dimensions:* Map Page • 6.1: Defining and Delimiting an Engineering Problem (ETS1.A) • 6.2: Developing and Optimizing Design Solutions (ETS1.B & ETS1.C)
VI. **Assessment Expectation**	**VIA:** *State Standards* • Examine your state's standards **VIB:** *Next Generation Science Standards:* Performance Expectations • K: K-PS2-2, p. 5; K-ESS3-2, p. 7; K-ESS3-3, p. 6 • Grade 2: 2-LS2-2, p. 14; 2-ESS2-1, p. 15 • K–2: K-2-ETS1-1, K-2-ETS1-2, K-2-ETS1-3, p. 16 • Grade 4: 4-PS3-4, p. 23; 4-PS4-3, p. 24; 4-ESS3-2, p. 26 • 3–5: 3-5-ETS1-1, 3-5-ETS1-2, 3-5-ETS1-3, p. 32 • MS: MS-PS1-6, p. 37; MS-PS3-3, p. 40; MS-LS2-5, p. 49; MS-ETS1-1, MS-ETS1-2, MS-ETS1-3, MS-ETS1-4, p. 63 • HS: HS-PS1-6, p. 70; HS-PS2-3, p. 72; HS-PS3-3, p. 74; HS-LS2-7, HS-LS4-6, p. 83; HS-ESS3-2, HS-ESS3-4, p. 99; HS-ETS1-1, HS-ETS1-2, HS-ETS1-3, HS-ETS1-4, p. 102 **VIC:** *NSTA Quick Reference Guide to the NGSS K–12:* Performance Expectations • K: K-ESS3-2, p. 92; K-ESS3-3, K-PS2-2, p. 93 • Grade 2: 2-LS2-2, 2-ESS2-1, p. 96 • K–2: K-2-ETS1-1, K-2-ETS1-2, K-2-ETS1-3, p. 98 • Grade 4: 4-ESS3-2, p. 108; 4-PS3-4, p. 109; 4-PS4-3, p. 110 • 3–5: 3-5-ETS1-1, 3-5-ETS1-2, 3-5-ETS1-3, p. 114 • MS: MS-PS1-6, p. 124; MS-PS3-3, p. 126; MS-LS2-5, p. 130; MS-ETS1-1, MS-ETS1-2, MS-ETS1-3, MS-ETS1-4, p. 138 • HS-PS1-6, p. 148; HS-PS2-3, p. 149; HS-PS3-3, p. 151, HS-LS2-7, p. 157, HS-LS4-6, p. 160, HS-ESS3-2, p. 165; HS-ESS3-4, p. 166; HS-ETS1-1, HS-ETS1-2, HS-ETS1-3, HS-ETS1-4, p. 167

Review Chapter 2 instructions on how to use this guide.
Visit curriculumtopicstudy2.org for more information about CTS and additional resources.

NOTE: *Atlas* page numbers have not been provided because *The NSTA Atlas of the Three Dimensions* was produced concurrently with this edition. Titles and map codes are accurate.

Additional Readings:

Available for download at www.curriculumtopicstudy2.org

Copyright © 2020 by Corwin Press, Inc. All rights reserved. Reprinted from *Science Curriculum Topic Study: Bridging the Gap Between Three-Dimensional Standards, Research, and Practice* (2nd ed.) by Page Keeley and Joyce Tugel. Thousand Oaks, CA: Corwin, www.corwin.com. Reproduction authorized for educational use by educators, local school sites, and/or noncommercial or nonprofit entities that have purchased the book.

Improving Proposed Designs
Grades K–12 Standards- and Research-Based Study of a Curricular Topic

Section and Outcome	Selected Sources and Readings for Study and Reflection Read and examine **related** parts of
I. Content Knowledge	**IA:** *Framework for K–12 Science Education:* Narrative • Ch. 8: ETS1.C: Optimizing the Design Solution, pp. 208–209 **IB:** *Disciplinary Core Ideas: Reshaping Teaching and Learning* • Ch. 13: ETS1.C: Optimizing the Design Solution, p. 251 **IC:** *The NSTA Atlas of the Three Dimensions:* Narrative Page • 6.2: Developing and Optimizing Design Solutions (ETS1.C)
II. Concepts, Core Ideas, or Practices	**IIA:** *Framework for K–12 Science Education:* Grade Band Endpoints • Ch. 8: ETS1.C: Optimizing the Design Solution, pp. 209–210 **IIB:** *Next Generation Science Standards:* Disciplinary Core Ideas Column • Grade 2: ETS1.C, p. 15 • K–2: K-2-ETS1.C, p. 16 • Grade 4: ETS1.C, p. 24 • 3–5: 3-5-ETS1.C, p. 32 • MS: ETS1.C, pp. 37, 63 • HS: ETS1.C, pp. 70, 72, 102 **IIC:** *NSTA Quick Reference Guide to the NGSS K–12:* Disciplinary Core Ideas • Grade 2: ETS1.C, p. 96 • K–2: K-2-ETS1.C, p. 98 • Grade 4: ETS1.C, p. 110 • 3–5: 3-5-ETS1.C, p. 114 • MS: ETS1.C, pp. 124, 138 • HS: ETS1.C, pp. 148, 149, 167
III. Curriculum, Instruction, and Formative Assessment	**IIIA:** *Disciplinary Core Ideas: Reshaping Teaching and Learning* • Ch. 13: Teaching Strategies for Developing Possible Solutions, pp. 258–259
IV. Research on Commonly Held Ideas	**IVB:** *Disciplinary Core Ideas: Reshaping Teaching and Learning* • Ch. 13: What Do Students Find Challenging About Engineering Design? pp. 255–256
V. K–12 Articulation and Connections	**VA:** *Next Generation Science Standards:* Appendices: Progression Graphics • Appendix I: Engineering Design, pp. 3–6 **VB:** *NSTA Quick Reference Guide to the NGSS K–12:* Progression Chart • ETS1: Optimizing the Design Solution, p. 81 **VC:** *The NSTA Atlas of the Three Dimensions:* Map Page • 6.2: Developing and Optimizing Design Solutions (ETS1.B & ETS1.C)

Section and Outcome	Selected Sources and Readings for Study and Reflection Read and examine **related** parts of
VI. **Assessment Expectation**	**VIA: *State Standards*** • Examine your state's standards **VIB: *Next Generation Science Standards:*** Performance Expectations • Grade 2: 2-ESS2-1, p. 15 • K–2: K-2-ETS1-3, p. 16 • Grade 4: 4-PS4-3, p. 24 • 3–5: 3-5-ETS1-3, p. 32 • MS: MS-PS1.6, p. 37; MS-ETS1-4, p. 63 • HS: HS-PS1-6, p. 70; HS-PS2-3, p. 72; HS-ETS1-2, p. 102 **VIC: *NSTA Quick Reference Guide to the NGSS K–12:*** Performance Expectations • Grade 2: 2-ESS2-1, p. 96 • K–2: K-2-ETS1-3, p. 98 • Grade 4: 4-PS4-3, p. 110 • 3–5: 3-5-ETS1-3, p. 114 • MS: MS-PS1-6, p. 124; MS-ETS1-4, p. 138 • HS-PS1-6, p. 148; HS-PS2-3, p. 149; HS-ETS1-2, p. 167

Review Chapter 2 instructions on how to use this guide.

Visit curriculumtopicstudy2.org for more information about CTS and additional resources.

NOTE: *Atlas* page numbers have not been provided because *The NSTA Atlas of the Three Dimensions* was produced concurrently with this edition. Titles and map codes are accurate.

Additional Readings:

Available for download at www.curriculumtopicstudy2.org

Copyright © 2020 by Corwin Press, Inc. All rights reserved. Reprinted from *Science Curriculum Topic Study: Bridging the Gap Between Three-Dimensional Standards, Research, and Practice* (2nd ed.) by Page Keeley and Joyce Tugel. Thousand Oaks, CA: Corwin, www.corwin.com. Reproduction authorized for educational use by educators, local school sites, and/or noncommercial or nonprofit entities that have purchased the book.

Influence of Science, Technology, and Engineering on Society and the Natural World

Grades K–12 Standards- and Research-Based Study of a Curricular Topic

Section and Outcome	Selected Sources and Readings for Study and Reflection Read and examine **related** parts of
I. Content Knowledge	**IA:** *Science for All Americans* • Ch. 3: Technology and Science, pp. 26–28 • Ch. 8: Agriculture, pp. 108–110 • Ch. 8: Materials and Manufacturing, pp. 111–114 • Ch. 8: Energy Sources and Use, pp. 114–116 • Ch. 8: Health Technology, pp. 123–126 **IB:** *Framework for K–12 Science Education:* Narrative • Ch. 8: ETS2.B: Influence of ETS on Society and the Natural World, pp. 212–213 **IC:** *Disciplinary Core Ideas: Reshaping Teaching and Learning* • Ch. 14: Influence of ETS on Society and the Natural World, pp. 265–268 **1D:** *Next Generation Science Standards:* Appendices Narrative • Appendix J: Influence of ETS on Society and the Natural World, pp. 2–3 **IE:** *The NSTA Atlas of the Three Dimensions:* Narrative Page • 8.2, Influence of Engineering, Technology, and Science on Society and the Natural World (INFLU)
II. Concepts, Core Ideas, or Practices	**IIA:** *Framework for K–12 Science Education: Grade Band Endpoints* • Ch. 8: ETS2.B: Influence of Engineering, Technology, and Science on Society and the Natural World, pp. 213–214 **IIB:** *Next Generation Science Standards:* Disciplinary Core Ideas—Connections to Engineering, Technology, and Applications of Science • K: Influence of ETS on Society and the Natural World, p. 7 • Grade 1: Influence of ETS on Society and the Natural World, pp. 9, 10 • Grade 2: Influence of ETS on Society and the Natural World, pp. 13, 15 • Grade 3: Influence of ETS on Society and the Natural World, p. 21 • Grade 4: Influence of ETS on Society and the Natural World, pp. 23, 26 • 3–5: Influence of ETS on Society and the Natural World, p. 32 • MS: Influence of ETS on Society and the Natural World, pp. 35, 38, 42, 49, 58, 60, 63 • HS: Influence of ETS on Society and the Natural World, pp. 74, 76, 96, 99, 129, 102 **IIC:** *NSTA Quick Reference Guide to the NGSS K–12:* Connections to Engineering, Technology, and Applications of Science Charts • K–2: Influence of ETS on Society and the Natural World, p. 90 • 3–5: Influence of ETS on Society and the Natural World, p. 102 • MS: Influence of ETS on Society and the Natural World, p. 120 • HS: Influence of ETS on Society and the Natural World, p. 144
III. Curriculum, Instruction, and Formative Assessment	**IIIA:** *Disciplinary Core Ideas: Reshaping Teaching and Learning* • Ch. 14: How Does Student Understanding of This Idea Develop Over Time? pp. 268–269 • Ch. 14: Elementary p. 273 • Ch. 14: Middle, p. 273 • Ch. 14: High School, pp. 273–274 • Ch. 14: Connecting the Two Components of the Disciplinary Core Idea, pp. 274–276
IV. Research on Commonly Held Ideas	**IVA:** *Benchmarks for Science Literacy:* Chapter 15 Research • 1C: The Scientific Enterprise, p. 333 • 3C: Issues in Technology, p. 335

Section and Outcome	Selected Sources and Readings for Study and Reflection Read and examine **related** parts of
V. **K–12 Articulation and Connections**	**VA:** *Next Generation Science Standards:* Appendices: Progression Chart • Appendix J: 2. Influence of ETS on Society and the Natural World, pp. 3–4 **VB: NSTA Quick Reference Guide to the NGSS K–12:** Progression • Influence of ETS on Society and the Natural World, p. 85 **VC:** *The NSTA Atlas of the Three Dimensions:* Map Page • 8.2, Influence of Engineering, Technology, and Science on Society and the Natural World (INFLU)
VI. **Assessment Expectation**	**VIA:** *State Standards* • Examine your state's standards **VIB: Next Generation Science Standards:** Performance Expectations • K: K-ESS3-2, p. 7 • Grade 1: 1-PS4-4, p. 9; 1-LS1-1, p. 10 • Grade 2: 2-PS1-2, p. 13; 2-ESS2-1, p. 15 • Grade 3: 3-ESS3-1, p. 21 • Grade 4: 4-PS3-4, 4-ESS3-1, p. 23; 4-ESS3-2, p. 26 • 3–5: 3-5-ETS1-1, 3-5-ETS1-2, p. 32 • MS: MS-PS1-3, p. 35; MS-PS2-1, p. 38; MS-PS4-3, p. 42; MS-LS2-2, p. 49; MS-ESS3-1, p. 58; MS-ESS3-2, MS-ESS3-3, MS-ESS3-4, p. 60; MS-ETS1-1, p. 63 • HS: HS-PS3-3, p. 74; HS-PS4-2, HS-PS4-5, p. 76; HS-ESS2-2, p. 96; HS-ESS3-1, HS-ESS3-2, HS-ESS3-3, HS-ESS3-4, p. 99; HS-ETS1-1, HS-ETS1-3, p. 102 **VIC: NSTA Quick Reference Guide to the NGSS K–12:** Performance Expectations • K: K-ESS3-2, p. 92 • Grade 1: 1-PS4-4, p. 95; 1-LS1-1, p. 94 • Grade 2: 2-PS1-2, p. 97; 2-ESS2-1, p. 96 • Grade 3: 3-ESS3-1, p. 105 • Grade 4: 4-PS3-4, p. 109; 4-ESS3-1, 4-ESS3-2, p. 108 • 3–5: 3-5-ETS1-1, 3-5-ETS1-2, p. 114 • MS: MS-PS1-3, p. 122; MS-PS2-1, p. 124; MS-PS4-3, p. 127; MS-LS2-2, p. 130; MS-ESS3-1, MS-ESS3-2, MS-ESS3-3, p. 136; MS-ESS3-4, p. 137; MS-ETS1-1, p. 138 • HS: HS-PS3-3, p. 151; HS-PS4-2, p. 152; HS-PS4-5, p. 153; HS-ESS3-1, HS-ESS3-2, p. 165; HS-ESS3-3, HS-ESS3-4, p. 166; HS-ETS1-1, HS-ETS1-3, p. 167

Review Chapter 2 instructions on how to use this guide.

Visit curriculumtopicstudy2.org for more information about CTS and additional resources.

NOTE: *Atlas* page numbers have not been provided because *The NSTA Atlas of the Three Dimensions* was produced concurrently with this edition. Titles and map codes are accurate.

Additional Readings:

Available for download at www.curriculumtopicstudy2.org

Copyright © 2020 by Corwin Press, Inc. All rights reserved. Reprinted from *Science Curriculum Topic Study: Bridging the Gap Between Three-Dimensional Standards, Research, and Practice* (2nd ed.) by Page Keeley and Joyce Tugel. Thousand Oaks, CA: Corwin, www.corwin.com. Reproduction authorized for educational use by educators, local school sites, and/or noncommercial or nonprofit entities that have purchased the book.

Interdependence of Science, Engineering, and Technology
Grades K–12 Standards- and Research-Based Study of a Curricular Topic

Section and Outcome	Selected Sources and Readings for Study and Reflection Read and examine **related** parts of
I. **Content Knowledge**	**IA:** *Science for All Americans* • Ch. 3: Technology and Science, pp. 26–28 **IB:** *Framework for K–12 Science Education:* Narrative • Ch. 8: ETS2.A: Interdependence of Science, Engineering, and Technology, pp. 210–211 **1C:** *Next Generation Science Standards:* Appendices Narrative • Appendix J: Interdependence of Science, Engineering, and Technology, pp. 1–2 **ID:** *Disciplinary Core Ideas: Reshaping Teaching and Learning* • Ch. 14: ETS2.A: Interdependence of Science, Engineering, and Technology, p. 264 **IE:** *The NSTA Atlas of the Three Dimensions:* Narrative Page • 8.1, Interdependence of Science, Engineering, and Technology (INTER)
II. **Concepts, Core Ideas, or Practices**	**IIA:** *Framework for K–12 Science Education:* Grade Band Endpoints • Ch. 8: ETS2.A: Interdependence of Science, Engineering, and Technology, pp. 211–212 **IIB:** *Next Generation Science Standards:* Disciplinary Core Ideas Column—Connections to Engineering, Technology, and Applications of Science • K: Interdependence of Science, Engineering, and Technology, p. 7 • Grade 3: Interdependence of Science, Engineering, and Technology, pp. 18, 19 • Grade 4: Interdependence of Science, Engineering, and Technology, pp. 22, 24 • MS: Interdependence of Science, Engineering, and Technology, pp. 35, 45, 50, 56 • HS: Interdependence of Science, Engineering, and Technology, pp. 75, 92, 96 **IIC:** *NSTA Quick Reference Guide to the NGSS K–12:* Connections to Engineering, Technology, and Applications of Science • K–2: Interdependence of Science, Engineering, and Technology, p. 90 • 3–5: Interdependence of Science, Engineering, and Technology, p. 102 • MS: Interdependence of Science, Engineering, and Technology, p. 120 • HS: Interdependence of Science, Engineering, and Technology, p. 144
III. **Curriculum, Instruction, and Formative Assessment**	**IIIA:** *Disciplinary Core Ideas: Reshaping Teaching and Learning* • Ch. 14: Teaching ETS2.A Interdependence of Science, Engineering, and Technology, pp. 269–272 • Ch. 14: Connecting the Two Components of the Disciplinary Core Idea, pp. 274–276
IV. **Research on Commonly Held Ideas**	**IVA:** *Benchmarks for Science Literacy:* Chapter 15 Research • 3A: Technology and Science, p. 334
V. **K–12 Articulation and Connections**	**VA:** *Next Generation Science Standards:* Appendices: Progression Chart • Appendix J: 1. Interdependence of Science, Engineering, and Technology, p. 3 **VB:** *NSTA Quick Reference Guide to the NGSS K–12:* Progression • Interdependence of Science, Engineering, and Technology, p. 85 **VC:** *The NSTA Atlas of the Three Dimensions:* Map Page • 8.1, Interdependence of Science, Engineering, and Technology (INTER)

Section and Outcome	Selected Sources and Readings for Study and Reflection Read and examine **related** parts of
VI. **Assessment Expectation**	**VIA:** *State Standards* • Examine your state's standards **VIB:** *Next Generation Science Standards:* Performance Expectations • K: K-ESS3-2, p. 7 • Grade 3: 3-PS2-4, p. 18; 3-LS4-3, p. 19 • Grade 4: 4-PS4-3, p. 24; 4-ESS3-1, p. 22 • MS: MS-PS1-3, p. 35; MS-LS1-1, p. 45; MS-LS4-5, p. 50; MS-ESS1-3, p. 56 • HS: HS-PS4-5, p. 75; HS-ESS1-2, HS-ESS1-4, p. 92; HS-ESS2-3, p. 96 **VIC:** *NSTA Quick Reference Guide to the NGSS K–12:* Performance Expectations • K: K-ESS3-2, p. 92 • Grade 3: 3-PS2-4, p. 106; 3-LS4-3, p. 105 • Grade 4: 4-PS4-3, p. 110; 4-ESS3-1, p. 108 • MS: MS-PS1-3, p. 122; MS-LS1-1, p. 128; MS-LS4-5, p. 132; MS-ESS1-3, p. 133 • HS: HS-PS4-5, p. 153; HS-ESS1-2, p. 161; HS-ESS1-4, p. 162; HS-ESS2-3, p. 163

Review Chapter 2 instructions on how to use this guide.

Visit curriculumtopicstudy2.org for more information about CTS and additional resources.

NOTE: *Atlas* page numbers have not been provided because *The NSTA Atlas of the Three Dimensions* was produced concurrently with this edition. Titles and map codes are accurate.

Additional Readings:

Available for download at **www.curriculumtopicstudy2.org**

Copyright © 2020 by Corwin Press, Inc. All rights reserved. Reprinted from *Science Curriculum Topic Study: Bridging the Gap Between Three-Dimensional Standards, Research, and Practice* (2nd ed.) by Page Keeley and Joyce Tugel. Thousand Oaks, CA: Corwin, www.corwin.com. Reproduction authorized for educational use by educators, local school sites, and/or noncommercial or nonprofit entities that have purchased the book.

Hypotheses, Theories, and Laws
Grades K–12 Standards- and Research-Based Study of a Curricular Topic

Section and Outcome	Selected Sources and Readings for Study and Reflection Read and examine **related** parts of
I. Content Knowledge	**IA:** *Science for All Americans* • Ch. 1: Science Is a Blend of Logic and Imagination, pp. 5–6 • Ch. 1: Science Explains and Predicts, p. 6 **IB:** *The NSTA Atlas of the Three Dimensions:* Narrative Page • 7.4: Science Models, Laws, Mechanisms, and Theories Explain Natural Phenomena (ENP)
II. Concepts, Core Ideas or Practices	**IIA:** *Next Generation Science Standards:* (Scientific and Engineering Practices Under Connections to Nature of Science) • Grade 2: Science Models, Laws, Mechanisms, and Theories Explain Natural Phenomena, p. 13 • Grade 5: Science Models, Laws, Mechanisms, and Theories Explain Natural Phenomena, p. 29 • MS: Science Models, Laws, Mechanisms, and Theories Explain Natural Phenomena, p. 37 • Science Models, Laws, Mechanisms, and Theories Explain Natural Phenomena, pp. 96, 102, 116, 121 **IIB:** *NSTA Quick Reference Guide to the NGSS K–12:* Connections to Nature of Science • K–12: Science Models, Laws, Mechanisms, and Theories Explain Natural Phenomena, p. 83 • K–2: Science Models, Laws, Mechanisms, and Theories Explain Natural Phenomena, p. 91 • 3–5 Science Models, Laws, Mechanisms, and Theories Explain Natural Phenomena, p. 103 • MS: Science Models, Laws, Mechanisms, and Theories Explain Natural Phenomena, p. 121 • HS: Science Models, Laws, Mechanisms, and Theories Explain Natural Phenomena, pp. 72, 76, 87, 92, 94
III. Curriculum, Instruction, and Formative Assessment	**IIIA:** *Next Generation Science Standards:* Appendices • Appendix H: Implementing Instruction to Facilitate Understanding of the Nature of Science, pp. 7–8 **IIIB:** *Uncovering Student Ideas:* Assessment Probe and Suggestions for Instruction and Assessment • USI.3: Is It a Theory? pp. 83, 89–90; What Is a Hypothesis? pp. 101, 106–107
IV. Research on Commonly Held Ideas	**IVA:** *Benchmarks for Science Literacy:* Chapter 15 Research • 1A: The Scientific World View, p. 332 • 1B: Theory (Explanation) and Evidence, p. 332 • 12E: Theory and Evidence, p. 361 **IVB:** *Uncovering Student Ideas:* Related Research • USI.3: Is It a Theory? pp. 88–89; What Is a Hypothesis? p. 106

Section and Outcome	Selected Sources and Readings for Study and Reflection Read and examine **related** parts of
V. **K–12 Articulation and Connections**	**VA:** *Next Generation Science Standards:* Appendices: Progression Chart • Appendix H: Science Models, Laws, Mechanisms, and Theories Explain Natural Phenomena, p. 5 **VB:** *NSTA Quick Reference Guide to the NGSS K–12:* Progression Chart • Scientific Investigations Use a Variety of Methods, p. 83 **VC:** *The NSTA Atlas of the Three Dimensions:* Map Page • 7.4: Science Models, Laws, Mechanisms, and Theories Explain Natural Phenomena (ENP)
VI. **Assessment Expectation**	**VIA:** *State Standards* • Examine your state's standards **VIB:** *Next Generation Science Standards:* Performance Expectations • Grade 2: 2-PS1-4, p. 13 • Grade 5: 5-LS2-1, p. 29 • MS: MS-PS1-5, p. 37 • HS-PS2-1, HS-PS2-4, p. 72, HS-PS4-3, p. 76; HS-LS4-1, p. 87; HS-ESS1-2, p. 92; HS-ESS1-6, p. 94 **VIC:** *NSTA Quick Reference Guide to the NGSS K–12:* Performance Expectations • Grade 2: 2-PS1-4, p. 97 • Grade 5: 5-LS2-1, p. 111 • MS: MS-PS1-5, p. 123 • HS-PS2-1, HS-PS2-4, p. 149; HS-PS4-3, p. 152; HS-LS4-1, p. 159; HS-ESS1-2, p. 161; HS-ESS1-6, 162

Review Chapter 2 instructions on how to use this guide.

Visit curriculumtopicstudy2.org for more information about CTS and additional resources.

NOTE: *Atlas* page numbers have not been provided because *The NSTA Atlas of the Three Dimensions* was produced concurrently with this edition. Titles and map codes are accurate.

Additional Readings:

Available for download at www.curriculumtopicstudy2.org

Copyright © 2020 by Corwin Press, Inc. All rights reserved. Reprinted from *Science Curriculum Topic Study: Bridging the Gap Between Three-Dimensional Standards, Research, and Practice* (2nd ed.) by Page Keeley and Joyce Tugel. Thousand Oaks, CA: Corwin, www.corwin.com. Reproduction authorized for educational use by educators, local school sites, and/or noncommercial or nonprofit entities that have purchased the book.

Methods of Scientific Investigations
Grades K–12 Standards- and Research-Based Study of a Curricular Topic

Section and Outcome	Selected Sources and Readings for Study and Reflection Read and examine **related** parts of
I. Content Knowledge	**IA:** *Science for All Americans* • Ch. 1: Scientific Inquiry, pp. 3–7 **IB:** *The NSTA Atlas of the Three Dimensions:* Narrative Page • 7.1: Scientific Investigations Use a Variety of Methods (VOM)
II. Concepts, Core Ideas, or Practices	**IIA:** *Next Generation Science Standards:* (Scientific and Engineering Practices Column—Under Connections to Nature of Science) • K: Scientific Investigations Use a Variety of Methods, pp. 5, 7 • Grade 1: Scientific Investigations Use a Variety of Methods, p. 9 • Grade 3: Scientific Investigations Use a Variety of Methods, p. 18 • HS: Scientific Investigations Use a Variety of Methods, pp. 80, 98 **IIB:** *NSTA Quick Reference Guide to the NGSS K–12:* **Connections to Nature of Science** • K–2: Scientific Investigations Use a Variety of Methods, p. 91 • 3–5 Scientific Investigations Use a Variety of Methods, p. 103 • MS: Scientific Investigations Use a Variety of Methods, p. 121 • HS: Scientific Investigations Use a Variety of Methods, p. 145
III. Curriculum, Instruction, and Formative Assessment	**IIIA:** *Next Generation Science Standards: Appendices* • Appendix H: Implementing Instruction to Facilitate Understanding of the Nature of Science, pp. 7–8 **IIIB:** *Uncovering Student Ideas:* Assessment Probe and Suggestions for Instruction and Assessment • USI.3: Doing Science, pp. 93, 98–99
IV. Research on Commonly Held Ideas	**IVA:** *Benchmarks for Science Literacy:* Chapter 15 Research • 1B: Experimentation, p. 332 **IVB:** *Uncovering Student Ideas:* Related Research • USI.3: Doing Science, p. 97
V. K–12 Articulation and Connections	**VA:** *Next Generation Science Standards:* Appendices: Progression Chart • Appendix H: Scientific Investigations Use a Variety of Methods, p. 5 **VB:** *NSTA Quick Reference Guide to the NGSS K–12:* Progression Chart • Scientific Investigations Use a Variety of Methods, p. 82 **VC:** *The NSTA Atlas of the Three Dimensions:* Map Page • 7.1: Scientific Investigations Use a Variety of Methods (VOM)
VI. Assessment Expectation	**VIA:** *State Standards* • Examine your state's standards **VIB:** *Next Generation Science Standards:* Performance Expectations • K: K-PS2-1, p. 5; K-PS3-1, p. 7 • Grade 1: 1-PS4-1, p. 9 • Grade 3: 3-PS2-1, p. 18 • HS: HS-LS1-3, p. 80; HS-ESS3-5, p. 98

Section and Outcome	Selected Sources and Readings for Study and Reflection Read and examine **related** parts of
	VIC: *NSTA Quick Reference Guide to the NGSS K–12:* Performance Expectations • K: K-PS2-1, K-PS3-1, p. 93 • Grade 1: 1-PS4-1, p. 95 • Grade 3: 3-PS2-1, p. 105 • HS: HS-LS1-3, p. 154, HS-ESS3-5, p. 166

Review Chapter 2 instructions on how to use this guide.

Visit curriculumtopicstudy2.org for more information about CTS and additional resources.

NOTE: *Atlas* page numbers have not been provided because *The NSTA Atlas of the Three Dimensions* was produced concurrently with this edition. Titles and map codes are accurate.

Additional Readings:

Available for download at **www.curriculumtopicstudy2.org**

Copyright © 2020 by Corwin Press, Inc. All rights reserved. Reprinted from *Science Curriculum Topic Study: Bridging the Gap Between Three-Dimensional Standards, Research, and Practice* (2nd ed.) by Page Keeley and Joyce Tugel. Thousand Oaks, CA: Corwin, www.corwin.com. Reproduction authorized for educational use by educators, local school sites, and/or noncommercial or nonprofit entities that have purchased the book.

Science Addresses Questions
Grades K–12 Standards- and Research-Based Study of a Curricular Topic

Section and Outcome	Selected Sources and Readings for Study and Reflection Read and examine **related** parts of
I. Content Knowledge	**IA:** *Science for All Americans* • Ch. 1: Science Cannot Provide Complete Answers to All Questions, p. 3 **IB:** *The NSTA Atlas of the Three Dimensions:* Narrative Page • 7.8: Science Addresses Questions About the Natural and Material World (AQAW)
II. Concepts, Core Ideas, or Practices	**IIA:** *Next Generation Science Standards:* (Scientific and Engineering Practices Column Under Connections to Nature of Science) • Grade 2: Science Addresses Questions About the Natural and Material World, p. 15 • Grade 5: Science Addresses Questions About the Natural and Material World, p. 30 • MS: Science Addresses Questions About the Natural and Material World, pp. 49, 50, 60 • HS: Science Addresses Questions About the Natural and Material World, p. 100 **IIB:** *NSTA Quick Reference Guide to the NGSS K–12:* Connections to Nature of Science • K–12: Science Addresses Questions About the Natural and Material World, p. 84 • K–2: Science Addresses Questions About the Natural and Material World, p. 91 • 3–5 Science Addresses Questions About the Natural and Material World, p. 103 • MS: Science Addresses Questions About the Natural and Material World, p. 121 • HS: Science Addresses Questions About the Natural and Material World, p. 146
III. Curriculum, Instruction, and Formative Assessment	**IIIA:** Next Generation Science Standards: Appendices • Appendix H: Implementing Instruction to Facilitate Understanding of the Nature of Science, pp. 7–8
IV. Research on Commonly Held Ideas	**IVA:** *Benchmarks for Science Literacy:* Chapter 15 Research • 1C: The Scientific Enterprise, p. 333
V. K–12 Articulation and Connections	**VA:** *Next Generation Science Standards:* Appendices: Progression Chart • Appendix H: Science Addresses Questions About the Natural and Material World, p. 6 **VB:** *NSTA Quick Reference Guide to the NGSS K–12:* Progression Chart • Science Addresses Questions About the Natural and Material World, p. 84 **VC:** *The NSTA Atlas of the Three Dimensions:* Map Page • 7.8: Science Addresses Questions About the Natural and Material World (AQAW)

Section and Outcome	Selected Sources and Readings for Study and Reflection Read and examine **related** parts of
VI. **Assessment** **Expectation**	**VIA:** *State Standards* • Examine your state's standards **VIB:** *Next Generation Science Standards:* Performance Expectations • Grade 2: 2-ESS2-1, p. 15 • Grade 5: 5-ESS3-1, p. 30 • MS: MS-LS2-5, p. 49; MS-LS4-5, p. 50; MS-ESS3-4, p. 60 • HS: HS-ESS3-2, p. 99 **VIC:** *NSTA Quick Reference Guide to the NGSS K–12:* Performance Expectations • Grade 2: 2-ESS2-1, p. 96 • Grade 5: 5-ESS3-1, p. 112 • MS: MS-LS2-5, p. 130; MS-LS4-5, p. 132; MS-ESS3-4, p. 137 • HS: HS-ESS3-2, p. 165

Review Chapter 2 instructions on how to use this guide.

Visit curriculumtopicstudy2.org for more information about CTS and additional resources.

NOTE: *Atlas* page numbers have not been provided because *The NSTA Atlas of the Three Dimensions* was produced concurrently with this edition. Titles and map codes are accurate.

Additional Readings:

 Available for download at www.curriculumtopicstudy2.org

Copyright © 2020 by Corwin Press, Inc. All rights reserved. Reprinted from *Science Curriculum Topic Study: Bridging the Gap Between Three-Dimensional Standards, Research, and Practice* (2nd ed.) by Page Keeley and Joyce Tugel. Thousand Oaks, CA: Corwin, www.corwin.com. Reproduction authorized for educational use by educators, local school sites, and/or noncommercial or nonprofit entities that have purchased the book.

Science as a Human Endeavor

Grades K–12 Standards- and Research-Based Study of a Curricular Topic

Section and Outcome	Selected Sources and Readings for Study and Reflection Read and examine **related** parts of
I. Content Knowledge	**IA:** *Science for All Americans* • Ch. 1: The Scientific Enterprise, pp. 8–12 **IB:** *The NSTA Atlas of the Three Dimensions:* Narrative Page • 7.7: Science Is a Human Endeavor (HE)
II. Concepts, Core Ideas, or Practices	**IIA:** *Next Generation Science Standards:* (Scientific and Engineering Practices Column—Under Connections to Nature of Science) • Grade 3: Science Is a Human Endeavor, p. 21 • Grade 4: Science Is a Human Endeavor, p. 23 • MS: Science Is a Human Endeavor, pp. 42, 45 • HS: Science Is a Human Endeavor, pp. 85, 99 **IIB:** *NSTA Quick Reference Guide to the NGSS K–12:* Connections to Nature of Science • K–2: Science Is a Human Endeavor, p. 91 • 3–5 Science Is a Human Endeavor, p. 103 • MS: Science Is a Human Endeavor, p. 121 • HS: Science Is a Human Endeavor, p. 146
III. Curriculum, Instruction, and Formative Assessment	**IIIA:** *Next Generation Science Standards: Appendices* • Appendix H: Implementing Instruction to Facilitate Understanding of the Nature of Science, pp. 7–8
IV. Research on Commonly Held Ideas	**IVA:** *Benchmarks for Science Literacy:* Chapter 15 Research • 1C: The Scientific Enterprise, p. 333
V. K–12 Articulation and Connections	**VA:** *Next Generation Science Standards:* Appendices: Progression Chart • Appendix H: Science Is a Human Endeavor, p. 6 **VB:** *NSTA Quick Reference Guide to the NGSS K–12:* Progression Chart • Science Is a Human Endeavor, p. 84 **VC:** *The NSTA Atlas of the Three Dimensions:* Map Page • 7.7: Science Is a Human Endeavor (HE)
VI. Assessment Expectation	**VIA:** *State Standards* • Examine your state's standards **VIB:** *Next Generation Science Standards:* Performance Expectations • Grade 3: 3-ESS3-1, p. 21 • Grade 4: 4-PS3-4, p. 23

Section and Outcome	Selected Sources and Readings for Study and Reflection Read and examine **related** parts of
	• MS: MS-PS4-3, p. 42; MS-LS1-3, p. 45 • HS: HS-LS3-3, p. 85; HS-ESS3-3, p. 100 **VIC: *NSTA Quick Reference Guide to the NGSS K–12:*** Performance Expectations • Grade 3: 3-LS4-3, 3-ESS3-1, p. 105 • Grade 4: 4-PS3-4, p. 109 • MS: MS-PS4-3, p. 127; MS-LS1-3, p. 128 • HS: HS-LS3-3, p. 158; HS-ESS3-3, p. 166

Review Chapter 2 instructions on how to use this guide.

Visit curriculumtopicstudy2.org for more information about CTS and additional resources.

NOTE: *Atlas* page numbers have not been provided because *The NSTA Atlas of the Three Dimensions* was produced concurrently with this edition. Titles and map codes are accurate.

Additional Readings:

Available for download at **www.curriculumtopicstudy2.org**

Copyright © 2020 by Corwin Press, Inc. All rights reserved. Reprinted from *Science Curriculum Topic Study: Bridging the Gap Between Three-Dimensional Standards, Research, and Practice* (2nd ed.) by Page Keeley and Joyce Tugel. Thousand Oaks, CA: Corwin, www.corwin.com. Reproduction authorized for educational use by educators, local school sites, and/or noncommercial or nonprofit entities that have purchased the book.

Science as a Way of Knowing
Grades K–12 Standards- and Research-Based Study of a Curricular Topic

Section and Outcome	Selected Sources and Readings for Study and Reflection Read and examine **related** parts of
I. **Content Knowledge**	**IA:** *Science for All Americans* • Ch. 1: The Scientific World View, pp. 2–3 • Ch. 1: Science Demands Evidence, pp. 4–5 • Ch. 1: Science Explains and Predicts, p. 6 **IB:** *The NSTA Atlas of the Three Dimensions:* Narrative Page • 7.5: Science Is a Way of Knowing (WOK) • 7.6: Scientific Knowledge Assumes an Order and Consistency in Natural Systems (AOC)
II. **Concepts, Core Ideas, or Practices**	**IIA:** *Next Generation Science Standards:* (Scientific and Engineering Practices Under Connections to Nature of Science) • Grade 1: Scientific Knowledge Assumes an Order and Consistency in Natural Systems, p. 11 • Grade 3: Scientific Knowledge Assumes an Order and Consistency in Natural Systems, p. 19 • Grade 4: Scientific Knowledge Assumes an Order and Consistency in Natural Systems, p. 26 • Grade 5: Scientific Knowledge Assumes an Order and Consistency in Natural Systems, p. 28 • MS: Scientific Knowledge Assumes an Order and Consistency in Natural Systems, pp. 47, 52, 56 • HS: Scientific Knowledge Assumes an Order and Consistency in Natural Systems, pp. 70, 74–75, 87, 92 **IIB:** *NSTA Quick Reference Guide to the NGSS K–12:* Connections to Nature of Science • K–2: Science Is a Way of Knowing, Scientific Knowledge Assumes an Order and Consistency in Natural Systems, p. 91 • 3–5 Science Is a Way of Knowing, Scientific Knowledge Assumes an Order and Consistency in Natural Systems, p. 103 • MS: Science Is a Way of Knowing, Scientific Knowledge Assumes an Order and Consistency in Natural Systems, p. 121 • HS: Science Is a Way of Knowing, Scientific Knowledge Assumes an Order and Consistency in Natural Systems, p. 146
III. **Curriculum, Instruction, and Formative Assessment**	**IIIA:** *Next Generation Science Standards: Appendices* • Appendix H: Implementing Instruction to Facilitate Understanding of the Nature of Science, pp. 7–8
IV. **Research on Commonly Held Ideas**	**IVA:** *Benchmarks for Science Literacy:* Chapter 15 Research • 1A: The Scientific World View, p. 332 • 1B: Nature of Knowledge, p. 333

Section and Outcome	Selected Sources and Readings for Study and Reflection Read and examine **related** parts of
V. **K–12 Articulation and Connections**	**VA:** *Next Generation Science Standards:* Appendices: Progression Chart • Appendix H: Science Is a Way of Knowing, p. 6 • Appendix H: Scientific Knowledge Assumes an Order and Consistency in Natural Systems, p. 6 **VB:** *NSTA Quick Reference Guide to the NGSS K–12:* Progression Chart • Science Is a Way of Knowing, p. 84 • Scientific Knowledge Assumes an Order and Consistency in Natural Systems, p. 84 **VC:** *The NSTA Atlas of the Three Dimensions:* Map Page • 7.5: Science Is a Way of Knowing (WOK) • 7.6: Scientific Knowledge Assumes an Order and Consistency in Natural Systems (AOC)
VI. **Assessment Expectation**	**VIA:** *State Standards* • Examine your state's standards **VIB:** *Next Generation Science Standards:* Performance Expectations • Grade 1: 1-ESS1-1, p. 11 • Grade 3: 3-LS4-1, p. 19 • Grade 4: 4-ESS1-1, p. 26 • Grade 5: 5-PS1-2, p. 28 • MS: MS-LS2-3, p. 47; MS-LS4-1, MS-LS4-2, p. 52; MS-ESS1-1, MS-ESS1-2, p. 56 • HS: HS-PS1-7, p. 70; HS-PS3-1, p. 74; HS-LS-4-1, HS-LS4-4, p. 87; HS-ESS1-2, p. 92 **VIC:** *NSTA Quick Reference Guide to the NGSS K–12:* Performance Expectations • Grade 1: 1-ESS1-1, p. 94 • Grade 3: 3-LS4-1, p. 104 • Grade 4: 4-ESS1-1, p. 107 • Grade 5: 5-PS1-2, p. 112 • MS: MS-LS2-3, p. 130; MS-LS4-1, MS-LS4-2, p. 131; MS-ESS1-1, MS-ESS1-2, p. 133 • HS: HS-PS1-7, p. 148; HS-PS3-1, p. 150; HS-LS-4-1, p. 159; HS-LS4-4, p. 160; HS-ESS1-2, p. 161

Review Chapter 2 instructions on how to use this guide.

Visit curriculumtopicstudy2.org for more information about CTS and additional resources.

NOTE: *Atlas* page numbers have not been provided because *The NSTA Atlas of the Three Dimensions* was produced concurrently with this edition. Titles and map codes are accurate.

Additional Readings:

 Available for download at www.curriculumtopicstudy2.org

Copyright © 2020 by Corwin Press, Inc. All rights reserved. Reprinted from *Science Curriculum Topic Study: Bridging the Gap Between Three-Dimensional Standards, Research, and Practice* (2nd ed.) by Page Keeley and Joyce Tugel. Thousand Oaks, CA: Corwin, www.corwin.com. Reproduction authorized for educational use by educators, local school sites, and/or noncommercial or nonprofit entities that have purchased the book.

Scientific Knowledge Demands Evidence
Grades K–12 Standards- and Research-Based Study of a Curricular Topic

Section and Outcome	Selected Sources and Readings for Study and Reflection Read and examine **related** parts of
I. Content Knowledge	**IA:** *Science for All Americans* • Ch. 1: Science Demands Evidence, pp. 4–5 **IB:** *The NSTA Atlas of the Three Dimensions:* Narrative Page • 7.2: Scientific Knowledge Is Based on Empirical Evidence (BEE) • 7.3: Scientific Knowledge Is Open to Revision in Light of New Evidence (OTR)
II. Concepts, Core Ideas, or Practices	**IIA:** *Next Generation Science Standards:* (Scientific and Engineering Practices Under Connections to Nature of Science) • K: Scientific Knowledge is Based on Empirical Evidence, pp. 6, 7 • Grade 1: Scientific Knowledge is Based on Empirical Evidence, p. 10 • Grade 2: Scientific Knowledge is Based on Empirical Evidence, p. 14 • Grade 3: Scientific Knowledge is Based on Empirical Evidence, pp. 18, 20 • Grade 4: Scientific Knowledge is Based on Empirical Evidence, p. 24 • MS: Scientific Knowledge is Based on Empirical Evidence, pp. 37, 38, 40, 42, 47, 52; Scientific Knowledge Is Open to Revision in Light of New Evidence, p. 57 • HS: Scientific Knowledge is Based on Empirical Evidence, pp. 97, 98; Scientific Knowledge Is Open to Revision in Light of New Evidence, p. 81, 83 **IIB:** *NSTA Quick Reference Guide to the NGSS K–12:* Connections to Nature of Science • K–2: Scientific Knowledge is Based on Empirical Evidence, Scientific Knowledge Is Open to Revision in Light of New Evidence, p. 91 • 3–5 Scientific Knowledge is Based on Empirical Evidence, Scientific Knowledge Is Open to Revision in Light of New Evidence, p. 103 • MS: Scientific Knowledge is Based on Empirical Evidence, Scientific Knowledge Is Open to Revision in Light of New Evidence, p. 121 • HS: Scientific Investigations Use a Variety of Methods, p. 145
III. Curriculum, Instruction, and Formative Assessment	**IIIA:** *Next Generation Science Standards: Appendices* • Appendix H: Implementing Instruction to Facilitate Understanding of the Nature of Science, pp. 7–8
IV. Research on Commonly Held Ideas	**IVA:** *Benchmarks for Science Literacy:* Chapter 15 Research • 1A: The Scientific World View, p. 332 • 1B: Theory (Explanation) and Evidence; Nature of Knowledge, pp. 332–333 • 12E: Theory and Evidence, p. 361
V. K–12 Articulation and Connections	**VA:** *Next Generation Science Standards:* Appendices: Progression Chart • Appendix H: Scientific Knowledge is Based on Empirical Evidence, p. 5 • Scientific Knowledge Is Open to Revision in Light of New Evidence, p. 5 **VB:** *NSTA Quick Reference Guide to the NGSS K–12:* Progression Chart • Scientific Knowledge is Based on Empirical Evidence, p. 82 • Scientific Knowledge Is Open to Revision in Light of New Evidence, p. 83

Section and Outcome	Selected Sources and Readings for Study and Reflection Read and examine **related** parts of
	VC: *The NSTA Atlas of the Three Dimensions:* Map Page • 7.2: Scientific Knowledge is Based on Empirical Evidence (BEE) • 7.3: Scientific Knowledge Is Open to Revision in Light of New Evidence (OTR)
VI. Assessment Expectation	**VIA:** *State Standards* • Examine your state's standards **VIB:** *Next Generation Science Standards:* Performance Expectations • K: K-LS1-1, p. 6; K-ESS2-1, p. 7 • Grade 1: 1-LS1-2, p. 10 • Grade 2: 2-LS4-1, p. 14 • Grade 3: 3-PS2-2, p. 18; 3-LS1-1, p. 20 • Grade 4: 4-PS4-1, p. 24 • MS: MS-PS1-2, p. 37; MS-PS2-2, MS-PS2-4, p. 38; MS-PS3-4, MS-PS3-5, p. 40; MS-MS-PS4-1, p. 42; MS-LS1-6, MS-LS2-4, p. 47; MS-LS4-1, p. 52; MS-ESS2-3, p. 57 • HS: HS-LS2-2, p. 83; HS-LS2-3, p. 81; HS-ESS2-3, p. 97; HS-ESS2-4, HS-ESS3-5, p. 98 **VIC:** *NSTA Quick Reference Guide to the NGSS K–12:* Performance Expectations • K: K-LS1-1, K-ESS2-1, p. 92 • Grade 1: 1-LS1-2, p. 94 • Grade 2: 2-LS4-1, p. 96 • Grade 3: 3-PS2-2, p. 106; 3-LS1-1, p. 104 • Grade 4: 4-PS4-1, p. 109 • MS: MS-PS1-2, p. 122; MS-PS2-2, p. 124; MS-PS2-4, p. 125; MS-PS3-4, MS-PS3-5, p. 126; MS-MS-PS4-1, p. 127; MS-LS1-6, p. 129; MS-LS2-4, p. 130; MS-LS4-1, p. 131; MS-ESS2-3, p. 134 • HS: HS-LS2-2, HS-LS2-3, p. 156; HS-ESS2-3, p. 163; HS-ESS2-4, p. 164; HS-ESS3-5, p. 166

Review Chapter 2 instructions on how to use this guide.

Visit curriculumtopicstudy2.org for more information about CTS and additional resources.

NOTE: *Atlas* page numbers have not been provided because *The NSTA Atlas of the Three Dimensions* was produced concurrently with this edition. Titles and map codes are accurate.

Additional Readings:

 Available for download at www.curriculumtopicstudy2.org

Copyright © 2020 by Corwin Press, Inc. All rights reserved. Reprinted from *Science Curriculum Topic Study: Bridging the Gap Between Three-Dimensional Standards, Research, and Practice* (2nd ed.) by Page Keeley and Joyce Tugel. Thousand Oaks, CA: Corwin, www.corwin.com. Reproduction authorized for educational use by educators, local school sites, and/or noncommercial or nonprofit entities that have purchased the book.

References

American Association for the Advancement of Science. (1990). *Science for all Americans*. New York, NY: Oxford University Press.

American Association for the Advancement of Science. (2001). *Atlas of science literacy* (Vol. 1). Arlington, VA: NSTA Press.

American Association for the Advancement of Science. (2009). Benchmarks for science literacy online. Retrieved from www.project2061.org/publications/bsl/online

Beane, J. (1995). Introduction: What is a coherent curriculum? In J. Beane (Ed.), *Toward a coherent curriculum: The 1995 ASCD yearbook*. Alexandria, VA: Association for Supervision and Curriculum Development.

Bransford, J., Brown, A., & Cocking, R. (2000). *How people learn*. Washington, DC: National Academy Press.

Chingos, M., & Whitehurst, G. (2012). *Choosing blindly: Instructional materials, teacher effectiveness, and the common core*. Washington, DC: Brookings Institution.

Driver, R., Squires, A., Rushworth, P., & Wood-Robinson, V. (1994). *Making sense of secondary science*. New York, NY: Routledge.

Duncan, R. C., Krajcik, J., & Rivet, A. E. (Eds.). (2017). *Disciplinary core ideas: Reshaping teaching and learning*. Arlington, VA: National Science Teachers Association.

Hazen, R., & Trefil, J. (2009). *Science matters: Achieving scientific literacy*. New York, NY: Anchor Books.

Heritage, M. (2010). *Formative assessment: Making it happen in the classroom*. Thousand Oaks, CA: Corwin.

Kahle, J. B. (1999, June). Testimony, U.S. House of Representatives, Washington, DC. Retrieved from http://archives-republicans-edlabor.house.gov/archive/hearings/106th/pet/teacher61099/kahle.htm

Keeley, P. (2005). *Science curriculum topic study: Bridging the gap between standards and practice*. Thousand Oaks, CA: Corwin.

Keeley, P. (2011). *Uncovering student ideas in life science: 25 new formative assessment probes*. Arlington, VA: National Science Teachers Association.

Keeley, P. (2013). *Uncovering student ideas in primary science: 25 new formative assessment probes for K–2*. Arlington, VA: National Science Teachers Association.

Keeley, P. (2015). *Science formative assessment: 50 more strategies linking assessment, instruction, and learning*. Thousand Oaks, CA: Corwin.

Keeley, P. (2018). *Uncovering student ideas in science: 25 formative assessment probes* (2nd ed.). Arlington, VA: National Science Teachers Association.

Keeley, P., & Cooper, S. (2019). *Uncovering student ideas in physical science (Vol. 3): 32 new matter and energy formative assessment probes*. Arlington, VA: National Science Teachers Association.

Keeley, P., Eberle, F., & Dorsey, C. (2008). *Uncovering student ideas in science: Another 25 formative assessment probes*. Arlington, VA: National Science Teachers Association.

Keeley, P., Eberle, F., & Tugel, J. (2007). *Uncovering student ideas in science: 25 more formative assessment probes*. Arlington, VA: National Science Teachers Association.

Keeley, P., & Harrington, R. (2010). *Uncovering student ideas in physical science (Vol. 1): 45 new force and motion formative assessment probes*. Arlington, VA: National Science Teachers Association.

Keeley, P., & Harrington, R. (2014). *Uncovering student ideas in physical science (Vol. 2): 39 new electricity and magnetism formative assessment probes*. Arlington, VA: National Science Teachers Association.

Keeley, P., & Sneider, C. (2012). *Uncovering student ideas in astronomy: 45 new formative assessment probes*. Arlington, VA: National Science Teachers Association.

Keeley, P., Sneider, C., & Ravel, M. (In press). *Uncovering student ideas about engineering and technology*. Arlington, VA: National Science Teachers Association.

Keeley, P., & Tucker, L. (2015). *Uncovering student ideas in earth and environmental science: 45 new formative assessment probes*. Arlington, VA: National Science Teachers Association.

Keeley, P., & Tugel, J. (2009). *Uncovering student ideas in science: 25 new formative assessment probes*. Arlington, VA: National Science Teachers Association.

National Academy of Engineering and National Research Council. (2014). *STEM integration in K–12 education: Status, prospects, and an agenda for research*. Washington, DC: National Academies Press.

National Research Council. (1996). *National science education standards*. Washington, DC: National Academies Press.

National Research Council. (2002). *Investigating the influence of standards: A framework for research in mathematics, science, and technology education*. Washington, DC: National Academies Press.

National Research Council. (2007). *Taking science to school: Learning and teaching science in grades K–8*. Washington, DC: National Academies Press.

National Research Council. (2012). *A framework for K–12 science education: Practices, crosscutting concepts, and core ideas*. Washington, DC: National Academies Press.

National Research Council. (2014). *Developing assessments for the next generation science standards*. Washington, DC: National Academies Press.

National Research Council. (2015). *Guide to implementing the Next Generation Science Standards*. Washington, DC: National Academies Press.

NGSS Lead States. (2013). *Next Generation Science Standards: For states by states*. Washington DC: Achieve. Retrieved from www.nextgenscience.org/next-generation-science-standards

Schwarz, C. V., Passmore, C., & Reiser, B. J. (Eds.). (2017). *Helping students make sense of the world using next generation science and engineering practices*. Arlington, VA: National Science Teachers Association.

Shulman, L. S. (1986). Those who understand: Knowledge growth in teaching. *Educational Researcher, 15*(2), 4–14.

Wiggins, G., & McTighe, J. (2005). *Understanding by design* (2nd ed.). Alexandria, VA: Association for Supervision and Curriculum Development.

Wiliam, C. (2011). *Embedded formative assessment*. Bloomington, IN: Solution Tree Press.

Willard, T. (Ed.). (2015). *The NSTA quick-reference guide to the NGSS K–12*. Arlington, VA: NSTA Press.

Willard, T. (2019). *The NSTA Atlas of the Three Dimensions*. Arlington, VA: National Science Teachers Association.

Index

Adaptation, 124–125
Ahlgreen, A., 23
Analyzing and Interpreting Data, 258–259
Argumentation, 266–267
Asking Questions in Science, 250–251
Atoms and Molecules, 146–147

Backwards design, 8, 68–69
Behavior, Senses, Feedback, and Response, 90–91
Behavior and Characteristics of Gases, 148–149
Benchmarks for Science Literacy-Chapter 15 Research Summaries, 5 (figure), 33–34, 33 (figure), 38, 39
   crosscutting concepts guides, 271
   Earth and space science guides, 200
   life science guides, 86
   physical science guides, 142
   scientific and engineering practices guides, 247
Biodiversity and Human Impact, 126–127
Biogeology, 220–221
Biological Evolution, 128–129
Biological structure and function guides
   Behavior, Senses, Feedback, and Response, 90–91
   Brain and the Nervous System, 92–93
   Cell Division and Differentiation, 94–95
   Cells and Biomolecules: Structure and Function, 96–97
   Characteristics of Living Things, 98–99
   Energy Extraction, Food, and Nutrition, 100–101
   Macroscopic Structure and Function of Organisms, 102–103
   Organs and Systems of the Human Body, 104–105
   Photosynthesis and Respiration at the Organism Level, 106–107
Brain and the Nervous System, 92–93
Bundling, 71

Cause and Effect, 276–277
Cell Division and Differentiation, 94–95
Cells and Biomolecules: Structure and Function, 96–97
Characteristics of Living Things, 98–99
Chemical Bonding, 150–151
Chemical Reactions, 152–153

Classroom assessment, 79
   using CTS to inform, 79–84
Coherent curriculum, 71–72
Committee on STEM Integration, 76
Concept of Energy, 168–169
Conservation of Matter, 154–155
Constructing Scientific Explanations, 20–21, 262–263
Content knowledge enhancement, 66–68
Crosscutting concepts guides, 50 (figure), 271–273
   Cause and Effect, 276–277
   Energy and Matter: Flows, Cycles, and Conservation, 282–283
   Patterns, 274–275
   Scale, Proportion, and Quantity, 278–279
   Stability and Change, 286–287
   Structure and Function, 284–285
   Systems and Systems Models, 280–281
Crosscutting Concepts-Strengthening Science Learning, 21
Curriculum
   coherence and articulation across grades of, 71–72
   CTS for informing, 68–76
   implementing, 73, 74 (figure)
   informed with CTS, 76–79
   selection of, 73
Curriculum Topic Study (CTS) guides, 2–6, 13
   adding supplementary material to, 20–21
   building a professional collection of, 21–22
   common resources used in, 5, 5–6 (figure), 21, 22–35
   crosscutting concepts. *See* Crosscutting concepts guides
   CTS website reference in, 21
   Earth and space science. *See* Earth and space science guides
   example of K-12, 3–4 (figure)
   impacting beliefs about teaching and learning, 11–12
   life science. *See* Life science guides
   organization of, 13–14 (figure)
   physical science. *See* Physical science guides
   reasons for studying curriculum topics using, 6–9
   scientific and engineering. *See* Scientific and engineering practices guides

sections of, 5, 17, 53
selected readings column, 18–19, 19–20 (figure)
selecting specific, 41, 41–52 (figure), 52
STEM connections. *See* STEM connections guides
website, 21
why focus on topics using, 9–10
Curriculum Topic Study (CTS) process
deciding which resources to use in, 37–38
gathering and managing CTS resources in, 38–40
getting started with, 53
group work in, 39–40
guiding questions for getting started with, 54–60
guiding the, 37
selecting a guide, defining your purpose, and choosing your outcomes in, 41, 41–52 (figure), 52
Curriculum Topic Study (CTS) use and application, 61–62
additional, 84
in classroom assessment, 79–84
for different purposes, 66
in enhancing content knowledge, 66–68
group, 64–66
individual, 62, 63–64 (figure)
in informing curriculum, 68–76
Cycling of Matter and Flow of Energy in Ecosystems, 108–109
Cycling of Matter in Ecosystems, 110–111

Decomposition and Decay, 112–113
Defining and Delimiting and Engineering Problem, 292–293
Defining Problems, 252–253
Designing Solutions (Engineering), 264–265
Developing and Using Models, 254–255
Developing Possible Design Solutions, 294–295
Diagnostic assessment, 79–80
Disciplinary Core Ideas: Reshaping Teaching and Learning, 6 (figure), 19, 27–28, 27 (figure)
additional questions, 59
Earth and space science guides, 200–201
life science guides, 87, 88
physical science guides, 143, 144
STEM connections guides, 289–291
Diversity of Species and Evidence of Common Ancestry, 130–131
DNA, Genes, and Proteins, 132–133
Driver, R., 34
Duncan, R. C., 27

Earth, Moon, Sun System, 234–235
Earth and Our Solar System, 232–233
Earth and space science guides, 46–48 (figure), 200–203
Biogeology, 220–221
Earth, Moon, Sun System, 234–235
Earth and Our Solar System, 232–233
Earthquakes and Volcanoes, 222–223
Earth's History, 224–225

Earth's Materials and Systems, 204–205
Earth's Natural Resources, 206–207
Earth-Sun System, 236–237
Formation of Earth, the Solar System, and the Universe, 238–239
Global Climate Change, 208–209
Human Impact on Earth Systems, 210–211
Natural Hazards, 226–227
Phases of the Moon, 240–241
Plate Tectonics, 228–229
Seasons and Seasonal Patterns in the Sky, 242–243
Stars and Galaxies, 244–245
Structure of the Solid Earth, 212–213
Water Cycle and Distribution, 214–215
Water in the Earth System, 216–217
Weather and Climate, 218–219
Weathering, Erosion, and Deposition, 230–231
Earth history and processes that change the Earth guides
Biogeology, 220–221
Earthquakes and Volcanoes, 222–223
Natural Hazards, 226–227
Plate Tectonics, 228–229
Weathering, Erosion, and Deposition, 230–231
Earth in space, solar system, and the universe guides
Earth, Moon, Sun System, 234–235
Earth and Our Solar System, 232–233
Earth-Sun System, 236–237
Formation of Earth, the Solar System, and the Universe, 238–239
Phases of the Moon, 240–241
Seasons and Seasonal Patterns in the Sky, 242–243
Stars and Galaxies, 244–245
Earthquakes and Volcanoes, 222–223
Earth's History, 224–225
Earth's Materials and Systems, 204–205
Earth's Natural Resources, 206–207
Earth structure, materials, and systems guides
Earth's Materials and Systems, 204–205
Earth's Natural Resources, 206–207
Global Climate Change, 208–209
Human Impact on Earth Systems, 210–211
Structure of the Solid Earth, 212–213
Water Cycle and Distribution, 214–215
Water in the Earth System, 216–217
Weather and Climate, 218–219
Earth-Sun System, 236–237
Ecosystems and ecological relationships guides
Cycling of Matter and Flow of Energy in Ecosystems, 108–109
Cycling of Matter in Ecosystems, 110–111
Decomposition and Decay, 112–113
Food Chains and Food Webs (K-8), 116–117
Group Behaviors and Social Interactions in Ecosystems, 118–119
Interdependency in Ecosystems, 120–121
Transfer of Energy in Ecosystems, 122–123
Ecosystem Stability, Disruptions, and Change, 114–115

Elements, Compounds, and the Periodic Table, 156–157
Energy and Matter: Flows, Cycles, and Conservation, 282–283
Energy Extraction, Food, and Nutrition, 100–101
Energy in Chemical Processes and Everyday Life, 174–175
Engineering, technology, and applications of science guides, 289
    Defining and Delimiting an Engineering Problem, 292–293
    Developing Possible Design Solutions, 294–295
    Engineering Design, 296–297
    Improving Proposed Designs, 298–299
    Influence of Science, Technology, and Engineering on Society and the Natural World, 300–301
    Interdependence of Science, Engineering, and Technology, 302–303
Engineering Design, 296–297

5E (engage, explore, explain, elaborate, elaborate) approach, 77
Food Chains and Food Webs (K-8), 116–117
Force, motion, and energy guides
    Concept of Energy, 168–169
    Conservation of Energy, 170–171
    Electric Charge and Current, 172–173
    Elements, Compounds, and the Periodic Table, 156–157
    Energy in Chemical Processes and Everyday Life, 174–175
    Force and Motion, 176–177
    Forces Between Objects, 178–179
    Gravitational Force, 180–181
    Kinetic and Potential Energy, 182–183
    Magnetism, 184–185
    Nuclear Energy, 186–187
    Relationship Between Energy and Forces, 188–189
    Sound, 190–191
    Transfer of Energy, 192–193
    Visible Light and Electromagnetic Radiation, 194–195
    Waves and Information Technologies, 196–197
    Waves and Wave Properties, 198–199
Force and Motion, 176–177
Forces Between Objects, 178–179
Formation of Earth, the Solar System, and the Universe, 238–239
Formative assessment, 80
Framework for K-12 Science Education: Practices, Crosscutting Concepts, and Core Ideas, 5 (figure), 6, 7, 10, 18, 21–22, 25–26, 25 (figure)
    additional questions, 58
    crosscutting concepts guides, 271–273
    in CTS process, 37, 38, 39
    Earth and space science guides, 200
    on enhancing content knowledge, 67–68
    life science guides and, 87, 88
    on modifying lessons, 76–77

    physical science guides, 142–143
    scientific and engineering practices guides, 247–249
    STEM connections guides, 289–291
    three-dimensional curriculum, 70–71

Global Climate Change, 208–209
Gravitational Force, 180–181
Group Behaviors and Social Interactions in Ecosystems, 118–119
Group CTS use, 64–66
Group work in CTS process, 39–40
Guiding questions for CTS, 54–60

Helping Students Make Sense of the World Using Next Generation Science and Engineering Practices, 6 (figure), 28–29, 28 (figure)
    additional questions, 59
    scientific and engineering practices guides, 247
Human Impact on Earth Systems, 210–211
Hypothesis, Theories, and Laws, 304–305

Improving Proposed Designs, 298–299
Individual CTS use, 62, 63–64 (figure)
Influence of Science, Technology, and Engineering on Society and the Natural World, 300–301
Informing curriculum with CTS, 68–76
Informing instruction with CTS, 76–79
Inheritance of Traits, 14–16 (figure), 18, 134–135
Inquiry-based teaching, 77–78
Integration and interdisciplinary connections, 75
Interdependence of Science, Engineering, and Technology, 302–303
Interdependency in Ecosystems, 120–121

Jigsaw readings, 65

KIDS (Kids Involved Doing Service learning) model, 75
Kinetic and Potential Energy, 182–183
Krajcik, J., 27

Large group discussion, 66
Learning intentions, 80–84
Lee, O., 21
Life science guides, 41–44 (figure), 86–89
    Adaptation, 124–125
    Behavior, Senses, Feedback, and Response, 90–91
    Biodiversity and Human Impact, 126–127
    Biological Evolution, 128–129
    Brain and the Nervous System, 92–93
    Cell Division and Differentiation, 94–95
    Cells and Biomolecules: Structure and Function, 96–97
    Characteristics of Living Things, 98–99
    Cycling of Matter and Flow of Energy in Ecosystems, 108–109
    Cycling of Matter in Ecosystems, 110–111
    Decomposition and Decay, 112–113

Diversity of Species and Evidence of Common Ancestry, 130–131
DNA, Genes, and Proteins, 132–133
Ecosystem Stability, Disruptions, and Change, 114–115
Energy Extraction, Food, and Nutrition, 100–101
Food Chains and Food Webs (K-8), 116–117
Group Behaviors and Social Interactions in Ecosystems, 118–119
Inheritance of Traits, 14–16 (figure), 18, 134–135
Interdependency in Ecosystems, 120–121
Macroscopic Structure and Function of Organisms, 102–103
Natural Selection, 136–137
Organs and Systems of the Human Body, 104–105
Photosynthesis and Respiration at the Organism Level, 106–107
Reproduction, Growth, and Development, 138–139
Transfer of Energy in Ecosystems, 122–123
Variation of Traits, 140–141
Life's continuity, change, and diversity guides
    Adaptation, 124–125
    Biodiversity and Human Impact, 126–127
    Biological Evolution, 128–129
    Diversity of Species and Evidence of Common Ancestry, 130–131
    DNA, Genes, and Proteins, 132–133
    Inheritance of Traits, 134–135
    Natural Selection, 136–137
    Reproduction, Growth, and Development, 138–139
    Variation of Traits, 140–141

Macroscopic Structure and Function of Organisms, 102–103
Magnetism, 184–185
Making Sense of Secondary Science: Research Into Children's Ideas, 5 (figure), 34, 34 (figure)
    Earth and space science guides, 200
    life science guides, 88
    physical science guides, 142
Matter guides
    Atoms and Molecules, 146–147
    Behavior and Characteristics of Gases, 148–149
    Chemical Bonding, 150–151
    Chemical Reactions, 152–153
    Concept of Energy, 168–169
    Conservation of Matter, 154–155
    Elements, Compounds, and the Periodic Table, 156–157
    Mixtures and Solutions, 158–159
    Nuclear Processes, 160–161
    Particulate Nature of Matter, 162–163
    Properties of Matter, 164–165
    States of Matter, 166–167
McTighe, J., 8, 68–69
Methods of Scientific Investigations, 306–307
Mixtures and Solutions, 158–159

National Research Council (NRC), 25
National Science Foundation (NSF), 7
National Youth Leadership Council, 75
Natural Hazards, 226–227
Natural Selection, 136–137
Nature of science guides, 289
    Hypothesis, Theories, and Laws, 304–305
    Methods of Scientific Investigations, 306–307
    Science Addresses Questions, 308–309
    Science as a Human Endeavor, 310–311
    Science as a Way of Knowing, 312–313
    Scientific Knowledge Demands Evidence, 314–315
Next Generation Science Standards (NGSS), 2, 5 (figure), 6, 7, 21–22, 26–27, 26 (figure)
    additional questions, 58
    crosscutting concepts guides, 271
    in CTS process, 37, 38, 39
    Earth and space science guides, 200–201
    life science guides, 88
    physical science guides, 143
    scientific and engineering practices guides, 247
    STEM connections guides, 289
    topic and disciplinary core ideas in, 9–10
    unpacking learning goals and standards in, 69–70
NGSS Bozeman Science, 21
Nordine, J., 21
NSTA Atlas of the Three Dimensions, The, 5 (figure), 34–35, 34 (figure)
    additional questions, 59–60
    crosscutting concepts guides, 271
    Earth and space science guides, 200–201
    life science guides, 87, 89
    physical science guides, 143, 145
    scientific and engineering practices guides, 247
    STEM connections guides, 289–291
NSTA Quick Reference Guide to the NGSS, K–12, The, 5 (figure), 18, 27, 27 (figure)
    additional questions, 58
    crosscutting concepts guides, 271–273
    in CTS process, 37, 39
    Earth and space science guides, 200–201
    life science guides, 86–87
    physical science guides, 142–145
    scientific and engineering practices guides, 247
    STEM connections guides, 289
Nuclear Energy, 186–187
Nuclear Processes, 160–161

Obtaining, Evaluating, and Communicating Information, 268–269
Organs and Systems of the Human Body, 104–105

Particulate Nature of Matter, 162–163
Passmore, C., 28
Patterns, 274–275

Pedagogical content knowledge, 9
Phases of the Moon guide, 240–241
Phenomena, selection of, 78–79
Photosynthesis and Respiration at the Organism Level, 106–107
Physical science guides, 44–46 (figure), 142–145
   Atoms and Molecules, 146–147
   Behavior and Characteristics of Gases, 148–149
   Chemical Bonding, 150–151
   Chemical Reactions, 152–153
   Concept of Energy, 168–169
   Conservation of Matter, 154–155
   Electric Charge and Current, 172–173
   Elements, Compounds, and the Periodic Table, 156–157
   Energy in Chemical Processes and Everyday Life, 174–175
   Force and Motion, 176–177
   Forces Between Objects, 178–179
   Gravitational Force, 180–181
   Kinetic and Potential Energy, 182–183
   Magnetism, 184–185
   Mixtures and Solutions, 158–159
   Nuclear Energy, 186–187
   Nuclear Processes, 160–161
   Particulate Nature of Matter, 162–163
   Properties of Matter, 164–165
   Sound, 190–191
   States of Matter, 166–167
   Transfer of Energy, 192–193
   Visible Light and Electromagnetic Radiation, 194–195
   Waves and Information Technologies, 196–197
   Waves and Wave Properties, 198–199
Planning and Carrying Out Investigations, 256–257
Plate Tectonics, 228–229
Professional learning community (PLC), 64, 71
Project based learning (PBL), 75
Properties of Matter, 164–165

Quinn, H., 27

Reiser, B., 28
Reproduction, Growth, and Development, 138–139
Rivet, A., 27
Rushworth, P., 34
Rutherford, J., 23

Scale, Proportion, and Quantity, 278–279
Schwarz, C., 28
Science Addresses Questions, 308–309
Science and Children, 21
Science as a Human Endeavor, 310–311
Science as a Way of Knowing, 312–313
Science for All Americans (SFAA), 5 (figure), 23, 23 (figure), 24 (figure)
   additional questions, 58
   in CTS process, 38, 39
   on enhancing content knowledge, 67–68

Science Matters: Achieving Scientific Literacy, 5 (figure), 24–25, 24 (figure), 58
   Earth and space science guides, 200
   life science guides, 87
   physical science guides, 143
Science Scope, 21
Science Teacher, The, 21
Scientific and engineering practices guides, 48–49 (figure), 77–78, 247–249
   Analyzing and Interpreting Data, 258–259
   Argumentation, 266–267
   Asking Questions in Science, 250–251
   Constructing Scientific Explanations, 262–263
   Defining Problems, 252–253
   Designing Solutions (Engineering), 264–265
   Developing and Using Models, 254–255
   Obtaining, Evaluating, and Communicating Information, 268–269
   Planning and Carrying Out Investigations, 256–257
   Using Mathematics and Computational Thinking, 260–261
Scientific Knowledge Demands Evidence, 314–315
Seasons and Seasonal Patterns in the Sky, 242–243
Sections and outcomes column of CTS guides, 5, 17
Selected readings column of CTS guides, 18–19, 19–20 (figure)
Service learning, 75
Shulman, 9
Small group discussion, 66
Sound, 190–191
Squires, A., 34
Stability and Change, 286–287
Stars and Galaxies, 244–245
States of Matter, 166–167
State standards, 6 (figure), 35
STEM connections guides, 51–52 (figure), 289
   Defining and Delimiting and Engineering Problem, 292–293
   Developing Possible Design Solutions, 294–295
   Engineering Design, 296–297
   Hypothesis, Theories, and Laws, 304–305
   Improving Proposed Designs, 298–299
   Influence of Science, Technology, and Engineering on Society and the Natural World, 300–301
   Interdependence of Science, Engineering, and Technology, 302–303
   Methods of Scientific Investigations, 306–307
   Science Addresses Questions, 308–309
   Science as a Human Endeavor, 310–311
   Science as a Way of Knowing, 312–313
   Scientific Knowledge Demands Evidence, 314–315
STEM education, 2, 7, 10
   CTS impacting beliefs about teaching and learning in, 11–12
   integration of, 75–76

Structure and Function, 284–285
Structure of the Solid Earth, 212–213
Study guides
    crosscutting concepts. *See* Crosscutting concepts guides
    Earth and space science. *See* Earth and space science guides
    life science. *See* Life science guides
    physical science. *See* Physical science guides
    scientific and engineering. *See* Scientific and engineering practices guides
    STEM connections. *See* STEM connections guides
Summative assessment, 80
Systems and Systems Models, 280–281

Table groups, 65
Three-dimensional classroom assessment tasks, 84
Three-dimensional curriculum, 70–71
Transfer of Energy, 122–123, 192–193

Uncovering Student Ideas in Science, 5 (figure), 18, 29, 30–33 (figure)
    on classroom assessment, 83
    Earth and space science guides, 200
    life science guides, 86
    physical science guides, 142
    scientific and engineering practices guides, 247
    STEM connections guides, 289
    on STEM integration, 75
Understanding by Design, 8
Unpacking learning goals or standards, 69–70
Using Mathematics and Computational Thinking, 260–261

Variation of Traits, 140–141
Visible Light and Electromagnetic Radiation, 194–195

Water Cycle and Distribution, 214–215
Water in the Earth System, 216–217
Waves and Information Technologies, 196–197
Waves and Wave Properties, 198–199
Weather and Climate, 218–219
Weathering, Erosion, and Deposition, 230–231
Website, CTS, 21
Wiggins, G., 8, 68–69
Wood-Robinson, V., 34

Helping educators make the greatest impact

CORWIN HAS ONE MISSION: to enhance education through intentional professional learning.

We build long-term relationships with our authors, educators, clients, and associations who partner with us to develop and continuously improve the best evidence-based practices that establish and support lifelong learning.

NSTA's mission is to promote excellence and innovation in science teaching and learning for all.

The Arlington, VA–based National Science Teachers Association is the largest professional organization in the world promoting excellence in science teaching and learning, preschool through college. NSTA's membership includes approximately 50,000 science teachers, science supervisors, administrators, scientists, business representatives, and others involved in science education.

Solutions YOU WANT | Experts YOU TRUST | Results YOU NEED

EVENTS

>>> **INSTITUTES**
Corwin Institutes provide large regional events where educators collaborate with peers and learn from industry experts. Prepare to be recharged and motivated!

corwin.com/institutes

ON-SITE PD

>>> **ON-SITE PROFESSIONAL LEARNING**
Corwin on-site PD is delivered through high-energy keynotes, practical workshops, and custom coaching services designed to support knowledge development and implementation.

corwin.com/pd

>>> **PROFESSIONAL DEVELOPMENT RESOURCE CENTER**
The PD Resource Center provides school and district PD facilitators with the tools and resources needed to deliver effective PD.

corwin.com/pdrc

ONLINE

>>> **ADVANCE**
Designed for K–12 teachers, Advance offers a range of online learning options that can qualify for graduate-level credit and apply toward license renewal.

corwin.com/advance

Contact a PD Advisor at (800) 831-6640 or visit www.corwin.com for more information

CORWIN